CW00995081

THE LIFE OF SIR JAMES BROOKE

Yours very sincerely
J Brooke

(From the painting by Sir Francis Grant, P.R.A)

THE LIFE OF SIR JAMES BROOKE RAJAH OF SARAWAK

FROM HIS PERSONAL PAPERS
AND CORRESPONDENCE

BY
SPENSER ST. JOHN

WITH AN INTRODUCTION BY
R. H. W. REECE

KUALA LUMPUR
OXFORD UNIVERSITY PRESS
OXFORD SINGAPORE NEW YORK
1994

Oxford University Press

Oxford New York Toronto
Delhi Bombay Calcutta Madras Karachi
Kuala Lumpur Singapore Hong Kong Tokyo
Nairobi Dar es Salaam Cape Town
Melbourne Auckland Madrid

and associated companies in
Berlin Ibadan

Oxford is a trade mark of Oxford University Press

First published in 1879 by William Blackwood and Sons, London
Reprinted in Oxford in Asia Hardback Reprints *in 1994*

ISBN 967 65 3036 0

Printed by Kyodo Printing Co. (S) Pte. Ltd., Singapore
Published by Oxford University Press,
19–25, Jalan Kuchai Lama, 58200 Kuala Lumpur, Malaysia .

Spenser St. John, amanuensis, confidant and adviser to Sir James Brooke for ten years. (Royal Geographical Society)

INTRODUCTION

IN early 1878, Her Britannic Majesty's Consul-General and Resident Minister in Peru, Spenser St. John, was completing the biography of his former employer, patron, and friend, Sir James Brooke. Situated 8,000 feet above sea-level, Lima was a cold and windswept place and a far cry from the tropical climate of Sarawak and Brunei where St. John had spent almost two decades. It was also a difficult place from which to communicate with possible publishers. Consequently, he was obliged to enlist the assistance of an old Sarawak Service contemporary, Charles Grant, then living in retirement at St. Andrew's in Scotland.

In their correspondence about the biography over the ensuing eighteen months, the two men relived much of the drama they had participated in during those earlier years, notably the first Rajah's disinheritance in February 1863 of his elder nephew, Captain John Brooke Johnson (better known as Brooke Brooke), who was also Charles Grant's brother-in-law. St. John had been a staunch supporter of the Rajah, while 'Charlie' Grant remained loyal to his late sister's husband. Twenty years on, St. John was anxious not to open up old wounds. By asking Grant to be the *de facto* editor of his manuscript, he was employing the diplomatic skills which had already taken him a long way in the foreign service.

The correspondence is significant for the light it throws on the process of editing and on the evolving relationship between author and editor. In his letters, St. John said things which he could not say in his book. He also recorded his response to its reviews and his ultimate disappointment with its reception. For his part, Grant demonstrated his good sense and fair-mindedness.

The third son of James Augustus St. John, a London journalist, Spenser Buckingham was born (appropriately enough) at St. John's Wood on 22 December 1825. He received a good private school education and his social connections were later to secure him an appointment as James Brooke's private secretary and subsequently a career in the diplomatic service. James St. John wrote articles about Borneo for the *Morning Chronicle* and various journals in 1846 and became friendly with Brooke's commercial agent, Henry Wise. Subsequently, he had a number of interviews with Lord Palmerston on a commercial concession for Wise and on the colonization of Labuan, which seems to have been his pet project. He also wrote the text to accompany a series of drawings by Captain Drinkwater Bethune, RN, and others, published in 1847 as *Views in the Indian Archipelago*. Another of his sons, Horace, produced a history of the Indian Archipelago in 1853 which was unflattering of Brooke and reflected Wise's bitter antagonism towards him after 1848.

Even before he met James Brooke, Spenser St. John read Captain Henry Keppel's 1846 edition of his journals and wrote 'innumerable articles' about Borneo, one of them a semi-fictionalized account of some of Keppel's adventures. Wise had persuaded him to try for an appointment in Borneo, rather than Persia which had been his earlier interest. Indeed, he was already

Charles Grant, *c.*1856. Although Grant had fallen out with St. John over the Rajah's disinheritance of Brooke Brooke (Grant's brother-in-law) in 1862, he served him well in editing his biography of the Rajah and arranging for its publication.

studying the Malay language. Introduced to the Rajah by his father at Mivart's Hotel in late 1847, he was 'struck with that winning manner which in those days made every one who approached him his friend'. Thirty years later, St. John paid tribute to the extraordinary personal charm which accounted for the Rajah's success in inspiring both loyalty and affection: 'He was so handsome, elegant in looks as well as in manner, fond of the lighter accomplishments of music and poetry, so winning in ways as to be beloved by all those he met, and full of ability, and with his friends brilliant in talk. Yet in general society he was reserved and rarely gave sign of the power which was in him.'

It was agreed by the Colonial Office that St. John would serve as Private Secretary to the Rajah in his newly appointed capacity as Governor of Labuan, a salary of £200 a year being allowed 'in a roundabout way' by Lord Palmerston. Accompanying the Rajah to Sarawak on HMS *Maeander* (Captain Keppel) in February 1848, St. John found himself closeted with a boisterous party of young men, including James's younger nephew, Charles Brooke Johnson, Charles Grant, and Hugh Low. A botanist who had gone out to Sarawak in 1845 to collect exotic plants for his father's London nursery, Low had stayed on as James's interpreter and general factotum and was now going out as Colonial Secretary to the government of newly acquired Labuan. He and St. John were to become good friends.

For the next ten years St. John served as the Rajah's amanuensis, confidant, and adviser. He also became close friends with Brooke Brooke who had come to Sarawak in September 1848 to be trained as the Rajah's successor, progressively assuming responsibility for the government during his uncle's many absences. They explored the hinterland of the Sarawak River together.

Possibly at the Rajah's instigation, St. John in 1849 began to contribute a series of articles on the history of piracy in the region to the *Journal of the Indian Archipelago and Eastern Asia*, published by the Singapore lawyer, newspaper owner, and antiquarian, J. R. Logan. During the Parliamentary Commission of Inquiry into Brooke's activities held in Singapore in 1854, St. John was also one of the principal witnesses providing support for the Rajah's policy towards piracy. Significantly, he later observed that it was the early parliamentary criticism of him by Joseph Hume in 1850 which was the turning-point in the Rajah's life: 'Sir James Brooke was of a very excitable and nervous temperament. The savage attacks to which he was subjected roused his anger, and did him permanent injury. He never was again that even-tempered gay companion of former days. He thought too much of these attacks, and longed to answer every petty insult and calumnious insinuation.' Other observers, notably Bishop McDougall and Brooke's old friend John Templer, believed that it was the smallpox he contracted in 1853 which brought about a mental imbalance bordering on insanity.

Urbane and articulate, St. John's unorthodox views on religion and other subjects entertained the Rajah and the bevy of young men with whom he surrounded himself but alarmed the head of the Anglican mission, the Revd (and later Bishop of Labuan and Borneo) F. T. McDougall. Harriette McDougall, being more intellectually sophisticated, appreciated St. John's conversational skills and his apparently sympathetic support for their work. However, she was also to be alienated by his actions.

From his arrival in Kuching in 1848, McDougall

Bishop Francis McDougall. His missionary efforts in Sarawak were strongly criticized by Spenser St. John in *Life in the Forests of the Far East* (1862).

fought a losing battle against the prevailing custom for young European officers to take native 'keeps' or mistresses. St. John himself established a relationship with a certain Dayang Kamariah, a Malay woman of aristocratic ancestry, and had three children by her. She may well have been introduced to him by Hugh Low who had a relationship with her sister. Apparently encouraging his companions to follow his example, St. John was subsequently accused by the Bishop of 'corrupting' Charles Fox, a catechist who had been sent over from the Anglican theological college in Calcutta, and two government officers, Harry Nicholetts and Henry Steele. Letters detailing St. John's 'immorality' were sent to St. John's women friends in London by one of the Bishop's women teachers, Elizabeth Woolley (later Mrs Chambers), and when relations between St. John and the Bishop eventually broke down in 1862, the accusations were made public in the Anglican newspaper, the *Guardian*. Spenser's younger brother, James, who was employed as Surveyor-General at Labuan, also had a Malay *nonya* and numerous children but was persuaded by the colonial chaplain to marry her.

In 1855, St. John was appointed as British Consul-General to the native states of Borneo, a position which Brooke himself had resigned in 1853. Establishing himself at Brunei, he left Dayang Kamariah and their two children at Labuan under the watchful eye of his friend Hugh Low, who had by then installed Kamariah's sister as his *nonya*. He quickly attuned himself to life in the tiny, run-down sultanate which James Brooke had once pledged himself to restore to its former glory. As there was little official business to negotiate, St. John had plenty of time to visit Labuan and to make extensive explorations of the hinterland of Brunei.

Apart from early expeditions up the Brunei, Limbang, and Baram Rivers, he also accompanied Hugh Low on the latter's second ascent of Mount Kinabalu in April 1858. Possibly out of deference to James Brooke, he failed to acknowledge the first exploration of the Kayan country in 1847 by Robert Burns, a Singapore-based trader whom the Rajah ruthlessly excluded from any commercial involvement in north-west Borneo. In his biography of the Rajah, St. John was content to dismiss Burns as 'a disreputable Scottish adventurer'.

It was on a visit to Kuching with Dayang Kamariah and their youngest child in late 1858 (the two older children had died of fever at Labuan while he was away with Low) that St. John alienated Harriette McDougall. By presenting his *nonya* to the missionary's wife, he was in her view publicly flaunting a relationship which she regarded as shameful and wicked. Conveying the shock of this unexpected and insulting confrontation, she told their mutual friend Brooke Brooke that she felt 'nothing but disgust for him ... he had not a particle of shame'.

Another allegation of St. John's 'immorality' was made in the following year by Sir Robert MacClure, a naval officer who had visited Brunei in HMS *Esk*. He apparently reported to Brooke Brooke that Her Majesty's Consul-General not only went about openly with native women but *dressed in native costume*. Responding in a way that reflected his extreme sensitivity to accusations of this kind, St. John told Brooke Brooke:

Sir Robert sent over to say he was not coming to Brunei, so I started off on a fishing excursion. When I was starving in the jungle, during my last expedition, my men began making vows, and I joining in the affair said, that if we all got safely home, I promise to give you a feast and you may invite your families. Well on the day appointed they assembled and about twenty pulled away in a big prahu and I followed by myself in my own boat. I met Sir Robert as I shoved off, and I was properly dressed for a boat: clean white silk pyjamas, clean shirt, and clean white silk jacket. I know I had that dress because it was a Sunday when I received no native visitors. I was about to go down to Muara with a bull for the feast and plenty of fishing tackle. Now Sir Robert never saw any native women but I told him of the promise I had made. We might as well have been accused of going away tubaing [using poison to catch fish] as pulling about with native women.

St. John returned to England with the Rajah in late 1860, where he completed and published at his own expense the two-volume *Life in the Forests of the Far East*.... Drawing on his time in Sarawak, Brunei, and Sulu, he also had access to his friend Hugh Low's journals of their 1858 expedition to Mount Kinabalu, which is one of the features of the book. Although it was not a great commercial success, *Life in the Forests* received more than thirty reviews, most of them very favourable, and went into a second, enlarged edition in 1863. No doubt it enhanced St. John's reputation, helping him to obtain his next posting as British chargé d'affaires for the new black republic of Haiti in January 1863, where he was later promoted to Resident Minister.

St. John used the book to exact revenge on McDougall and some of the other missionaries. His final chapter on the Borneo Mission was highly critical of the Bishop and his methods, accusing him of high-handed treatment of his subordinates and maladministration. He was also critical of the Bishop's actions during the rebellion by the Chinese goldminers of Bau in February 1857, although the Rajah's desperate escape at that time had left the resourceful and courageous

McDougall as the hero of the hour. When the Anglican newspaper, the *Guardian*, published a letter from McDougall's principal assistant, Walter Chambers accusing St. John of keeping a native woman, the Rajah himself wrote a savage attack on McDougall. This was published as a pamphlet, *The Bishop of Borneo*, under St. John's name.

Writing to James Brooke's brother-in-law, the Revd Edmund Evelyn, in March 1863 after the publication of the pamphlet, McDougall described what he regarded as St. John's malign influence:

> The attack upon me in St. John's book was as unexpected as it was foul and false.... That man has had a sad and baleful influence in this place not only by prejudicing the Rajah's mind against Brooke [Brooke] and myself which I am sure he has done (I had words with him about Brooke as I had also with the Rajah when I was in England) but by a positive active infidel propagandaism he was ever carrying [on] amongst younger men whom he thought he could influence ... Poor Fox told me last time I saw him (he always came to me as his old friend when he was here) that not only had St. John excited his doubts about H.[oly] S.[cripture] and our Lord's Divinity, which made him leave the Mission, but that he afterwards never let him alone with his taunts and sneers, until he followed his bad example and kept a native woman....

St. John told Brooke Brooke that 'Fox left because the Bishop called him a fool and a presumptious schoolboy etc.'. McDougall's forthright and tactless manner certainly antagonized people but his absolute honesty could always be relied upon—which is more than could be said for St. John and the Rajah. To put it another way, they were politicians and the Bishop was not.

Harriette McDougall was bitter about what she

regarded as St. John's betrayal of the trust which she and the Bishop had placed in him during the early days of the Mission when there were so many problems—not the least being the unsuitable people sent out to assist. In May 1862, she told Brooke Brooke:

> At the time Mr St. John came in and out of this house like a familiar spirit, used to hear all the troubles and difficulties as they arose, and no doubt Frank often spoke to him as if he had been speaking aloud of their foibles in his disappointment. I don't say Frank was worldlywise in doing this but you must remember how very intimate St. John was with us, and his peculiar way of seeking confidences. How ungenerous is the advantage he has taken of it I can better feel than say.

Chambers' response was more pointed: 'The Arab abstains from the salt of the man he would strike', he told McDougall. '*He* ate your salt that his stab might go deeper. He warmed himself with your confidence [in order] to betray it.'

Although he did not come out in public support of St. John, Charles Brooke's sympathies were naturally with him in his quarrel with the Bishop. In November 1862, he wrote to Brooke Brooke:

> The mildest part of the business is that he [St. John] *only kept one* [mistress]—he [Chambers] may bring to light some of my little affairs some of these days—but he [St. John] made [an] allowance [to his family] for many years and I believe that was the right thing to do in a legitimate monogamy. These are the weak points a parson snatches at. It can't be helped—for society will excuse such natural indulgences. After all a Harem will be the proper thing to keep when one gets richer.

At some point before his death in England in 1868, possibly when they were together there in 1861, James Brooke entrusted his personal papers to St. John on the understanding that he would act as his literary

James Brooke, *c.*1864. St. John described him as 'one of the noblest Englishmen that ever lived'.

executor and official biographer. These documents St. John subsequently carried with him in tin trunks to Haiti and Peru, although they proved to be of limited value to his enterprise. As he told Charles Grant, most of the Rajah's interesting papers had been burnt during the Chinese Rebellion of 1857 and those that survived 'principally refer to those interminable and unfortunate negotiations with foreign powers which so hurt and wearied his friends...'. That St. John had assisted the Rajah in these negotiations, even to the point of being described as his 'alter ego', was something he preferred to forget.

In March 1871, St. John approached the Rajah's brother-in-law, the Revd Edmund Evelyn, for access to the papers of his nephew, Brooke Brooke, who had died not long after the Rajah. 'My object in writing the Raja's life', he told Evelyn, 'is to present him as I believe he deserved to be presented to the public, not a panegyric, but a true account.' He was also at pains to emphasize that he would not give a partisan version of the disastrous quarrel between the Rajah and his elder nephew which led to the latter's disinheritance and public humiliation:

> ... it is not my intention to touch in any way on family matters, to avoid every subject of controversy which would tend to bring in the name of anyone of his family now passed away. With regard to Captain Brooke himself, you might rely on my treating his memory tenderly, as he was once my most intimate and dearest friend ... you may trust to my fairness and my feelings not to make any use of [Captain Brooke's papers], that could in any way annoy those who hold his name in respect. I would be silent on the whole subject rather than to do so....

For some reason, St. John did not seek access to the hundreds of letters James Brooke had written to

Baroness Angela Burdett-Coutts, which were located by Owen Rutter and published by him in 1935. Nor did he make any mention of the Rajah's eccentric millionaire patron in the biography beyond a passing reference to 'a generous friend'. This was remarkable because her financial support had been crucial to the Rajah's survival and at one point he named her as his heir. Sir Steven Runciman has suggested it was typical of the desire for anonymity by 'A Lady Unknown', but there is surely more to be said about such an extraordinary omission. Perhaps St. John was disappointed that Baroness Coutts had not supported his grand scheme of establishing a new chartered company for the whole of northern Borneo, with himself as the managing director.

More importantly, the Baroness played a key part in ensuring that the Rajah did not transfer entire authority over Sarawak to Brooke Brooke before the disinheritance crisis of 1863. Reflecting on this after the event, the Rajah wrote to her: 'I tried about May or June 1860, to release myself and impose the responsibility of Government upon Mr. Brooke, but in vain as the Missus always was always opposed to it, and thus saved Sarawak from falling into the hands of a man who has proved himself to be half rogue, half fool.'

'The Missus', as the Rajah playfully called Britain's wealthiest woman, had also intervened to prevent a reconciliation between uncle and nephew. As St. John wrote in his biography, 'had mutual friends both in England and Sarawak been more conciliatory, the estrangement between uncle and nephew would never have gone as far as it did'. St. John, too, was one of those mutual friends.

Hearing of St. John's enterprise, Charles Brooke, who was proclaimed Rajah shortly after his uncle's

Baroness Burdett-Coutts (left) and her companion, Mrs Brown, 1864. Through her financial support for the Rajah, Britain's wealthiest woman played a decisive part in Sarawak's dynastic politics in the 1860s.

death in June 1868, had given a collection of his papers to Miss Gertrude Jacob with the view of producing a rival biography. No doubt he feared that St. John's book would be less than flattering to the Rajah and himself. These papers included James's original journals and his letters to his old friend John Templer, most of which had already been published by Keppel, Captain Rodney Mundy, and Templer. The latter had refused to co-operate with St. John, probably because he hoped to write a book about the Rajah himself.

Gertrude Jacob was the niece of Major-General Sir George Le Grand Jacob who had visited Sarawak at the Rajah's invitation in 1853 and formed a highly favourable impression of his achievements: 'During my stay in Borneo [Sir George wrote later] I observed that the Raja was regarded by [the Dayaks] as almost a superhuman being sent for their deliverance from the oppression of their fellow-men, while the Malays looked up to him as to a great chief fit to rule and guide them.'

It seems that Sir George persuaded his niece to commemorate the Rajah in a publication of some kind and that her efforts attracted the attention of Templer, who had been unable to fulfil his own ambition of writing the Rajah's life. He was able to revise the last chapters of her manuscript before his death, after which his widow gave her access to his papers.

St. John first became aware of this rival enterprise when Miss Jacob's work was published as a series of articles in the *Monthly Packet* in 1873–4 under the title 'The English Raja'. Unperturbed by this 'very slight compilation', he told Grant in December 1874 that he was not impressed by Miss Jacob's literary skills. When her book was advertised in 1876 he wrote: 'I am glad it is coming out, as it may give me some information about the early life [of the Rajah]. I have no fear of its interfering in any way with my book, as mine will be a real life of the Rajah in Borneo and not a series of extracts connected by intervening paragraphs....' As it happened, the book revealed very little of Brooke's early life and this remained a problem for St. John. On the positive side, Miss Jacob's extensive use of quotations from Brooke's writings persuaded him that this was something he should avoid. Her book was 'too long and wordy and too full of extracts to be readable'.

In May 1877, an event occurred which tested St. John's loyalty to the Rajah. In a speech in the House of Commons about the atrocities recently committed in Bulgaria, William Gladstone compared them with the 'battle' of Beting Marau at the mouth of the Saribas River in July 1849. On that occasion, more than 500 lightly armed Dayaks were killed by a combined Sarawak and Royal Navy force led by Admiral Arthur Farquhar and a staggering £20,700 in pirate 'head money' was subsequently paid out to the sailors by the Admiralty Court in Singapore. Writing to Charles Grant after reading a report of Gladstone's speech, Bishop McDougall complained of St. John's failure to speak up publicly in support of his former patron: 'I have been expecting you to open fire.... You know more about the thrice confuted lies against the old Rajah than any man, except St. John perhaps, though one can expect no good from him, and I s[houl]d not wonder if Gladstone got hold of him. Though he owed every thing to the Rajah he had not the pluck or the honesty to come forward like a man and vindicate his old patron's memory which he c[oul]d do so well if he chose.' St. John had been present at this overwhelmingly one-sided 'battle' and had given evidence about it to the Singapore Commission.

In July, McDougall wrote again to Grant after

Gladstone had elaborated on his charges in the columns of the *Contemporary Review*: 'It seems that St. John has been in communication with Gladstone. Why does he not stick up for his old Master who *made him & spoilt him*? ... I cannot help feeling savage with a varmint who first calumniated a man and then makes money out of his lies by defending them in a review which pays him well.' Ironically, it was the Bishop himself who came to the Rajah's defence in a letter to *The Times*, thus earning Charles Brooke's gratitude. If St. John communicated with Gladstone privately, as he had done during the Commission crisis in 1854, there is no evidence of it.

By May 1878, St. John had completed his manuscript. Writing to Charles Grant, he asked him to arrange for its publication by William Blackwood and Sons. The advantage of using the Edinburgh publishers was that they 'would not mind any attack on Gladstone' and were associated with the Conservatives whose foreign policy St. John strongly supported at that point. Leaving it to Grant to negotiate the terms of publishing the manuscript, he told him: 'I am not satisfied with it, but it is the best I can manage under the circumstances...'. Blind worshippers of the Rajah would not be satisfied, of course, but he had tried to 'present him as he appeared to me'. He would have preferred a rewrite, but 'another ten years might pass without it being completed'.

On the two main areas of sensitivity—the disinheritance of Brooke Brooke and the controversy over the Borneo Mission—he felt that he had written 'in a Christian spirit', although even here his reference to McDougall was hardly in that vein:

I have treated the Brooke incident in as delicate a manner as I could and I do not think that any one will be pleased or displeased with it. I could not avoid touching on it, after Miss Jacob's reference to it: in fact complete silence would have done more harm that good. You will not like my reference to the Bishop—but then I must have a hit at one of my enemies and he is the fattest and being alive he can defend himself.

While seeking Grant's assistance with its publication, the only changes he authorized him to make to the manuscript at this point were the rearrangement of chapters and the correction of proper names. 'I don't want any statements of mine altered', he told Grant, 'because I am responsible for them.' He hoped that Grant would be able to arrange for the publication of British Consul-General H. T. Usher's official report on Sarawak as an appendix. Some time earlier he had written to their old contemporary, Hugh Low, now Resident of Perak in Malaya, asking him to contribute a background chapter but had received no reply. Also to be appended was 'Hints to Young Out-station Officers from the Rajah', a brief set of notes by James Brooke which provide some useful insights into his philosophy of native government. Finally, St. John asked Grant to locate the portrait of the Rajah as a young man which had previously been in the possession of his sister, Mrs Savage, and arrange for 'an exact reproduction' to be made as an illustration.

The frontispiece of the book was to be a reproduction of the original portrait by Grant's uncle, Sir Francis Grant, RA, which was purchased by St. John in July 1877 and subsequently donated by him to the National Portrait Gallery. 'If I can bring my book out', St. John told Grant, 'your uncle's picture would add to its chance of success.' Painted in late 1847 when Brooke made his triumphant visit to London, it contributed greatly to his popular image as a dashing adventurer and patriot in the maritime tradition of Raleigh and Drake. An engraving based on the portrait was published in the *Illustrated London News* and was displayed in shop windows. Indeed, it remains one of the icons of British imperialism in its romantic mid-nineteenth century phase.

That Sir Francis Grant exercised his artistic licence to a considerable extent can be gathered from the remarks of one disinterested contemporary, J. H. Williamson, who met the Rajah at Mivart's Hotel in January 1848. On the fly leaf of his copy of Mundy's book (which also used the engraving), he wrote: 'The portrait in this book is not at all like him. He appeared about Fifty Years of Age, rather sligh[t] figure with most intelligent expression of countenance.' Brooke never properly acknowledged his debt to Grant, the leading portrait artist of his day, who had made him a present of the painting.

Charles Grant duly wrote to William Blackwood submitting the manuscript and telling him that there were some passages 'which would require to be modified. I mean some which are rather too penned [i.e., barbed].' Writing to St. John to request his permission to make the necessary changes, he told him:

> I think (with the reservations which follow) the biography is excellent, at any rate it interested me as a witness of the Rajah's work from 1845 till his death. I cannot judge so well whether it will interest the general public, though I *think* it will. But notwithstanding the omission of several paragraphs at the suggestion of a judicious friend, there are still several [references to] personalities which would require to be deleted or modified before I could undertake what you require. These might not *harm* the persons reflected upon, but they would give needless offence and assuredly mar an otherwise good biography.

John Templer ought not to be referred to as a 'clever, fussy little man', nor the first managing director of the Borneo Company Ltd as an 'irredeemable blockhead'; the Revd Walter Chambers' wife ought not to be ridiculed. However, 'the one almost insurmountable difficulty' for Grant in preparing the manuscript for publication was its treatment of Bishop McDougall who had sided with Brooke Brooke during his dispute with the Rajah: 'It is in fact not part of the Biography [he told St. John] but rather a defence of the chapter on missions in your former work and as it stands is an illustration of the truth that those who cast the first stone are the most aggrieved when paid back in their own coin.'

St. John agreed to most of Grant's suggestions with good grace but there were problems in accepting those relating to McDougall: 'If you read over your observations calmly, I think you will acknowledge that you have treated me with scant courtesy. I do not particularly care to revive old scores, but I have reasons for being dissatisfied with the Bishop of which you know nothing.'

There was also the problem of Mrs Chambers: 'If you saw [James] Brooke's letters about Mrs. Chambers you will find he calls her the impersonation of "envy, hatred and malice and all uncharitableness" and she did nothing but mischief.' Nevertheless, he gave Grant permission to make the changes: 'I always looked upon you as a man of the strictest honour, and I could not give a greater proof than trusting you with the revision of *my* view of the Rajah's life. In striking out anything you may think offensive to your friends, be careful not to weaken the impression I wish to convey—that the Rajah was one of the noblest Englishmen that ever lived.'

St. John also indicated that 'one judicious friend had advised me to say nothing disagreeable about Templer and the young Rajah: I would carry out that wish as far as possible.' Templer, who sailed with Brooke to China in 1830, had been his closest friend and confidant, and the enigmatic reference may suggest that their relationship was at least latently homosexual. Although James proposed marriage to a young woman at Bath and subsequently claimed an illegitimate son, his real affinity was for the young men with whom he surrounded himself in Sarawak. This is not a subject which a nineteenth-century writer could have been expected to broach, but it is a pity that James Brooke's most recent biographer, Professor Nicholas Tarling, has not seriously addressed it. What he describes as the 'strangeness' of the Rajah's relationship with Baroness Coutts (we have already noticed that he liked to call her 'Missus') has been attributed by Ronald Hyam to their prurient interest in young men. It seems more plausible, however, that she was both a wife and a mother figure for him. 'Missus' was also a way of acknowledging his financial dependence on her.

In mid-March 1879, Grant told St. John that after a delay of almost a year Blackwood's had agreed to publish the revised manuscript and that he would now have to consult their correspondence and his critical notes 'as I feel in having undertaken the revision a sense of considerable responsibility and difficulty. Anyhow you must not hold me responsible or ever name me, for if I use the pruning hook in the manner you have authorized, I shall be careful not to introduce anything. All that will be necessary in this respect I fancy will be some slight alterations where connecting links are necessary between expunged passages and modifications here and there.' Grant went about his work in a careful and judicious way, sending St. John many pages of detailed notes on the latter's responses to his original comments and on the way he had dealt with these in his editing.

An example of how Grant toned down St. John's treatment of McDougall can be seen in a comparison of the two following passages:

> The poor Bishop who though a splendid surgeon and physician had little useful knowledge of anything else, thought himself in another world and was speedily lost when he attempted to give us, for example, a popular idea of free will and predestination and tried to reconcile the two. Finding that we were not silenced by his authority he *growled*, so that we had to suspend our talks when he was there.

> Bishop McD (of whom I have already spoken) when he joined in the discussions did not by his arguments or his tone give encouragement to the enquirers.

Grant and St. John also had some correspondence about references to the second Rajah, Charles Brooke, with whom St. John had once been on friendly terms. Describing him as 'but a poor stick, and incapable of developing anything', St. John had earlier written that due to Charles's 'insufferable conduct and stupidity', he himself would not mind taking over the government of Sarawak. He was also critical of Charles's shabby treatment of Brooke Brooke's only surviving child, Hope Brooke, whom he had refused to support. In response to Grant's urgings, he allowed him to modify his remarks about the second Rajah, while continuing to damn him with faint praise: 'K[nox] very justly remarked in a note he wrote to me—that the Rajah would not have been pleased at my running down the present Rajah in any way, and as it would do no good, and might do harm, I wish you would take the sting out of any thrust at that gentleman. There can be no doubt, but that Sarawak is a success, and that it owes something to him.'

When St. John subsequently wrote: 'As we are all in a Christian frame of mind, soften anything disagreeable in my references to the present Rajah, or omit it as it would render future negotiations more difficult,' Grant noted that this was 'about the most difficult job that a man in South America could give to another at the antipodes'. Grant himself had cut off all communication with Charles Brooke in 1869 over the latter's refusal to support Hope Brooke, his brother's only surviving child.

One of Grant's excisions had more to do with the proprieties governing relationships between men during the late Victorian era than other sensitivities. In his description of his parting with the Rajah at Torquay in April 1867, St. John had written: '... as I bent over and kissed him I felt that it was for the last time. As I reached the door he called me back, kissed me again and I saw the tears falling, and I could see that he also felt that it was our last adieu.'

This might have been acceptable behaviour in the early Victorian era but an increasingly censorious attitude to men's expression of emotions meant that by the late 1870s it was *infra dig*. 'I fear your critics will show it to be contrary to British taste,' he told St. John. Scoring out the references to kissing as 'too sensational Nelsonic', and 'best left out', Grant was referring to an incident (familiar to every English school child) when the mortally wounded hero of Trafalgar called on the captain of the *Victory* to 'kiss me, Hardy'. He was also removing evidence of a genuine affection between St. John and the Rajah which was surely significant.

By 29 March 1878 Grant had completed his task.

Writing to St. John with details of his final revisions he concluded:

> Now I have done. Without having conceded your views regarding your many enemies, I have endeavoured to modify them that you may be saved the irritation of reviewers who would be quick enough to detect personal animosity—to the prejudice of biography and Author—and you must not abuse me if in doing my best, I have either failed in the task I undertook, or have used too plain language … the alterations may at first sight appear considerable, but in reality compared with the size of the task they are not so.… I have just read all your letters over again and see that no one man could have given another greater carte blanche than you have done—to alter, amend or omit—and I only hope I have not abused the confidence.

The one tinge of regret that Grant felt was about his treatment of St. John's references to Charles Brooke. He had since learnt that the Rajah, who had been staying for some time with his wife and two sons at his newly acquired property in Wiltshire, had seen 'little or nothing' of his relatives. 'Had I known sooner I might have been induced to "soften" less,' Grant confessed. 'But no harm is ever done in that direction.'

When the first proofs of the book became available in early July, Grant sent copies to St. John but there was insufficient time for the latter to make corrections. One amusing mistake consequent on this was the typesetter's misreading of 'Mrs' as 'Mr' so that St. John's warm description of Harriette McDougall was lavished instead on the Bishop: 'Mr McDougall exercised more influence over the European Inhabitants than any one before or since, it was a happy party until many years later when the demon of discord entered there.' To do him justice, St. John saw the humour of this. He was also properly appreciative of Grant's editorial efforts. After receiving the last proofs, he wrote to him: 'I cannot help repeating how obliged I am at the great trouble you have taken with the book—it must have been a labour of love. Your alterations have been very judicious, and no one will I think be offended, at least not much offended.'

Writing to Grant from Perak in November 1879 when the book had already appeared, Hugh Low complained that 'nearly all thoughts about Borneo [have been] driven out of my head by the very hard work and responsibilities and [illeg.] which have fallen upon me'. Consequently, he had been unable to contribute the background chapter St. John had requested, a response which hardly warranted the latter's sharp comment on the forwarded letter that it was 'what might be expected from so selfish a man'.

In January 1880, William Blackwood reported to Grant that the book was selling 'steadily but not in large numbers'. He thought that there was a possibility of a second edition and St. John subsequently welcomed this as an opportunity to show off his knighthood with a new title page. He had prepared a preface some time earlier against such a contingency. However, the second edition never eventuated.

Predictably enough, the book was the subject of a long and generally favourable review article by Alexander Knox in *Blackwood's Magazine*. A former leader writer on *The Times*, Knox had been an intimate friend and supporter of James Brooke, attending his funeral at Sheepstor in Devon in June 1868. Needless to say, he took a highly personal interest in the way the Rajah's life was portrayed. He observed that while St. John possibly knew him better than anyone else, this had resulted in a somewhat overcritical assessment of his achievement:

> He was deeply attached to the Rajah—nobody was better acquainted with his history, public and private—and no doubt it would have been impossible to find amongst his followers a more qualified man. We notice, as a characteristic of the book, and as showing how honestly Mr St. John has endeavoured to perform his task, that whilst full and ample justice is done to his noble qualities, the smallest foible of the Rajah is duly registered. The poor man was not clever at keeping accounts and double entry—he is not excused an error in vulgar fractions. He was over-indulgent to the middies and youngsters about him: one would almost imagine at times in these pages that we were reading the life of the great Mr Midshipman Easy.
>
> When the chief and the secretary differ on more important matters, the chief is in the wrong. After laying the book down we are really not sure which could give the other checkmate at chess. In a word, the Rajah was the Rajah, but Mr St. John was the 'guide, philosopher and friend'. We are glad to notice these little points, for they afford a fair presumption that Mr St. John has honestly endeavoured to give us a true picture of the man—weaknesses, errors and all. He has, in a word, painted the Rajah as Oliver Cromwell wished to be painted, but he has paid great attention to the wart.

According to Knox, the least reliable section of the book dealt with Brooke's time in Sarawak before 1848, notably his role in the Sarawak Malays' rebellion against Brunei and his operations against piracy. Dismissing St. John's account of the rebellion as reading 'like the story of a Christmas pantomime', Knox also rejected his criticism that the Rajah should have supported the rebels against Brunei rather than assisting Brunei against them. Glossing over the disputes with Brooke Brooke and Bishop McDougall and dismissing

Joseph Hume, Richard Cobden, and other parliamentary critics as 'two or three of the most persevering bawlers in the country', Knox was yet another victim of the Rajah's powerful charm: 'There was in the man to the last a something so gracious and "winning", that, in the old Roman phrase, he seemed to "play around your heart". You could have no harsh or unkindly thoughts in the presence of one who appeared not to know the meaning of the word.'

Although he was in a much better position to be aware of them, St. John failed to acknowledge the guileful and manipulative qualities of the Rajah and his capacity for cold vindictiveness. John Grant, laird of Kilgraston and Brooke Brooke's father-in-law, was understandably loyal to Brooke Brooke but his description of the Rajah as 'an unscrupulous liar' was not without some foundation.

The *Saturday Review* liked the book, observing that St. John had only attempted to 'give a general idea of Brooke's life ... a task rendered sufficiently arduous by many of the topics with which he had to deal'. Altogether, he had 'related with admirable simplicity, clearness, and vigour, the story of a life on which Englishmen of future generations will dwell with unalloyed satisfaction and justifiable pride'. Predictably, however, the topic which dominated the review was Brooke's acquisition of power and his campaign against the Borneo pirates. It was this that he was remembered for. The rest of his career, after all, was of progressively diminishing interest after the early heroic phase. As a freelance imperialist, his exploits had given rise to the use of the term 'Sarawaking' or 'Sarawhacking' to describe similarly unorthodox acquisition of territory by European freebooters in Asia, Africa, and the Pacific. They inspired Rudyard Kipling's parable of imperialism in his story 'The Man Who Would Be King' and

Kilgraston, Scotland, home of the Grant family. The son of a Scottish laird, Charles Grant's social background was more elevated than St. John's journalistic antecdents.

later novels by Owen Rutter, Nicholas Monsarrat, and George MacDonald Fraser in his Flashman series. Generations of English children came to know of Rajah Brooke through such classics as *The Boy's Book of Pioneers*.

St. John discounted the suggestion by his old enemy, the *Guardian*, that Gladstone might be upset by his criticisms:

> The *Guardian* is angry at my attacking Gladstone, but he is not angry, as he has written to me by this packet about the book. He remarks that 'I have myself written words about Sir James Brooke which may serve to show that the difference between us is not so wide as might be supposed, and I freely admit that what I have questioned in his acts has been accepted by his legitimate superiors—the Government and the Parliament.' He promises to examine the work with care; if he does so, I fancy that he will not be over pleased with some of the remarks.

He was also dismissive of the personal reactions reported to him by Grant, including those of the second Rajah who was said to have observed that 'it was written for the glorification of Mr St. John and made the hero of the Memoir contemptible': 'As I did not write my book to please Charles Brooke, I do not care for his opinion. You knew that he refused me all help, and pushed Miss Jacob to bring out her book before mine.'

He was more sensitive to criticism of his ability as a writer, telling Grant in January 1880 that the notice in the *Saturday Review* 'would comfort me under twenty such notices as that in the *Scotsman*'. The *Scotsman* thought that the book was

> ... undoubtedly the most complete and authoritative record of the Rajah's public career that has yet been given to the world.... From actual personal experience and observation, and from documentary evidence, he has been able to compile a

narrative of the Rajah's public career which may be accepted as accurate. It deserves, also, to be called impartial, for though Mr St. John is a warm admirer of Brooke's character and work, and though he enters into an elaborate vindication of those phases of his policy which aroused hostile criticism in this country, he never attempts to distort or colour facts, and, moreover, he never pretends that the Rajah was faultless.

The sting in the review was its reference to his lack of literary artistry:

It might have been thought a difficult feat to make the story of Rajah Brooke—a story as full of strange adventure, sudden turns of fortune, and thrilling incidents as any ever conceived by the most imaginative of fictionists—dry, prosaic, and uninteresting in the telling: but this feat Mr St. John has successfully accomplished. The art of word painting is not his. His book reads like a long Foreign Office despatch. He seldom attempts description, and when he does, though the reader may be sure that there is accuracy of detail, there is nothing more. Consequently it is not at all likely that this will be a popular or widely-read book, but people who are wishful to be thoroughly informed respecting the early history and present circumstances of Sarawak will hold it in deserved respect.

The Times, too, was less than enthusiastic:

We can hardly say that Mr Spencer [sic] St. John has made the most of a romance of modern adventure, but unquestionably, from the ample materials at his disposal, he has written an interesting life of a most remarkable man; and though he was the protégé and intimate friend of the Rajah, he has written it with very creditable impartiality.

Both newspapers took the opportunity to reflect on the bitter debate of the early 1850s about Brooke's policies and to conclude that right had generally been on his side. For *The Times*, he was another Robert Clive born after his time.

A little surprisingly, the book was ignored by other leading journals, such as the *Edinburgh Review* and the *Quarterly Review*, which might have been expected to take an interest, and St. John must have been privately disappointed. In his letter to Grant of January 1880, he rejected the suggestion made in two or three reviews that the book had been designed for his own self-enhancement: 'The reviewers are no doubt right when they say that the style is cold, but that I ever intended to place myself on a higher pedestal than the Rajah is absurd. I look upon him as the bravest, best and most noble man I have ever known familiarly, and certainly I do not place myself above the first footstep around his pedestal.'

Perhaps the relative dearth of reviews reflected disappointment of an expectation that St. John would glorify the Rajah. However, he thought it was more a matter of Sarawak having dropped out of the news: 'It must be confessed that my book was rather a failure. In fact after the death of the Rajah all interest in Sarawak ceased, as Charles Johnson is not a man to excite any sympathy. However, it was a duty done.'

As the *Saturday Review* pointed out, St. John's *Life of Sir James Brooke* is uniquely authoritative because it is the only biography written by someone who knew him and shared in his work. For the most part, it is a judicious and dispassionate account of his official career and achievements. As a protégé and friend, however, St. John drew a veil over the least creditable parts of the story—the Rajah's actions during the Chinese Rebellion of 1857, his disinheritance of Brooke Brooke, and his negotiations with a number of European governments for the sale of Sarawak. There are times, too, when St. John is at best disingenuous about his own crucial role, as in his reference to the memorandum he wrote to the Governor of Singapore, Colonel Oliver Cavenagh, about the Rajah's efforts to sell Sarawak. This was shown to Brooke Brooke by Cavenagh and precipitated his break with the Rajah. When St. John subsequently accused Cavenagh of 'treachery', he was only attempting to obscure his own duplicity.

No other reason is given by St. John to explain the failed relationship between uncle and nephew, although he and Grant were both aware of events as early as 1858 which led Brooke Brooke to distrust the Rajah. While giving Brooke's father-in-law, John Grant, a written assurance that Brooke was his heir, at the same time he employed Brooke to trace the man he claimed to be his illegitimate son and subsequently spoke of taking him out to Sarawak. Charles Grant himself put this down to the mental imbalance which the Rajah had first demonstrated during the Chinese Rebellion. Another example of St. John's economy with the truth is his account of the Mukah affair of July 1860 when Governor G. W. Edwardes of Labuan failed to prevent the Brookes from seizing the prosperous sago-producing area from Brunei.

One of the people best qualified to criticize the book was Paul F. Tidman, a Borneo Company employee who had served in Sarawak from early 1857. Together with his superior, Ludwig Helms, he was one of the few independently employed and minded Europeans in Sarawak who could write about the Rajah 'without fear or favour'. In a letter to Charles Grant in February 1883, he described St. John's book as 'untruthful as regards the [Chinese] Insurrection and the quarrel with Brooke [Brooke] in the first instance, by deliberate affirmation, in the second, by suppression'. However, in spite of encouragement to do so, he could not see the point of 'disposing of these falsehoods' in any easily forgotten journal article. He hoped to write his own book about Sarawak 'from the abdication of James Brooke', but nothing more appeared in print than his account of the Chinese Rebellion in Helms's *Pioneering in the Far East* published in the previous year.

Although *The Life of Sir James Brooke* was not a commercial success, it no doubt added a certain gloss to St. John's successful diplomatic career. In March 1881, he was knighted for his intervention in the war between Peru and Chile, and in May 1883 the Foreign Office sent him to Mexico to negotiate the re-establishment of diplomatic relations with Britain. In the following year, he was appointed Envoy Extraordinary and Minister Plenipotentiary to Mexico, serving there until 1893 and subsequently in Stockholm until his retirement to Camberley, Surrey, in 1896. In 1894, he had been made GCMG. Little is known about the last fourteen years of his life but it seems likely that poor health prevented him from travelling. He did not, for example, become an active member of the Royal Geographical Society which he had first joined in 1862. Instead, he occupied himself in editing some essays on Shakespeare produced by a relative and in writing two books of loosely fictionalized Borneo adventures under the *nom de plume* of Captain Charles Hunter, RN. He never could get Sarawak out of his blood.

In April 1899, St. John married Mary, the daughter of an Indian army officer, Lieutenant-Colonel Frederick Macnaghten Armstrong, CB, who survived him when he died on 2 January 1910. There were only brief obituaries in *The Times* and the *Morning Post*, and even the *Sarawak Gazette*, which should have made an effort, was content to reprint the latter of these. Not even the *Journal* of the Royal Geographical Society noted his passing. Altogether, for reasons which may by now be apparent, St. John seems to have had very few friends.

Like Hugh Low, however, whose personal life paralleled his own in many ways, he maintained links with his earlier family. After returning to England in late 1860, he apparently sent money to Labuan for their support. Dayang Kamariah subsequently went to Singapore to live with another European but returned to Labuan where she was the victim of a disgraceful persecution by the odious Governor John Pope-Hennessy in his vendetta against Hugh Low. According to Pope-Hennessy's successor, Governor Henry Bulwer, Kamariah 'was not a virtuous woman, but certainly not a bad woman'. Possibly as a result of Pope-Hennessy's efforts, she went back to Singapore and died there not long afterwards. Their one surviving child, Sulong, had been born in Kuching during the Chinese Rebellion and attended the missionary W. D. Gomes's school at Lundu. In 1866, St. John arranged for him to go to England for training as a civil engineer and then persuaded Hugh Low to find him a government job in Perak. What happened to him subsequently is not known.

St. John was an intelligent and determined man whose powerful ambition was tempered by an awareness of what he himself called his 'deficiencies in education and manner'. Keeping his emotions under tight rein for the most part, his odd demeanour led many people to regard him as a 'cold fish', although this would have been a much more accurate description of his former friend Charles Brooke. Censorious and humourless to a fault, St. John could never suffer fools gladly. Charles Grant, with typical generosity, told his mother in December 1860 before the disinheritance crisis that St. John had 'a very bad manner and is not what we may call "one of our set" but he is a true colleague of mine, a true friend with a good heart, kind in sickness and ready in need...'. Nothing demonstrated this better than St. John's devoted nursing of the Rajah and other colleagues in 1848 during a fever epidemic and again in 1853 when the Rajah almost died of smallpox. On the other hand, Grant described St. John as having a 'meagre knowledge of what is what in good society—i.e., he expects to be run after when he has taken the trouble to pay his respects'. Grant was, after all, the son of a Scottish laird and St. John the son of a London hack.

It was probably St. John's very manner which endeared him to the Malays, although his fluency in the Malay language must have helped a great deal. He was candid enough to say later that 'the good treatment I received in most parts of Borneo arose from the Rajah having called me his adopted son, and not because I was Consul-General'. From all accounts, he was highly regarded in Brunei, where he was on close terms with both the Temenggong (later Sultan Hashim) and James Brooke's old adversary, Indera Mahkota Shahbandar Mohammed Salleh ibn Pengiran Sharmayuda ('the Mahkota'), whom he described as 'the most talented man I met in Borneo'. Nevertheless, in his book he cast Mahkota as arch-villain in the drama which resulted in Brooke becoming Rajah in 1841.

St. John had a profound admiration for James Brooke, who seems to have been the one person ever to inspire his loyalty and love. He also acknowledged the Rajah's enormous influence over him when he admitted in 1863: 'I am very determined and confirmed in my opinions when alone, but when the Rajah is present I have little will of my own.' Despite all this, he wrote his biography in the measured style of a diplomatic acquaintance rather than a personal friend. St. John was aware of this coolness, telling Charles Grant in January 1880 that his official work had left its mark on him: 'The habit of despatch writing is not conducive to ornamental literary work, as we are expected to explain a subject in the fewest words possible. I wrote last year between 300 and 400 despatches, besides keeping up a correspondence with neighbouring Legations and my nine Vice Consuls. No wonder my hand refuses to write more.'

From his letters to Grant, however, it is clear that St. John was capable of expressing himself with some warmth when powerful negative emotions were dominant, as with his vindictive references to the Bishop and Mrs Chambers. He also wrote waspishly of the Brooke family who no doubt resented his intimacy with James: 'The Rajah has an unfair share of great and good qualities, as he appears to have deprived all his family of their portions, and left them ill-mannered, ignorant, selfish, grasping and ungrateful.'

From the viewpoint of patriotic and politically conservative contemporaries, St. John was not well cast in the role of biographer to the romantic hero of early Victorian Britain. Something more Carlylean purple was wanted. For later generations, his careful and measured opinions have proved more useful than the hagiography some of his critics would no doubt have preferred. It was in response to this expectation that he produced an abbreviated version of the book for Fisher and Unwin's 'Builders of Greater Britain' series in 1899. Nevertheless, his *Life of James Brooke* must itself be recognized as belonging to the genre of 'court history', enhancing as it does the historical reputation of the first White Rajah.

Whether it is possible to write a biography of James

Brooke which is anything other than an account of his official career seems doubtful. St. John knew him better than anyone else and yet was unable to penetrate his deeper personal motives. And even after his own prodigious work on the original sources over many years, Professor Tarling felt obliged to admit that 'hundreds, perhaps thousands of letters later, I find Sir James still elusive, still baffling, and there are many parts of his life which remain obscure'.

Regardless of historians' efforts, the popular legend of James Brooke which St. John helped to foster lives on, notably in Sarawak where this eulogy was recently penned by Ooi Keat Gin:

James Brooke arrived in '39
Chaos and bloodshed all around
Pangeran's misrule in country and town
In Brooke a saviour was found
With diplomacy and skillful tact
Restored order with peace intact
Brooke made an impressive impact
Sultan made him Sarawak's Prefect

Brooke faced troubled times ahead
Piracy and rebellion besieged his state
Took initiative 'fore it was too late
On his side stood God and Fate

Lanun roamed the Borneon seas
Plundering, raping and killing persist
Disrupting trade and many lives amiss
Shaking the young state with such 'disease'

The Royal Navy assisted in the fight
Piratical activities were soon put to flight
But back in Parliament trouble was in sight
Brooke was accused of using enormous might

Chinese rebels from Bau added to trouble
Kuching town was sacked and reduced to rubble

Brooke's Sarawak staggered and nearly crumbled
Timely rescue by brave Iban and Charles the able

A Malay plot threatened to overthrow the Raj
Lack of popular support did not achieve such
In spite of coup trusted Malay just as much
Through Malay counsel Rajah and the people in close touch

Solvency problems beset Brooke throughout his reign
Assistance from home came in trickles with appeals in vain
True friend in the Baroness alleviated this recurring 'pain'
Sheltered the young Sarawak from the financial 'rain'

Shattered spirit, tired and old
Health and age increasingly took their toll
The Rajah left for England bade farewell to all
Sadness befell Sarawak when the Rajah breath no more

Fremantle,
Western Australia,
December 1993

R. H. W. REECE

References

Baring-Gould, S. and Bampfylde, C. A., *A History of Sarawak under Its Two White Rajahs 1839–1908*, London, Henry Sotheran, 1909; reprinted Singapore, Oxford University Press, 1989, with an introduction by Nicholas Tarling.

[Brooke, J.], *The Bishop of Labuan: A Vindication of the Statements respecting the Bornean Mission, contained in the last chapter of 'Life in the Forests of the Far East'*, London, Ridgeway, 1862.

Cranbrook, Lord, 'A Note on the Appearance of Rajah James Brooke at the Age of 44', *Sarawak Museum Journal*, Vol. XXIX, No. 50 (December 1981), pp. 143–5.

Dictionary of National Biography, Supplement January 1901–December 1911, Oxford University Press, 1969, pp. 249–50.

Harrisson, T. H., 'Robert Burns: The First Ethnologist and Explorer of Interior Sarawak', *Sarawak Museum Journal*, Vol. V, Pt. 3 (November 1951), pp. 463–94.

Helms, L., *Pioneering in the Far East, and Journeys to California in 1849, and to the White Sea in 1878*, London, W. H. Allen, 1882.

Hyam, R., *Britain's Imperial Century 1815–1914: A Study of Empire and Expansion*, London, Batsford, 1976.

Irwin, G., *Nineteenth Century Borneo: A Study in Diplomatic Rivalry*, 's-Gravenhage, Martinus Nijhoff, 1955; reprinted Singapore, Donald Moore, 1965.

Jacob, G. L., *The Raja of Sarawak: An Account of Sir James Brooke, K.C.B, LL.D., given chiefly through letters and journals*, 2 vols., London, Macmillan, 1876.

Keppel, H., *The Expedition to Borneo of H.M.S. Dido for the Suppression of Piracy: with extracts from the journal of James Brooke, Esq. of Sarawak....*, 2 vols., London, Chapman and Hall, 1846; reprinted Singapore, Oxford University Press, 1991, with an introduction by R. H. W. Reece.

McDougall, H., *Sketches of Our Life at Sarawak*, London, S.P.C.K., 1882; reprinted Singapore, Oxford University Press, 1992, with an introduction by R. H. W. Reece and A. J. M. Saint.

Monsarrat, N., *The White Rajah*, London, Cassell and Co., 1961.

Mundy, R., *A Narrative of Events in Borneo and Celebes, down to the Occupation of Labuan: from the Journals of James Brooke Esq.*, 2 vols., London, John Murray, 1848.

Ooi Keat Gin, 'Sir James Brooke The First White Rajah of Sarawak', *Sarawak Gazette*, December 1991, p. 42.

Pope-Hennessy, J., *Verandah: Some Episodes in the Crown Colonies 1867–1889*, London, Allen and Unwin, 1964.

Reece, B., 'Two Accounts of the Chinese Rebellion', *Sarawak Museum Journal*, Vol. XLIII, No. 64 (December 1992), pp. 265–89.

Reece, R. H. W., 'European-Indigenous Miscegenation and Social Status in Nineteenth Century Borneo', in V. Sutlive, ed., *Female and Male in Borneo: Contributions and Chal-*

lenges to Gender Studies, Borneo Research Council Monograph Series, Vol. 1, Williamsburg (Va.), 1990, pp. 455–88.

Runciman, S., *The White Rajahs: A History of Sarawak from 1841 to 1946*, Cambridge, Cambridge University Press, 1960.

Rutter, O., *The White Rajah*, London, Hutchinson and Co., 1931.

_____, ed., *Rajah Brooke & Baroness Burdett Coutts, Consisting of Letters from Sir James Brooke first White Rajah of Sarawak, to Miss Angela (afterwards Baroness) Burdett Coutts*, London, Hutchinson and Co., 1935.

Saint, M., *A Flourish for the Bishop and Brooke's Friend Grant: Two Studies in Sarawak History 1848–1868*, Braunton (Devon), Merlin Books, 1985.

St. John, H., *The Indian Archipelago, its History and Present State*, 2 vols., London, Longman, 1853.

St. John, J. A., *Views in the Indian Archipelago: Borneo, Sarawak, Labuan &C, &C from drawings made on the spot by Captain Drinkwater Bethune, R.N.C.B., Commander L. G. Heath, R.N., and others*, London, Thomas McLean, 1847.

Saunders, G., *Bishops and Brookes: The Anglican Mission and the Brooke Raj in Sarawak 1848–1941*, Singapore, Oxford University Press, 1992.

Tarling, N., *Britain, the Brookes and Brunei,* Kuala Lumpur, Oxford University Press, 1971.

_____, 'Spenser St. John and his "Life in the Forests of the Far East", *Sarawak Museum Journal*, Vol. XXIII, No. 44 (July–December 1975), pp. 193–305.

_____, 'Sir James Brooke: a new biography. Some comments', *Sarawak Museum Journal*, Vol. XXIX, No. 50 (December 1981), pp. 137–41.

_____, *The Burthen, the Risk, and the Glory: A Biography of Sir James Brooke*, Kuala Lumpur, Oxford University Press, 1982.

_____, 'St. John's Biography of Sir James Brooke', *Sarawak Museum Journal*, Vol. XLI, No. 62 (December 1990), pp. 255–70.

Templer, J. C., ed., *The Private Letters of Sir James Brooke, K.C.B, Rajah of Sarawak, Narrating the Events of His Life, from 1838 to the Present Time*, 3 vols., London, Richard Bentley, 1853.

Wood, E., *The Boy's Book of Pioneers*, London, Cassell and Co., n.d.

Obituaries

The Times, 4 January 1910

Morning Post, 4 January 1910

Works by Spenser St. John

(i) *Published Works*

1847 'Tom Hunter's First Cruize', *Mirror Monthly Magazine*, No. 16 (November 1847).

1849 'Piracy in the Indian Archipelago', *Journal of the Indian Archipelago and Eastern Asia*, Vol. III, No. 3, pp. 251–60.

1849 'The Population of the Indian Archipelago', *Journal of the Indian Archipelago and Eastern Asia*, Vol. III, pp. 379–84.

1849 'Piracy and Slave Trade of the Indian Archipelago', *Journal of the Indian Archipelago and Eastern Asia*, Vol. III, pp. 581–8.

1849 'Piracy and Slave Trade of the Indian Archipelago', *Journal of the Indian Archipelago and Eastern Asia*, Vol. III, pp. 629–36.

1850 'The Piracy and Slave Trade of the Indian Archipelago', *Journal of the Indian Archipelago and Eastern Asia*, Vol. IV, pp. 45–52.

1850 'The Piracy and Slave Trade of the Indian Archipelago', *Journal of the Indian Archipelago and Eastern Asia*, Vol. IV, pp. 144–62.

1850 'The Piracy and Slave Trade of the Indian Archipelago', *Journal of the Indian Archipelago and Eastern Asia*, Vol. IV, pp. 400–10.

1850 'The Piracy and Slave Trade of the Indian Archipelago', *Journal of the Indian Archipelago and Eastern Asia*, Vol. IV, pp. 617–28.

1850 'The Piracy and Slave Trade of the Indian Archipelago', *Journal of the Indian Archipelago and Eastern Asia*, Vol. IV, pp. 734–46.

1851 'The Piracy and Slave Trade of the Indian Archipelago, *Journal of the Indian Archipelago and Eastern Asia*, Vol. V, pp. 374–82.

1854–5 'Reports of Brooke Inquiry', *Parliamentary Papers*, Vol. XXIX.

1862 'Observations on the Physical and Political Geography of the N.W. Coast of Borneo', *Journal of the Royal Geographical Society*, Vol. 32, pp. 217–33.

1862 *Life in the Forests of the Far East*, 2 vols., London, Smith, Elder and Co; reprinted Singapore, Oxford University Press, 1986, with an introduction by T. H. Harrisson.

1879 *The Life of James Brooke, Rajah of Sarawak, from his personal papers and correspondence*, Edinburgh and London, William Blackwood and Sons.

1884 *Hayti; or the Black Republic*, London, Smith Elder and Co.

1899 *Rajah Brooke: The Englishman as Ruler of an Eastern State*, London, Fisher Unwin.

1908 *Essays on Shakespeare and His Works*, London.

Under the pseudonym of Captain Charles Hunter, RN:

1905 *Adventures of a Naval Officer, edited by Sir Spenser St. John*, London, Digby, Long and Co.

1906 *Earlier Adventures of a Naval Officer*, London, Digby, Long and Co.

(ii) *Manuscript Letters and Dispatches:*

Brooke Papers, Rhodes House Library, Oxford, MSS. Pac. s. 90.
National Library of Scotland, Edinburgh.
Public Record Office, London, Foreign Office official dispatches.
Royal Geographical Society, London, correspondence.

Reviews of *The Life of Sir James Brooke*

Blackwood's Edinburgh Magazine, Vol. 127 (February 1880), pp. 193–210.
Saturday Review, 22 November 1879, pp. 633–4.
Scotsman, 9 October 1879.
The Times, 28 October 1879.

THE

LIFE OF SIR JAMES BROOKE

RAJAH OF SARĀWAK

FROM HIS PERSONAL PAPERS AND CORRESPONDENCE

BY

SPENSER ST JOHN, F.R.G.S.

FORMERLY SECRETARY TO THE RAJAH

LATE H.M. CONSUL-GENERAL IN BORNEO; NOW H.M. MINISTER-RESIDENT TO THE REPUBLIC OF PERU; AND AUTHOR OF 'LIFE IN THE FORESTS OF THE FAR EAST'

WILLIAM BLACKWOOD AND SONS
EDINBURGH AND LONDON
MDCCCLXXIX

"THE DYAKS HAD HEARD, THE WHOLE WORLD HAD HEARD, THAT THE SON OF EUROPE WAS THE FRIEND OF THE DYAK."

PREFACE.

Eleven years have passed since the death of Sir James Brooke, and it is only now that I am enabled partially to fulfil the promise I made him to write his life in Borneo. Long and wearisome illness, as well as continued active service in tropical countries, are the only excuses I can offer for this delay. I wish that I could feel satisfied with my performance of this friendly duty; but I have been somewhat oppressed by the fear of wearying the reader by repetitions of a tale that has been many times told.

I have endeavoured to be very exact in my relation of the facts contained in this biography, and if I have dwelt on many points that might have been but lightly touched on, recent discussions have proved the necessity, and shown how profound is the ignorance of even enlightened statesmen, when they venture beyond the beaten track. Notwithstanding all that has been written, but little is yet popularly known of Borneo.

I could readily have increased the bulk of this work by inserting the unpublished correspondence of Sir James Brooke, of which I possess sufficient to fill volumes, but I have thought it unnecessary. Documents and letters and journals have appeared, until the Sarāwak literature would now form a library in itself; and to these I must refer any reader who may desire to study in detail the work which Sir James Brooke did in Borneo. I propose only to give a general idea of his life.

One of the Rajah's oldest friends, Mr Hugh Low, promised to write me a chapter giving an account of the Court of Borneo as he found it in 1846, and another as he left that same Court in 1876, after thirty years of arduous and almost continuous service there; but a sudden summons to return to the East, to take up his new appointment as H.M. Resident in Perak, has deprived me of this important addition to my book. I hope, however, to be able to include it later on, should the public call for a second edition of this biography.

I have not attempted to sum up the character of the Rajah. I have described him as he appeared to me, and I leave the reader to form his own opinion: but as I have also formed mine, I may say that I consider him to have been one of the noblest and best of men.

CONTENTS.

INTRODUCTION.

EARLY LIFE OF JAMES BROOKE.

1803–1839.

CHAPTER I.

FIRST VISIT TO SARĀWAK AND TO CELEBES.

1839–1840.

CHAPTER II.

SECOND VISIT TO SARĀWAK—CIVIL WAR.

1840.

CHAPTER III.

WRETCHED STATE OF THE COUNTRY—BROOKE ASSUMES THE GOVERNMENT OF SARĀWAK.

1840–1843.

CHAPTER IV.

FIRST EXPEDITIONS AGAINST THE SERIBAS AND SAKARANG PIRATES—CAPTAIN KEPPEL.

1843-1844.

CHAPTER V.

EVENTS IN BRUNEI AND ON THE NORTH-WEST COAST—MURDER OF MUDA HASSIM AND HIS FAMILY—CAPTURE OF BRUNEI.

1844-1847.

CHAPTER VI.

MR BROOKE VISITS ENGLAND.

1847–1849.

CHAPTER VII.

RETURN TO SARĀWAK.

1848-1849.

CHAPTER VIII.

THE SERIBAS AND SAKARANG PIRATES.

1849.

CHAPTER IX.

THE BATTLE OF BATANG MARAU.

1849.

CHAPTER X.

MR GLADSTONE AND THE RAJAH.

1849.

CHAPTER XI.

PEACE MEASURES—ATTACKS ON SIR JAMES BROOKE'S POLICY—MISSION TO SIAM.

1849–1851.

CHAPTER XII.

SECOND VISIT TO ENGLAND.

1851–1853.

CHAPTER XIII.

RETURN TO BORNEO.

1853–1854.

CHAPTER XIV.

THE ROYAL COMMISSION—EXPEDITION AGAINST RENTAB.

1854–1856.

CHAPTER XV.

THE CHINESE INSURRECTION.

1857.

CHAPTER XVI.

THIRD VISIT TO ENGLAND.

1857–1860.

CHAPTER XVII.

LAST VISIT TO BORNEO—RETURN TO ENGLAND—QUIET LIFE—DEATH IN 1868.

1860–1868.

LIFE OF SIR JAMES BROOKE,

RAJAH OF SARĀWAK.

INTRODUCTION.

EARLY LIFE OF JAMES BROOKE.

1803–1839.

VERY little of interest is known of the early life of Sir James Brooke, beyond the facts that he was born in Benares in 1803, came early to England, and was confided to a fond though injudicious grandmother, with whom he had completely his own way. He went to Norwich school, then kept by Mr Valpy, but on his second essay there, got tired of the restraint and ran away. When he reached home, however, he felt somewhat ashamed of his exploit, and wandered about the neighbourhood and the garden till nightfall, when his grandmother, hearing of his presence, sent the servants for him. This incident is worth mentioning, as it accounts for many things in Sir James's after-career. Then a private tutor was engaged, who appears to have had little success with the wayward lad. On the return of his parents from India, another school was tried, but he would not stay there. This want of regular training was of infinite disadvantage to young Brooke, who thus started in life with little knowledge, and with no idea of self-control. He had been indulged and petted by all around him, as he had a lovable disposition; but in after-life he had often cause to regret his neglected education, and the imperfect discipline to which he had been subjected. He would often endeavour to defend this system of education, and argue that boys should not be thwarted; and certainly he carried his system into practice with all the lads that came under his control, and certainly also with very markedly bad results. Occasionally a bright lad, like James Brooke, will rise superior to a neglected education, but in the mass the want of culture must tell.

In May 1819 he received his first commission, and soon left for Bengal to join his regiment, the 6th Native Infantry. In 1822 he became Sub-Assistant Commissary General, a post for which he was totally unfitted. He now began to feel his neglected education, and from that time appears to have devoted himself to reading, and in later life it would have been difficult to have met with a better read man. He did not confine himself to one branch, but was familiar with the great works on theology, history, natural history; and after he entered the diplomatic service he was unwearied in the study of public law.

Had the choice of the branch of the army in which he was to serve been left to himself, he would have chosen the Cavalry; but the Commissariat gave him abundant leisure, and he spent a fair amount of his time in pig-sticking and shooting, and many a story I have heard him tell of his adventures.

Sir James was very fond of relating the following incidents of a hunt in which he was engaged. One day a native brought into camp the news that a huge bear had taken up his quarters in a neighbouring ravine, and a party was made up to go out and attack the monster. "Seven of us started; among others, one who was so blind that, without his glasses, he could not see ten yards. Our guide brought us to the almost dried bed of a stream, filled with boulders, and with huge rocks jutting from either bank. Presently he made a sign, and we all mounted on the top of a boulder, and then saw, about fifty yards up the ravine, Bruin lying on the bank. When we came in sight he rose, and seeing retreat cut off by the steep hill, immediately rushed at us, and began to clamber up the boulder. He was not ten yards off, and all but the blind man fired their guns at him. Scared by the volley, he turned and fled, when, the glasses being adjusted, the purblind let fly at him, and the bear rolled over, just as he was half-way up the left bank. 'His skin must be spoilt,' said one. 'I put two balls into him,' said another. But when they came to examine the body, they found but a single shot-hole—all but the purblind had been too flurried to aim correctly." He used to tell this as a warning to us not to engage in dangerous hunts without proper preparation.

At another time he went out with his brother to shoot tigers from the back of an elephant. There was a large party, and the beaters reported a fine Bengal tiger to be in the grassy plain before them. The elephants were arranged to advance as a crescent, the two points gradually joining so as to surround the beast.[1] Then, with a tremendous roar, the tiger arose and sprang upon the neck of his brother's elephant. The powerful beast, with one ponderous shake, sent the attacker flying in the air. Again and again the tiger returned to the charge, selecting always the same elephant, but ever with the same result. At one time his brother was in great danger; but at length an opportunity was given, and young Brooke sent a shot through its brain, and laid the royal beast dead at their feet.

[1] Compare with this the Zulu formation of which we have lately heard so much.

James Brooke was full of fun, and one of his pranks had a very useful result. In the encampment there arrived a major and his wife—the former a fine-looking, gentleman-like fellow, the picture of good-nature; the latter a pretty, little, demure woman. After a few days whispers went abroad that this good-natured major was in the habit of beating his wife, and people who went in the evening near to their house could hear the cries of the little woman, and "What a brute you are to use me so!" The major began to be treated rather coolly by all in the cantonment, but he took no notice. At last Mr Brooke said to a friend—afterwards Colonel Goldney—"I cannot understand Major So-and-so beating his wife; let us go and see." So one evening after sunset these two went down to the bungalow, and, hearing the cries, Brooke mounted on the shoulders of his companion and looked into the window. There he saw the major seated in a chair, good-humouredly defending himself from the blows which the little vixen was aiming at him, while she kept shrieking, "What a brute you are to use me so!" This was too much for Brooke's gravity: he burst out laughing, jumped down, and they both ran away. The story spread through the cantonment, and whenever the little woman passed she would hear, "What a brute you are to use me so!" till she became so ashamed of the whole affair that the cries ceased, and, let us hope, the beatings. It reminds one of the late Lord Derby's tale—"It pleases she, and it don't hurt I, so I lets her do it."

On the breaking out of the Burmese war in 1825, Brooke was sent to join the invading army which was to operate in Assam. His commissary duties did not occupy him much, so he mixed a good deal with all that was going on; and one day hearing the general in command complain that they had no light cavalry to act as scouts, Lieutenant Brooke immediately offered to raise a troop if he might have his own way. Leave was given him, and he promptly called for volunteers, and his call was responded to by those native infantry soldiers who could ride. He soon had as many as he could mount, and as they were all disciplined men, he had little trouble in drilling them into fairly efficient irregular cavalry. Being of an adventurous and daring temper, he used to ride ahead of the advancing column, work round the stockades, send information, and then push on. His force would too often scatter, so that it was generally very late when he returned to camp, but his services were fully appreciated. One day, while the infantry were attacking a stockade, the Burmese were observed collecting in a body in an opening to the left, and Lieutenant Brooke asked permission to charge them. Leave being given, he put himself at the head of his men, and, after a few inspiriting words, dashed on at the enemy. They fired a volley, but as this did not check the cavalry rush, they were bewildered, and after a brief resistance turned and fled, pursued by Brooke's irregulars until they were entirely dispersed. For this brilliant little affair he was publicly thanked.

A few days after, the general in command heard of a strong stockade being in front, and sent out Lieutenant Brooke to reconnoitre, but he was not able to return in time to prevent the advance-guard from falling into a sort of ambuscade. As the foremost company turned a corner in the road, they were received by a volley which knocked over a number of men. In the midst of the confusion, Brooke came galloping up, and putting himself at the head of the men, charged, and "foremost, fighting, fell." When the affair was over, and the enemy driven from their stockades, Lieutenant-Colonel Richards asked after Lieutenant Brooke, whom he had seen fall, and he was reported dead. "Take me to his body," was his reply, and they rode to the spot. "Poor Brooke!" said the Colonel, getting off his horse to have a last look at him; and kneeling over him he took his hand. "He is not dead!" he cried, and instantly had him carefully removed to camp. Not dead, it was true, but he hovered for weeks between life and death, as the bullet lodged in one lung could not then be removed. In these long dreary days, too weak to speak, to read, to do anything, he used, he said, to be absorbed in soft melancholy thoughts, and he often longed to be at rest.

At last he was considered to be strong enough to be removed, but he could not bear the movement of a carriage or a horse, so he was put into a canoe and gently paddled down the stream. Though other scenes had faded away, that one dwelt upon his memory—the gliding gently down a magnificent river, whose smooth stream carried him swiftly but gently to his destination. He could not speak much, but he could observe, and though in pain, every object he passed seemed to fix itself in his mind's eye.

On his arrival in Calcutta, it was soon found that Lieutenant Brooke's only chance of recovery would be from a voyage and a return home. He therefore obtained a long furlough, and a pension for his wound of about £70 a-year, which rendered him, as he said, independent for life. He left for home in August 1825.

The voyage, however, and a long furlough, did not restore him completely, so that his leave had to be extended until four years and a half had passed, and after five years' absence he would cease to be a member of the Bengal Army. His first effort to return to India was delayed by shipwreck; the second was a slow vessel, the Castle Huntley, which brought him to Madras July 18, 1830, just leaving him twelve days to reach Bengal. This was not possible in the days of tubs of sailing-vessels, so that he made this his excuse for throwing up the service. The fact was that he had formed many friendships among the officers of the ship, and they had fired his imagination with descriptions of the Eastern Archipelago and of China, and he determined to have a look at these countries. Among the officers were Sir J. Dalrymple Elphinstone, M.P., Mr James Templer, and Captain Jolly whom subsequently I had the pleasure of meeting.

James Brooke did not really care for the East India Company's service, and in fact was not suited to any post which required steady methodical work, and his position in the Commissariat he particularly disliked. What appeared to be a rash act was perhaps the best thing he could have done, for it opened a new career, for which his abilities admirably suited him.

Although occasionally inclined to have a light joke about the East India Company's Government in India, no one had a greater respect for those grand old officers who founded and carried on our rule in that country. Though formerly of a different opinion, of one thing Brooke lived to be thoroughly convinced, and that is, that no Asiatic is fitted to govern a country: under European guidance, yes—but alone, no.

Having thrown up the service, Mr Brooke sailed in the Castle Huntley for China, and visited Penang, Malacca, and Singapore, committing enough imprudences in his expeditions at these places to have cut off his career for ever. He escaped, however, with a few attacks of fever, which he accounted for by the climate in his letters to his mother.[1]

In China he had to suffer, as every man of honour suffered, from the sight of the bullying to which the English submitted, and had to submit, in order to pre-

[1] Some very interesting letters of this period of his career have been published by Miss Jacob in her 'Raja of Sarāwak.' (London, Macmillan & Co.: 1876.)

serve their trade; and this lasted until the first Chinese war taught, or rather ought to have taught, a lesson to that unteachable people.

I know of no man of the talent of Sir James Brooke who more enjoyed boyish fun than he did, and during his youth this propensity led him into many a scrape. I am afraid, however, that I do not remember sufficiently well the details of his more than usually madcap exploit, when he and a party entered Canton in disguise.

In January 1831 the Castle Huntley was on its way home, and without adventure it safely arrived in England in June. During this voyage Mr Brooke had been forming a plan as to his future. No longer in the Company's service, he felt that he must do something, and though formed by nature to shine in the most refined society, yet over-sensitiveness made him shun it. At this time of his life James Brooke had every qualification to fit him to take a brilliant social position; he was handsome, elegant in look as well as in manner, fond of the lighter accomplishments of music and poetry, so winning in his ways as to be beloved by all those he met, and full of ability, and with his friends brilliant in talk. Yet in general society he was reserved, and rarely gave sign of the power that was in him. This over-sensitiveness and shyness, which, though often concealed, were never lost, made him delight in the thought of wild adventure; and his companions being of congenial tastes, a plan was formed that, as soon as money could be found, a schooner was to be bought, and he and his friends were to embark in some wild schemes, which appeared however to them to be full of promise, and even of profit. Mr Thomas Brooke, the father, though far from being a clever man, was sensible enough to see the wild side of his son's project, and man of the world enough to understand the folly of his imagining that he could succeed as a trader.

Sometimes Mr Brooke thought of going into Parliament, but had not enough money to contest Bath; for, strange to say, with all his contempt for civilised life, he had become an ardent reformer, not only of Parliament but of poor-laws. In fact, he would have dived deep into the causes of national distress, and thought that an earnest inquiry would find a remedy—for every great cause which could advance the happiness of the people had always his hearty sympathy.

About this time he passed through that ordeal which punishes most men—he fell in love, and became engaged. What were the causes which induced the lady or her family to break off the engagement I do not know, but it was broken off, and Mr Brooke appeared to look upon it as final; and he from that time seems to have withdrawn from all female blandishments. He never spoke of it to us, though an occasional allusion made us think that his thoughts often reverted to this episode in his history. I notice in one letter it is stated that this young lady, whose name is given, died shortly afterwards.

I have said that at bottom James Brooke was full of fun; he delighted in telling us the following story. When mesmerism was occupying the thoughts of most men, and gaining many converts, Mr Brooke and a friend determined to be present at a *séance*, and accordingly went. The mesmeriser, a certain Dr ——, after a few preliminary experiments, informed the assembled company that there was a woman present so susceptible to the influence of mesmerism that when under it you might place her in any posture, and in that she would remain until he had freed her from its power. Forthwith he called for a woman who, under his influence, did various things, but at length became completely insensible, and was placed in the corner of the room in some fantastic attitude. The lecture went on, but the two young men continued hovering about this woman, until one of them said, "I am sure she is shamming." They pinched her, but she gave no sign. At length one of them gave her a tap, a strong one, on the top of her heavy bonnet, which, regularly bonneting her, so startled the woman, that she gave a shriek and darted at the offender, who rapidly mixed with the crowd. The whole humbug of the thing was thus so clearly shown that the lecturer was only too glad to escape into another room and allow his audience to disperse.

The disappointment occasioned by the breaking off of his marriage engagement made young Brooke the more anxious to be doing something; he therefore began to urge his father to lend him the money necessary to start his schooner scheme. His father was too sensible a man to encourage him, yet too loving a parent to disappoint him, so the money was found, and the brig Findlay bought, laden with merchandise, and Mr Brooke set sail for China with his old friends Kennedy and Harry Wright. Different views and different systems of management soon brought about disagreements. Kennedy evidently was a strict disciplinarian—Brooke was lax, and I can readily imagine, interfered with the captain. The end was that the brig and cargo were sold at a loss, and Brooke returned to England.

Here he whiled away his time in summer evenings in a little yacht, and later in the year in vigorous fox-hunting.

In the autumn of 1835 his father died, leaving him about £30,000. Within a few months he returned to his project of buying a vessel, but this time there were to be no partners—no one to interfere with and mar his projects. He soon found a yacht—the Royalist, a vessel of 142 tons burden—which pleased him, and he purchased her. To test the vessel and the crew before entering on any great enterprise, he undertook a voyage to the Mediterranean, and visited Spain, Malta, and the Bosphorus, enjoying his trip thoroughly, for he had read much concerning those countries, and wished to verify what he had read. He particularly loved to dwell upon his visit to the Alhambra; and in after-years I heard him describing to the Malays, in full detail, this wonder of wonders.

Mr Brooke was well satisfied with his vessel, and in the autumn of 1838 everything was to be ready for his departure to the scene of his hopes, the Eastern Archipelago. What were his objects? They were in reality uncertain. He was longing for adventure—longing to visit wild countries, and see what no civilised man had yet seen. Geographical discovery was one of his principal motives, and in search of that he could satisfy the yearning of his soul for the wildest liberty of action. The Royalist set sail from Devonport in December 1838, and had a prosperous voyage to Singapore, as far as weather was concerned; but discord set in on board, and, one after the other, the officers were dismissed. The discipline of the vessel was no doubt faulty, but it is an undoubted fact that there could be no discipline in any vessel in which Mr Brooke had sway. He was too kind-hearted,—could never endure to see men punished, however guilty they may have been; too ready to encourage the escapades of the younger officers by so readily excusing them. But at Singapore new arrangements were made, and everything was soon ready for a voyage of discovery.

At that time but a very superficial knowledge of Borneo existed, even in the best works, and the Admiralty maps were but guess-work. It was known that a Sultan of Brunei existed on the north-west coast, but who he was, and what were his territories, were not known. A kind act towards some shipwrecked seamen on the part of a Rajah Muda Hassim, living at Sarāwak, first drew general attention in Singapore to this place; and Mr Brooke arriving at this time with a very undefined plan of a voyage of discovery, was requested to convey to this Rajah letters and some presents. Here then was something definite to do, and a good reason for visiting an unknown country. In the next chapter I commence the life of Sir James Brooke in Borneo.

CHAPTER I.

FIRST VISIT TO SARĀWAK AND TO CELEBES.

1839–1840.

On the 27th of July 1839 the Royalist sailed from Singapore, and at length Mr Brooke commenced that voyage of exploration on which his heart had long been set, and which was destined to have so great an effect on his future career. The wind was favourable, and in a few days he found himself off the coast of Borneo, where he was received in the true fashion of the country, with rain, thunder, and lightning; but with the morning sun the clouds dispersed, and he saw before him, in all its beauty, the wondrous island of Borneo.

To the world at that moment Borneo was the land of the unknown,—in fact, the north-east coast had not been visited since the end of the last century, and that portion lying between Sambas and Maludu Bay had scarcely ever been visited by any European. All that was known of it was unpropitious: it was a favourite cruising-ground of the Lanun and Balagñini pirates; it was infested by Dyak savages, who, both pirates and head-hunters, destroyed the native trade, and rendered life insecure; and even the Malays were looked upon with some reason as not only lawless but inimical to foreigners.

Mr Brooke used to say that, as he leant on the taffrail,

he thought of all this; but he had confidence in his vessel, confidence in his crew, and still more he had that without which there is no success,—he had confidence in himself.

After a few days' surveying between the points of Api and Datu, he passed the latter, and then for the first time he saw the territory of Sarāwak. The bay that lies between Capes Datu and Sipang is indeed a lovely one. To the right rises the splendid range of Poé, overtopping the lower but equally beautiful Gading hills; then the fantastic-shaped mountains of the interior, including Matang, Singhi, and Paninjow; while to the left the range of Santubong, end on towards you, looks like a solitary peak, rising as an island from the sea, as Teneriffe once appeared to me when sailing by in the Meander. From these hills flow many streams which add to the beauty of the view, from the Lundu with its broad shallow entrance, to the Sarāwak river, which emerges into the sea below the lofty peak of Santubong. But the gems of the scene are the little emerald isles that are scattered over the surface of the bay, presenting their pretty beaches of glistering sand, or their lovely foliage, drooping to kiss the rippling waves; Talang-Talang, noted for their turtles; Sumpadien and the Satangs, in their wild, new, rich, cocoa-nut groves.

From Talang-Talang came the first Malays whom Mr Brooke met in Borneo. Their chief, Bandari Daud, was polite, and dressed, as usual with these chiefs, in dark cloth trousers, a dark-green velvet jacket, and a *sarong* round his waist, thrown gracefully over the handles of two *krises* which he wore at his girdle. Having conciliated his goodwill by trifling presents and much polite attention, Mr Brooke continued his surveys to the 11th of August, when he set sail and bore away for the mouth of the Sarāwak, which he entered without difficulty, and soon found himself anchored under the peak of Santubong in deep water and in a snug anchorage.

There is no prettier spot: on the right bank rises this splendid peak, over 2000 feet in height, clothed from its summit to its base with noble vegetation, its magnificent buttresses covered with lofty trees showing over 100 feet of stem without a branch, and at its base a broad beach of white sand, fringed by graceful *casuarinas*, waving and trembling under the influence of the faintest breeze, and at that time thronged by wild hogs.

From the entrance of the river Mr Brooke despatched a boat to announce his arrival to Rajah Muda Hassim, who, on the following day, sent down some of his chiefs to welcome the stranger, and invite him to visit his town. The river of Sarāwak presents few features of interest after passing the entrance: a low forest of mangroves, whose tangled roots are submerged during every high tide—clusters of the humble but useful Nipa palm, or uninteresting jungle; but as you approach the town of Kuching the distant hills become visible, and add variety to the scene.

On the 15th of August 1839 the Royalist reached Kuching, and the same morning Mr Brooke had his first interview with Muda Hassim. After the usual salutes Mr Brooke landed, and entered a huge shed erected on piles, which, though from the outside of a very rough and uninviting appearance, was, in the inside, of a somewhat different aspect. Most Malay rulers have one of these reception-halls attached to their houses, differing only in size and ornament according to their wealth and rank. They are built either of wood or of the hard stems of the Nibong palm; the walls and partitions are of mats, held together by strips of bamboo, while the floors are of the split stem of the same palm, covered with rattan matting. These houses are sometimes ornamented with a few painted boards, or a few hangings of red cloth; but in general they are very simple, and, when new, very neat. Such reception-halls are necessary to men of rank, as custom does not permit the stranger to enter the dwelling-house.

As Mr Brooke was the first foreign gentleman whom Muda Hassim had ever met, he welcomed him in state, surrounded by his chiefs and his thirteen brothers. The Borneon Rajahs dress on these occasions in a very becoming manner, though the jacket too often gives their costume a formal appearance, with its stiff unbending collar, embroidered heavily with gold thread; but the material is generally rich—either fine cloth, velvet, or silk; the trousers of the same material, gold-edged; and the *sarong* —often of the heaviest gold-cloth, or sometimes of the simple dark plaid so highly prized by the Malays—is a manufacture of the Lanun women.

This visit of ceremony was soon over; but Mr Brooke was much struck with the intelligent though plain countenance of Muda Hassim, with his elegant and easy manners; and on that day a friendship was commenced which endured until the tragic death of the Malay prince. As their intercourse became more familiar, greater confidence was shown, and Muda Hassim was very anxious to know some details about the Dutch, and, as between England and Holland, "which was the cat and which the rat."

In order to understand the following narrative, a few words of explanation are necessary. The Sultan of Borneo proper, Omar Ali, was at the period of Mr Brooke's visit the nominal ruler of the coast which extends from Cape Datu to Maludu Bay, though many districts had ceased to obey the central government. Sarāwak, the most southern province, had been goaded into rebellion by the ill-treatment of Pangeran Makota, the governor of Sarāwak; and Muda Hassim, uncle to the Sultan and heir-presumptive to the throne, had been sent down from the capital to restore order. He had hitherto failed in his efforts; and, on the arrival of Mr Brooke, the insurgents were fortified and in force about twenty miles above the town of Kuching.

Kuching, in the year 1839, was a very small place, and consisted of a few decent mat or wooden houses, erected on piles, surrounded by inferior huts inhabited by the poorer classes. The population was estimated at 1500; but these consisted principally of the followers of the Borneon Rajahs, and people attracted there by the little trade that was left. The real Sarāwak population did not consist of above 500.

After spending a week in Kuching, Mr Brooke obtained from the Rajah permission to visit some of the Dyak tribes; and his first expedition was up the neighbouring river of Samarahan. As no particular incident occurred during this trip, it is not necessary to dwell on its details. But on his return from Samarahan, Mr Brooke was enabled to visit the district I have mentioned, to the west of the bay, lying under the Gading hills, and named Lundu, where there then resided a Sibuyow chief, who remained a faithful follower of his until his death. He was called Sijugah, or the Orang Kaya Tumanggong, and ruled over a small section of his tribe, about fifty families, who had fled from their native country, on the river Sibuyow, to escape the attacks of the piratical Dyaks. However, Mr Brooke found that even here they were not completely safe, as a boom stretched across the river to stop the first rush of enemies' boats, and stockades of a very primitive character, were erected to defend their village.

I do not consider it necessary to describe in detail the countries or the tribes visited by Mr Brooke; but it will add a little completeness to this biography to give a very short general sketch of the Dyaks, as it was among them that so many years of his life were passed.

The Dyaks may be divided into two distinct sections—the sea and the land Dyaks—the former using large boats, and frequenting the salt water; the latter confined at that time to the interior, and seldom venturing within the influence of the tides, or, as they expressed it, to where the rivers ran back towards their sources. The latter were generally a peaceful, timid race; while the former were divided into piratical and honest tribes, brave, energetic, and generally independent of Malay control.

Tunggong, the residence of Sijugah, consisted of a single village-house, about 600 feet in length, built on piles, 12 feet high, with a broad enclosed veranda in front, and rooms in the rear for the married people. In this broad veranda most of the work is done; and here also are hung the skulls taken in war—unsightly trophies of their prowess. This description will apply, more or less, to every sea-Dyak village-house, although they vary in size and finish; but those of the land Dyaks are of a different form, and much inferior. Sijugah was a quiet little man, modest and reserved, noted for his courage, and surrounded by sons, of whom he had reason to be proud. The Dyaks are a small race, few passing the height of 5 feet 4 inches —in fact they are generally shorter. But they are very active and enduring, without being athletic; and although their noses are flat, mouths large, and their eyes small, yet the general expression is not unpleasing, though too often melancholy and subdued.

I have mentioned the boom across the river: it had lately been of infinite service to them. A short time before Mr Brooke's arrival, the Orang Kaya had been visiting a village on the sea-shore, in a small but swift war-boat, when, after sunset, on returning up his river, he became aware of the presence of numerous *bangkongs*, or Dyak war-boats, ahead of him, which were pulling quietly up the stream. He instantly suspected that they were the Seribas, come to surprise his village; so he advanced cautiously, and as his boat and men could not be distinguished at night, passed in among the enemy, and gradually worked his way to the front. He was not hailed, as no one spoke in the fleet. He counted 73 *bangkongs*, which would contain about 2000 men. When he had cleared the leading boat, his people could not be restrained: they rose, gave a yell of defiance, and darted off for home. The sound of the paddles and the loud shouts roused the village, the opening in the boom was made ready, and the gallant chief dashed through with his boat, just in time to escape the enemy, who, furious at being thus outwitted, had pulled after him with their swiftest boats. To rush into the little battery that protected the boom was the work of an instant; and a discharge into the mass of the enemy, who were already trying to sever the fastenings, checked the advance. The sound of guns alarmed the neighbourhood, the few Malays hurried to the defence with their muskets, and the enemy drew off, as, their surprise having failed, they could not risk remaining to attack the place in force. This was the sort of life our brave Orang Kaya Tumanggong led.

Mr Brooke was exceedingly well received by this tribe, and it ever remained a favourite with him, their loyal chief meriting every confidence.

After a sojourn of ten days among these Sibuyows, Mr Brooke returned to Kuching, and entered into an unrestrained intercourse with the Rajah Muda Hassim and the other Borneon chiefs. Malays of high rank are generally very gentleman-like companions, whose manners never offend; but, as a rule, they are too ignorant to be entertaining, and visits are rarely enlivened by any pleasant conversation. Muda Hassim, though polished and kind, was no exception to this rule; but he had near him a chief who quite made up for his shortcomings, and who was one of the most companionable of men. This was Makota, the ruler of Sarāwak, whose tyranny and mismanagement had brought about that rebellion which prevented Mr Brooke from visiting the interior. He could both read and write, was the author of some pieces of poetry, and had also composed several of those poetic sayings called *pantuns,* so popular among the Malays.

Makota was a short, stout man, with an ugly but not unpleasant look; he was cheerful, talkative, ready to conform to the society in which he found himself, whether that of gentlemen or of rough Durham miners. He was the most talented man I met in Borneo; but his talents proved of little service to himself or to those who employed him, as he either ruined the country he attempted to administer or drove it into rebellion.

Makota now regularly visited the Royalist, partly to talk over the future trade of the country, partly to discover the real object of Mr Brooke's visit to Borneo. It was natural for the Malay chiefs to doubt whether any man would give himself the trouble to make so long a voyage, at so great an expense, merely to explore a country, survey its coasts, and collect specimens of natural history. They expected every moment to hear that Mr Brooke was the agent of the British Government, or at least the chosen envoy of the Governor of Singapore. As the acknowledged clever adviser, Makota was sent on board to worm out the secret, and when he could discover nothing, instead of believing Mr Brooke's assurances that his was a private voyage, he only acknowledged that he had discovered a cleverer diplomatist than himself.

However, no jealousy was shown by the Rajahs, and permission was now freely given to their English visitor to go wherever he pleased, and he was pressed to return again to the country, after having filled up provisions in Singapore. Before his departure, however, he visited the river of Sadong to the east of Sarāwak, and there made the acquaintance of Sirib Sahib, an Arab adventurer, who was afterwards an important actor in many events in which Mr Brooke was concerned. Among those whom he also met here was Datu Jembrang, an old Lanun pirate chief, who, having made a small fortune, had retired to spend his days in dignified idleness. He said to Mr Brooke, "I once met your countrymen before." "Where?" "Oh," he answered laughingly, "not very near. In 1814 I had put into Sambas with a squadron in which we were out cruising, in order to do a little trade with the Sultan of Sambas, when it was announced that the English fleet was off the river preparing to attack. Your ships were too large to come in, so you only could send your boats. Our Lanun squadron offered to aid in the defence, and we drew up in line behind the boom, protected by the batteries on shore. On came your boats; we all opened fire on them, and after a severe action we drove them off. I never saw men fight so gallantly as your countrymen, but they were overmatched." This story was true of our first attack: in the second we took Sambas, but in the meantime the Lanuns had left. Datu Jembrang added: "The officer who commanded the attack was so humiliated by his defeat that he committed suicide."

Mr Brooke had heard during his stay in Borneo of the savage tribe of pirates and head-hunters, the Seribas, but he little imagined that before his departure he should witness a specimen of their handicraft.

The Royalist was lying at the Moratabas entrance of the Sarāwak river, and half a mile from a rock, a lovely natural watering-place, where a tiny but pure stream trickles over the surface of the moss-covered stone, in quantity sufficient for a delicious bath, and enough to supply the native shipping. A Malay chief had anchored his boat inshore, waiting the morning ebb-tide to pilot the English vessel out of the river; it was a dark night, when the watch on board of the Royalist heard loud shouts on shore, "Dyak! Dyak!" from a dozen voices. Mr Brooke rushed on deck, guessed immediately that the pirates were attacking the Malays, ordered a blue light to be burnt, a gun to be loaded and fired; and the boat being launched, he sprang into it with a willing crew, and pulled off for the rock. The moment the blue light was burned and the gun fired all sounds ceased; the Dyaks fled, but not before they had severely wounded half the Malays in the pilot-boat. The dark night prevented anything being seen, and Mr Brooke could now only bring the wounded on board and place them under the surgeon's care. This circumstance delayed his departure, as he was forced, by the kindness of the messages from Muda Hassim, to return to Kuching in his gig; and he did so with more pleasure, as it afforded him an opportunity of seeing something of the interior life of a Malay Rajah. He was received with every hospitality, and, after some delay, served with what was meant to be an English dinner—very good in its way, for Malay cookery is often very tasty—an incident only worthy of being mentioned on account of the conduct of Muda Hassim, who insisted himself on superintending the feast, changing the plates with his own royal hands, uncorking bottles, proud to show his guests how well he could receive them—moved in fact by a genuine, hospitable, kindly feeling, which on similar occasions I have often observed among Malays of every rank.

In the evening Mr Brooke returned to the Royalist and set sail for Singapore, where he arrived without an accident. The merchants were delighted with his success, as it promised to open out a new field for their exertions; but the authorities were not so well pleased, being fearful of occasioning some complications with the Dutch—a natural though not a dignified fear.

Mr Brooke had every reason to be satisfied with the results which attended his first voyage of exploration.

He had roughly surveyed about 150 miles of coast with sufficient accuracy to show that the received maps were utterly incorrect. He had visited many rivers, some before unknown even by name; he had established a friendly intercourse with the Malay chiefs of the coast; and had been able to spend ten days among a Dyak tribe. He had obtained much information, which, on the whole, subsequent inquiry proved to be fairly correct, though no man was more convinced than himself that first impressions and notes made by travellers who only spend a few days in each place are rarely valuable, except so far as they relate to outward objects, and treat simply of the appearance of the country.

Though fully decided to visit Sarāwak a second time, Mr Brooke considered it more prudent to defer his voyage until the setting in of the fine monsoon, when it might be hoped that the termination of the insurrection in that province would afford him the desired opportunity of studying the interior of the country. But not to have to idle away his time in Singapore, he determined to carry out another object of his expedition, which was to visit the fantastically-shaped island of Celebes. The inhabitants, generally called Bugis, are noted for two qualities—their love of trade and their great personal bravery,—qualities which cause them to be respected and welcomed in every country they visit or in which they settle.

After a six weeks' stay in Singapore to refit and recruit his men, he set sail on the 30th November 1839 for Celebes, and, after a tedious voyage of upwards of three weeks, cast anchor in Bonthian Bay. Here, with his characteristic energy, he ascended the mountain of Lampu Batang, being the first European to set foot on its summit. He then continued his voyage, entering the deep but little known Bay of Boni. Among the states which then occupied its shores were those of Boni, Waju, Lawu, and some lesser ones of wild natives.

These proved a highly interesting people, brave, but wasting their energies in local quarrels and civil wars, instead of uniting against their common enemy, the Dutch. Here Mr Brooke spent his time in congenial pursuits, devoting it to the study of the people, their institutions, and their country, but filling up his leisure hours in the lighter pursuits of hunting and shooting, more for specimens than for sport. As a good horseman, accustomed to follow the hounds in England, he was ever in front, dashing after the deer with the noosed spear in hand, with the nerve and the dexterity of the most noted native hunter; or joining in their shooting matches, and astonishing the Bugis by the rapidity and precision of his fire. As a practical shot he was the best I ever met.

Mr Brooke being the first Englishman who had come in his own yacht to visit them, was treated with great distinction, and, notwithstanding his steady denials, was looked upon as a sort of observer for the English Government. His pleasant, genial manner gained on his hosts, and he was soon treated as an old friend. The ladies of rank crowded around him, talking and laughing with the most perfect ease, indulging in many an innocent freedom, and examining with scrupulous attention his various articles of toilet,—and Mr Brooke was particularly natty and elegant in all his personal surroundings. The chiefs soon grew confidential, and begged for advice as to how they were to act to settle their quarrels, and what policy they should pursue towards the Dutch. His answer to the latter was invariably the same, that if they desired to preserve their present comparative independence, they should avoid all cause of quarrel with their powerful neighbours, as a war with them could only end in one way, and that was in their absolute humiliation. As to the former he gave them much good advice, and although it was not rigidly followed in subsequent years, it used to please him to receive news from his Bugis hosts, and to feel that his voyage there had not been without benefit to them.

These Bugis states afforded curious studies: Waju, with its self-elected aristocratic council, tempered by powerful tribunes chosen by the people; Boni, nominally the same, but really under the despotic rule of one chief; Lawu, given up to disorder. All these Bugis states were, however, invariably in decay; and although one regrets to see independent communities extinguished, they are so utterly demoralised, so useless, that nothing but the guiding hand of a civilised state can restore them to prosperity. And this holds good whether the scene be laid in Western Asia, in India, or in Africa: no native state contains in itself the elements of regeneration.

Mr Brooke, new to the Archipelago, was not then convinced of this truth, and thought that a little notice from English authorities, and a little gentle pressure, would draw forth the resources of those rich countries.

An account of his visit to the King of Boni will give an idea of the power that these people still retained over their followers. And these were no slaves, but brave men, who had gallantly fought the Dutch, and won the respect of their civilised enemies.

Mr Brooke landed at Bajuè with ten companions, and mounted the horses they found ready for them: then winding their way over the cultivated plain, in about an hour they found themselves at the nominal capital. It had formerly been a large town; but during the last war with the Dutch it had been burnt to the ground, and although some good houses had been erected, it still was a capital but in name.

On approaching the palace, several thousand Bugis were perceived armed at the gates, dressed simply in trousers, the *sarong*, and a skull-cap,—in fact, being naked to the waist; a quiet orderly crowd, which remained silent while the strangers passed. At the gate of the courtyard stood ten spearmen, clad in coats of shining mail, or rather bright chain-armour, similar to that worn by some of the Lanun pirate chiefs, perhaps imitated from some specimens captured during their early wars with Europeans.

The King of Boni sat at the head of a table in a moderate-sized hall, surrounded by his nobles and body-guard, except on the left, where seats were placed for the strangers. He was plainly dressed in a long robe of English chintz, with gold buttons, a *kris*, and black skull-cap. A good-tempered, cheerful man, he entered readily into conversation, and was deeply interested when he found that the Englishman had visited Roum, or Constantinople, the Sultan of which is considered by most Mohammedans as their spiritual, almost their temporal, chief. They have an exaggerated idea of his power, and imagine him the greatest sovereign in Europe—indeed, in the world.

These chiefs, like other Asiatics, have but a vague notion of history and geography; but all have heard of Napoleon, and they often made inquiries of Mr Brooke about this great chief. In fact, when I lived in Borneo, I was often surprised to hear how interested the natives appeared to be in his life and victories, and one day I asked the Sultan of Brunei why he delighted to talk so much on that subject. He smilingly pointed to a very old man, and said that he had been in Egypt when the French army was there. On inquiry, I found that this Haji had been with the Turkish hosts at the defeat of Heliopolis, and was in Egypt also at the surrender of the French army to Abercromby. Thus the name of Napoleon was familiar to these Rajahs, who eagerly listened to all the old man's stories, which, in great part, I could readily confirm. It indeed appeared like living in another time to talk with a Malay who had fought at Heliopolis.

CHAPTER II.

SECOND VISIT TO SARĀWAK—CIVIL WAR.

1840.

THE second visit made by Mr Brooke to Borneo is, to my mind, not only one of the most important events, but one of the most romantic, even, in his romantic career. I wish I could render it in his own bright, energetic, and brilliant manner, as I have heard him tell the story during the twenty years of our familiar intercourse. I know the ground well; there is not a spot that I have not visited, and there have I heard the wondrous story in all its curious details.

When Mr Brooke arrived at Kuching in the month of August 1840, he found affairs in very nearly the same position as when he had left it in September 1839. Something had been done, but not much. The Orang-Kaya de Gading, a hard old man when I knew him, had been sent down from Brunei by the Sultan, nominally to urge Muda Hassim to exert himself, but really to watch what was occurring in Sarāwak, as the visit of an English ship to that province had somewhat disturbed the authorities in the capital. The consequence, however, had been that most of the wretched land Dyaks had come over to the Borneon party, thus lessening the opportunities of the insurgents for obtaining provisions. However, it was confidently believed that the Dutch-protected Sultan of Sambas favoured the rebels, and sent them small supplies of arms and ammunition. In fact, wearied of the struggle, and despairing of success, the native Sarāwak Malay chiefs had offered to renounce allegiance to the Sultan of Borneo, and to place themselves under the protection of their neighbour.

The insurrection, however, by the defection of the Dyaks, was now circumscribed, and only really occupied the ground held by its forces. Their fortified position was called Siniāwan, situated about twenty miles by river from Kuching. Aided by the Dyaks, the place would have proved impregnable to the Borneon forces; but deserted by them, it was exposed on the land side to constant attacks and surprises. Ascending the river, there lay to the right a little knoll, on which was erected a fort or strongly stockaded house, called Balidah; on the left rose the picturesque and steep mountain of Serambo, inhabited by three Dyak tribes, who had lately declared against the insurgents; at its base lay the village held by the Sarāwak Malays, whose fighting men were estimated at between 400 and 500.

After passing about a month at Kuching studying the situation of affairs, Mr Brooke at last acceded to the petitions of Muda Hassim, and resolved to aid him in bringing to an end this wretched warfare which was desolating the country. He felt that without the active aid of Muda Hassim the cherished object of this voyage—that of exploring Borneo—would be completely defeated. Muda Hassim had decided not to leave Sarāwak until the rebellion was over, otherwise his power and influence at the capital were lost. He had active enemies there, and even the half-witted Sultan viewed him with jealousy, being then completely under the influence of his clever half-brother, Pangeran Usup, who was naturally hostile to the legitimate branch of the family of the ancient Sultans of Borneo.

Whilst Mr Brooke was still weighing the reasons which now urged him to continue his voyage, now to stay for the solicited aid of the Rajah, the former for the moment gained the mastery, and he intimated to the Malay chief his resolution to depart, and in the evening he went to the palace to take his formal leave; but his heart smote him when he looked at the mournful countenance of his new friend, and he told him that he would remain if he saw the slightest chance of this war coming to an end. Now Muda Hassim's countenance brightened, and he hastened to assure him that he had the best of news to give; that their united fortunes would overcome all obstacles; that the land Dyaks had sent to him to entreat his forgiveness, with full promises of joining with all their forces when assured of his pardon. In fact, he continued, "the end is approaching; I have promised forgetfulness of the past, and now, with a little aid from you, all will go well. The Dyaks are rapidly coming in; their messengers are already with the army, and to-morrow the last of the chiefs will join us."

This determined Mr Brooke, and he promised to stay a few days longer, and even to go as far as Ledah Tanah, about fifteen miles above Kuching, to obtain positive information on the subject. He felt that the situation of the insurgents, if this news was true, must now be desperate. The Dyaks, deserting their cause, would cut them off from all supplies; the three tribes of the Serambo mountain looked down upon the fortified Malay village of Siniāwan; the numerous tribe of Singhi prevented them foraging the country on the left bank; the tribes in the further interior were either neutral, or lent some assistance to their enemies. They must now either surrender or cut their way through their opponents, and retire to foreign territory.

Fortune seemed to smile on Mr Brooke's resolve to

stay and aid Muda Hassim: detachments of Malays and Dyaks arrived from the coast; 200 Chinese came over from Sambas, animated by a desire to revenge the cruel treatment which they themselves had received at the hands of the Sarāwak Malays.

On the 18th of October Mr Brooke started for Ledah Tanah, or the tongue of land where the two branches of the Sarāwak river meet, and found the grand army encamped there under the protection of a stockade, erected to guard against surprise. At the moment of his arrival a column was starting to confirm the allegiance of the Serambo Dyaks, hold the villages on that mountain, and secure them against the Sarāwak Malays, who, pressed by hunger, would, it was feared, make desperate attacks on them in order to procure food.

Mr Brooke began now to appreciate the difficulties of the task he had half promised to undertake. He found Makota, the commander-in-chief, as he ever was, full of the most elaborate and clever schemes to circumvent the enemy, but as unwilling as incapable of taking a decisive resolution. When urged to make even a false attack, in order to divert the attention of the enemy from the column that was advancing on Serambo, to secure the most important posts, he would do nothing—he would wait: and even when a white flag, the preconcerted signal, was seen flying on the summit of the mountain—no, he would do nothing; he feared that the column was surrounded by the enemy.

The news that Muda Hassim had communicated to Mr Brooke as to the desertion of the Dyaks from the Siniāwan Malays now received practical confirmation. The chiefs of the tribes, surrounded by their warriors, came in looking worn out with hard work and want of food; indeed they frankly confessed that it was starvation that had induced them to abandon their former allies. The conditions they demanded were forgiveness of the past, and an assurance that none of the Seribas or Sakarran Dyaks should be employed in subduing Siniāwan, as they were, they said, "hateful in their eyes." These terms were readily granted, particularly the latter, as Makota was well aware that the piratical Dyaks were then fully engaged in their attack upon the district of Lingga.

As the new arrivals were not only thoroughly acquainted with the country, but also with the enemies' defences, their information was really valuable, and a council of war was now held, at which Mr Brooke assisted, to determine what was to be done. Makota was the greatest speaker, and full of projects, always impracticable; the other Malay chiefs—lazy, stupid, utterly uninterested in the war—were most unwilling to expose their persons to a chance shot. But the resolute counsels of the Englishman had their effect, and a movement in advance was determined on for the morrow. The stockade was to be pulled down, and transported to a position as close as possible to the enemy's chief fort of Balidah.

After this first specimen of a council of war, Mr Brooke thus reasoned with himself as to his motives in becoming a spectator of, or even a participator in, this scene: "In the first place, I must confess that curiosity strongly prompts me—since to witness the Malays, Chinese, and Dyaks in warfare is so new, that the novelty alone might plead as an excuse for this desire. But it is not the only motive; for my presence is a stimulus to our own party, and will probably depress the other in proportion. I look upon the cause of the Rajah as most just and righteous, and the speedy close of the war would be rendering a service to humanity, especially if brought about by treaty. At any rate, much may be done to ameliorate the condition of the rebels in case of their defeat; for though I cannot, perhaps ought not to, save the lives of the three leaders, yet all the others, I believe, will be forgiven on a slight intercession. At our arrival, too, I stated that if they wished me to remain, no barbarities must be committed, and especially that the women and children must not be fired upon. To counterbalance these motives is the danger, whatever it may amount to, but which does not weigh heavily on my mind. So much for reasons, which, after all, are poor and weak when we determine on doing anything, be it right or be it wrong. If evil befall, I trust the penalty may be on me rather than on my followers."

This reasoning would have been better had the cause of the Rajah Muda Hassim been just and righteous: on the contrary, the Sarāwak Malays had been driven into rebellion by the rapacity, cruelty, and extortions of Makota, the previous Governor of Sarāwak. The province, once comparatively flourishing, had been reduced to the lowest ebb by his infamous government. Had Muda Hassim really sought the truth he would have learnt it, and then banished Makota for ever from the dominions of the Sultan of Borneo. But such conduct is utterly unknown to Asiatic rulers. The poor are ever in the wrong, and the nobles right.

At daylight every one was astir; the stockade was pulled down, hauled to the river, and quickly formed into rafts; and then every one went to breakfast and to wait for the flood-tide. When that came no one was ready. Mr Brooke remonstrated in vain—procrastination was the order of the day; so that when they did start, the fleet only managed to move some four miles up the river, Makota showing that day an invariable repugnance to approach nearer to Siniāwan. However, next morning things went on better; a thick fog hid all movements from the enemy; the rafts were moved up to within a mile of Balidah Fort, the wood landed, and, all working with a will, the jungle was cleared; piles driven, forming a

15 yards' square ; the earth was scooped out in the centre, and thrown against the stockade to about 5 feet in height, and a Chinese garrison established in a small house in the centre. There were small watch-towers at each corner, and a few noisy but harmless Chinese swivels were mounted. The Dyaks worked at an outer fence, a sort of *chevaux-de-frise,* and around the whole were planted thousands of *ranjows* or pointed pieces of bamboo, dangerous to naked feet, and inflicting even a nasty wound about the ankles of shod men. In about eight hours the whole work had been finished.

Above the fort was a hill which completely commanded it. Mr Brooke mounted to the top to judge of the enemy's position, as their forts and the village of Siniāwan were visible from thence. The chief among the former was the position of Balidah, a hillock jutting out into the river encompassed by triple stockades, and separated from the mainland by a deep dry ditch. Thousands of *ranjows* were planted around the fort, pitfalls with pointed stakes, and every other contrivance of Malay and Dyak ingenuity to prevent their defences from being approached.

At a distance these defences seemed insignificant ; but they were not to be contemned, any more than those Burmese stockades against which our soldiers too often ran their heads, despising what appeared but a poor defence. The resources of the grand army were not more than sufficient to cope with the enemy's defences ; and without the energetic influence of the stranger, the attacking forces would have been ignominiously defeated.

The army that Muda Hassim had collected was but a motley crowd. In the first place were 200 Chinese under their captain,—fine muscular men, admirable at work but poor at fighting—wretchedly armed with fantastic spears, swords, and shields, a few muskets, and a certain number of curious weapons, consisting of long, thin, iron tubes with the bore of a musket, and carrying slugs. "These primitive weapons[1] were each managed by two men, one being the carrier of the ordnance, the other the gunner; for whilst one holds the tube over his shoulder, the other takes aim, turns away his head, applies his match, and is pleased with the sound." Nothing could better prove the little intercourse that Borneo had held with the outside world, than that so awkward and inefficient a weapon should have held its own to the middle of the nineteenth century.

The Malays, who numbered about 250, were the mainstay of the force: 150 had been sent to occupy the villages of Serambo, while the remainder formed Makota's bodyguard; half were armed with muskets, the rest with very serviceable spears and swords. As a defence against the cut of sword or the thrust of spear the Borneons wear a quilted jacket, which reaches over the hips; the arms are left bare to give freedom for the use of their weapons; and although the appearance of such a body of men is at first sight ludicrous, I have not noticed that their dress affected the activity of their movements.

To these may be added about 200 Dyaks of the various tribes in the neighbourhood, without muskets, although in other respects armed much like the Malays, and courageous enough when not exposed to the sound of musketry.

The history of this warfare is so curious that it is worthy of being followed in detail, and almost in Mr Brooke's original words. These natives, as a rule, rarely fought in the open, except when in very superior force; they preferred being sheltered by stockades. When, therefore, Mr Brooke pointed out to them that although the enemy were

[1] In the account of the Kashgar Mission, published in the 'Times' of March 28, 1874, a description is given of a similar weapon still in use in that country.

nearly as numerous as themselves they were saved the necessity of defending many detached posts and forts on different sides of the river, and that therefore now was the opportunity to assault these in detail, this proposal was hailed as an extreme of rashness, almost amounting to insanity. At a council of war it was subsequently decided that advances should be made from the hill behind the stockade to Balidah, by a chain of forts, the distance being a short mile, and that when the proper spot was reached, a battery should be erected, and a bombardment commenced—with their guns and gunners, likely to be very noisy, but perfectly harmless.

During the day there was comparative quiet, interrupted occasionally by the beating of gongs, shouts, and now and then a shot, to give life to the scene. With a spy-glass one could see the detachment of the Borneon forces, gathered well up the side of the Serambo hill, on a jutting rock called Paninjow, or the "look-out," a spot shaded by magnificent palms, and from whence a splendid view of the interior of Sarāwak can be obtained. No doubt these Malays were well pleased to see their comrades diverting from themselves the attention of the enemy. At night loud shouts and firing from the rebels caused preparations to be made for an attack, but it proved to be nothing but lights moving about the hillside, with what intent was not understood. The jungle on the left bank having been cleared, the enemies' skirmishers kept aloof; but a few spies approached the boats. With this exception no further disturbance took place, though the rebels kept up an incessant beating of gongs, and from time to time fired a few stray shots, whether against an enemy or not was doubtful.

Oct. 25th.—The grand army was lazy, and did not take the field until after the breakfast and the bath, when it moved and took possession of two eminences nearer the enemy, and commenced forts on each. About 11 A.M. intelligence arrived that the enemy was collecting on the left bank, as they had been heard by the scouts shouting one to another to gather together in order to attack the stockades in the course of building. Mr Brooke thereupon went up to one of the forts, when a universal shout from the rebels and a simultaneous beating of the gongs announced what appeared intended for a general action. But though the shouts continued loud and furious from both sides, and a gun or two was discharged in the air to refresh their courage, the enemy did not attack, and a heavy shower damped the ardour of the approaching armies, and reduced all to inaction. Like the heroes of old, however, the advance parties spoke to each other. "We are coming! we are coming!" exclaimed the rebels; "lay aside your muskets and fight us with swords." "Come on!" was the reply; "we are building a stockade, and want to fight you." And so the heroes continued to talk, but forgot to fight, except that the rebels opened a fire from Balidah with swivels, all the shot from which went over the tops of the trees. Peace, or rather rest, having been restored, the Borneons succeeded in intrenching themselves, and thus gained a field which had been obstinately assaulted by big words and louder cries. The distance of one fort from Balidah was about 800 yards; it was manned by sixty Malays, whilst a party of Chinese garrisoned the other. Evening fell upon this innocent warfare. The Borneons in this manner contend with vociferous shouts, and, preceding each shout, the leader of the party offers up a prayer aloud to the Almighty, the response being made by the soldiery, "Allah! Allah il-à hu!" The besiegers kept up firing and holloaing till midnight to disguise the advance of a party that was to seize an eminence and build a stockade within a shorter distance of Balidah. When they reached the spot, however, the night being dark, the troops sleepy, and the leaders of different opinions, they returned without effecting anything.

26*th.*—The advance of the party during the night was, as has been said, disguised by firing, drumming, and shouting from the fleet and forts; and in the deep stillness of the fine night the booming of the guns, the clangour of the gongs, and the outcries raised from time to time, fell on the ear like the spirit of discord breaking loose on a fair and peaceful paradise. About 6 A.M. Mr Brooke visited the three forts. The Chinese, Malays, and Dyaks were taking their morning meal, consisting of half a cocoa-nut-shell-full of boiled rice, with salt. The Dyaks were served in tribes; for as many of them were then at war among themselves, it was necessary to keep them separate, and though they would not fight the enemy, they would have had no objection to fall out with one another, and the slightest cause might have given rise to an instant renewal of intertribal hostilities.

About 9 A. M. a party proceeded to the elevation previously marked out within 300 yards of Balidah, and worked quickly till 2 P.M., by which time they had made considerable progress; and being then reinforced, they soon finished this new stockade, with a strong face towards their adversaries, and an outer fence.

This erection, however, being below the brow of the hill, was useless as a post whence to assault Balidah, but was meant to cover the working party that was preparing to erect another on the summit, from whence the rebel fort could be bombarded. The enemy, discovering at length the Borneon advance, opened fire for about half an hour, but finding it ineffectual, they sank into their usual apathy. The fact was that, deserted by their Dyak allies, the Sarāwak Malays were unable to skirmish effectively in the woods, and hands were wanting to oppose stockade to stockade. The Borneons by their successful advances appeared to gain confidence. To encourage them, and to make the issue favourable, Mr Brooke sent for two six-pounder carronades, together with a small addition to his European force.

Surprised at the little enterprise shown by the Sarāwak Malays, Mr Brooke asked Makota about the progress of his former campaign, when he had 1000 Malays and only a few Dyaks. He represented the enemy as active and daring then, very different from the want of spirit they showed during this campaign. He declared with an animated voice that they had had combats by sea and combats by land; stockade was opposed to stockade, and the fighting was constant and severe, but he never lost a man killed during the whole time, and only boasted of having killed five of the enemy. I have often heard him myself dwell on this theme, though very unwilling to touch on his subsequent career.

27th.—The night passed quietly, as usual. About 6 A.M. Mr Brooke started for the hills, and inspected each fort in turn. They were about commencing the fort previously referred to from which Balidah was to be bombarded; but while Mr Brooke was reconnoitring it, he was perceived by the enemy, who immediately opened fire upon the party. They shot wretchedly ill; but under cover of the smoke about thirty or forty men crept out and advanced stealthily to interrupt the work. The Malays, however, received them steadily, whilst the Chinese in the other fort placed them between two fires, and by a discharge from one of their famous tubes knocked down one man, the only one who had as yet fallen during these attacks. The enemy showed anxiety to possess themselves of their comrade, whilst the opposite party shouted, "Cut off his head!" But he was carried off, and the enemy, when they had saved his body, fled in all directions. Some fierce alarms were given of an attack by water, but they came to nothing, though both sides kept up a desultory firing until evening.

28th.—The stockade was completed during the night with *ranjows* stuck round the outer defence: it was admirably situated for battering the opposing fort.

During that day about 150 of the Sow and Singhi Dyaks joined, increasing the working portion of the army to about 500. Most of these men showed all the characteristics of a wild people,—never openly resisting their masters, but so obstinate that they nearly always got their own way, opposing threats and entreaties by a determined and immovable silence.

On the 29th the English guns arrived, but it was not till the 31st that they were dragged up to the fort. When once in position, however, they soon silenced the fire of the enemy, and struck down the stockade in such a way that a breach was opened by which several men could enter abreast. Seeing the effect of the fire, Mr Brooke proposed to Makota to storm the place with 150 Chinese and Malays, as the way from one fort to the other was protected. The enemy dared not show themselves on account of the showers of grape and canister, and nothing would have been easier than to take the place by storm; but Mr Brooke's proposition caused a commotion which it was difficult to forget, and still more difficult to describe. The Chinese consented, and Makota, the commander-in-chief, was apparently willing, but his inferiors were backward, and there arose a discussion which showed the violence of Malay passions, and their infuriated madness when once roused. This scene let Mr Brooke describe in his own words. "Pangeran Usman urged with energy the advantage of the proposal, and in the course of a speech lashed himself to a state of fury: he jumped to his feet, and with demoniac gestures stamped round and round, dancing a war-dance after the most approved fashion. His countenance grew livid, his eyes glared, his features were inflamed, and, for my part, not being able to interpret the torrent of his oratory, I thought the man possessed of a devil, or about to 'run amuck;'[1] but after a minute or two of this dance he resumed his seat, furious and panting, but silent. In reply, Subtu urged some objections to my plan, which, however, was warmly supported by Illudin, who apparently hurt Subtu's feelings, for the indolent placid Subtu leapt from his seat, seized his spear, and marched to the entrance of the stockade with his passions and his pride desperately aroused. I never saw finer action than when, with spear in hand, pointing to the enemy's fort, he challenged any one to rush on with him. Usman and Sirudeen, the bravest of the brave, like madmen seized their swords to inflame the courage of the rest: it was a scene of fiends: but in vain, —for though they appeared ready enough to quarrel and fight among themselves, there was no move to attack the enemy. All was confusion; the demon of discord and madness was among them, and I was glad to see them cool down, when the dissentients to the assault proposed making a road to-night, and attacking to-morrow. In the meantime our six-pounders were ready in battery, and it is certain that the assailants might walk nearly to the fort without any of the rebels daring to show themselves in opposition to our fire."

Pangeran Usman was really a brave man, and in 1854 I saw him perform an action worthy of the Victoria Cross. To this I shall have to refer at the proper moment.

November 1*st.*—The guns were ready to open their fiery mouths, and their masters ready to attend on them, but both had to wait till mid-day, when the chiefs of the grand army, having sufficiently slept, breakfasted, and bathed, lounged up with their straggling followers. The

[1] In Malay, *amok.*

instant the main division and headquarters of the army arrived at the battery, Mr Brooke renewed his proposal for an assault, which was variously received. If the Malays would go the Chinese agreed to fight, but the Malays had grown colder and colder. In order to encourage them, Mr Brooke opened fire on the rebel fort to show the effect of his guns; and having got a good range, every ball, as well as grape and canister, rattled against and through the wood. Mr Brooke then urged them again and again, but in vain. The coward chief displayed that dogged resolution which is invincible, an invincible resolution to do nothing; and the cold dumb looks of the others at once told the amount of their bravery. A council of war was called; grave faces covered timid hearts and fainting spirits. The Chinese contended, with justice, that in fairness they could not be expected to assault unless the Malays did the same. However, the latter would do nothing, and one of the chiefs delivered himself of a wise harangue to the effect that, "during the last campaign, when they had a fort, how had the enemy fired them, stabbed them, speared them?—and without a fort assaulting them! how could it be expected they should succeed? how unreasonable that they should go at all!" But even his stolid head seemed to comprehend the sarcasm, when Mr Brooke asked him how many men had been killed during all this severe fighting.

After waiting a couple of days, and finding that there was no serious intention to do anything, Mr Brooke saw that it was useless for him to stay further with the army, and therefore intimated his resolve to return to his ship. This caused the deepest surprise and vexation; but Mr Brooke was firm, and having embarked his guns and followers, he started for Kuching. His arrival was greatly felt by the Rajah Muda Hassim, as he now saw that the leader on whose assistance he had placed the greatest reliance was about to leave him. He begged and prayed him to remain, but Mr Brooke fully explained to him the situation of affairs, the uselessness of his stopping with the army as long as the present system of procrastination was allowed,—that, in fact, he saw no end to this devastating civil war under Makota's management. Muda Hassim was deeply moved by the thought of the disgrace that would attach to him should he fail in putting down the rebels: he knew that the prestige he had acquired by having a white ally was one of the reasons why other aid was coming in. He therefore again begged and entreated Mr Brooke not to desert him, and offered him at last the government of the province of Sarāwak if he would stay and aid him to suppress the insurgents. For the moment Mr Brooke declined to accept the grant, as he knew that it would be impossible for him to govern the country unless he had the Rajah's undoubted and spontaneous support. He remained at Kuching about a month, in daily communication with the Malay chief, and believing at length that the latter was really sincere and showed signs of true friendship, Mr Brooke no longer refused to return to Siniāwan, particularly as Pangeran Bedrudin, Muda Hassim's favourite brother, was now with the army, and he was reported to be as brave as he was undoubtedly intelligent.

Having prepared everything, Mr Brooke again started for the interior at the head of his sixteen gallant followers, with his guns and ammunition, and soon found himself with the grand army. The guns were mounted in their old positions, and everything was made ready for serious work.

Mr Brooke found that during his absence nothing had been done. The Borneons were in a state of torpor, eating, drinking, bathing, and walking up to the forts daily; but having built those imposing structures, and their appearance not having driven the enemy away, they were at a loss what to do next, or how to proceed.

On Mr Brooke's arrival he renewed his proposal to assault Balidah under cover of the fire of his guns. As Bedrudin warmly seconded the proposal, Makota was forced to express his assent, and the assaulting party was arranged as follows: Mr Brooke and ten of his men to lead, the rest of the Europeans to remain in the battery to serve the guns; Bedrudin, Makota, and all their chiefs and warriors were to attack by one path, while the Chinese were to push on by another. The Dyaks in the meantime were to remove the obstructions that might delay the advance.

At 8 A.M. the next morning Mr Brooke and his followers were in the battery, and at ten they opened fire and kept it up for an hour. The effect was severe: every shot told upon the thin defences of wood, which fell in many places, leaving practicable storming-breaches. Part of the roof was cut away and tumbled down, and the shower of grape and canister rattled so as to prevent the enemy from returning the fire except from a stray rifle. At mid-day the main forces came up to the fort, and it was then discovered that Makota had neglected to make a road, because it had rained during the night! It was evident that the rebels had gained intelligence of the projected assault, as they had erected a *chevaux-de-frise* of bamboo along their defences on the very spot that the besiegers had agreed to mount. Makota fancied that the want of a road would delay the attack; but Mr Brooke knew well that delay was equivalent to failure, and so it was at once agreed that all should advance without any path. The poor man's cunning and resources were now apparently nearly at an end, but he proved equal to the occasion. He could not refuse to accompany the assaulting party, but his courage could not be brought to the point, and pale and embarrassed he returned. Everything was ready, Bedrudin, the Capitan China, and Mr Brooke were at the head of the men, when Makota once more appeared, and raised a point of etiquette which answered his purpose. He represented to Bedrudin that the Malays were unanimously of opinion that the Rajah's brother could not expose himself in an assault; that their dread of the Rajah's indignation far exceeded their dread of death; and in case any accident happened to him, his brother's fury would fall on them. They stated their readiness to assault the place, but in case Bedrudin insisted on leading in person, they must decline accompanying him. Bedrudin was angry, all were angry, but anger was unavailing: it was clear that they did not intend to do anything in earnest, and after much discussion, in which Bedrudin insisted that if Mr Brooke went he would go likewise, and the Malays insisted that if he went they would not go, it was resolved that the English should serve the guns, whilst the Malays and Chinese should proceed to the assault. But Makota had gained his object, for neither he nor Subtu thought of exposing himself to a single shot. The artillery opened and was beautifully served, only three rifles answering from the fort. Two-thirds of the way the storming-party proceeded without the enemy being aware of their advance, and they might have reached the very foot of the hill without being discovered had not one of the Malay chiefs, from excess of piety and rashness, or rather cowardice, began most loudly to say his prayers. The three rifles commenced then to play on them: one Chinaman was killed; the whole party halted; the prayers were more vehement than ever; and after squatting under cover of the jungles for some time, they all returned. In this affair one of the Englishmen was wounded.

As Mr Brooke and his men could not leave the battery,

they had a house erected near it. It was a rough affair, about 20 feet long, with a loose floor of reeds and a roof of palm-leaf mats. After a time different attempts at theft induced Mr Brooke to have it fenced in, and divided into apartments: one at the end was used by Mr Brooke and his officers and servants, while the centre served as a store and a hospital, and in the room at the other end were the seamen. The unevenness of the reed floor was so uncomfortable, that at length they had it covered with the bark of trees, and then it appeared quite a luxurious abode.

Days and days were now wasted in building useless stockades, but at length it was resolved to push on to Sekundis, a spot from which the enemy would be out-flanked, and the command of the upper reach of the river secured. It was a most important position, as, if once properly protected, the Sarāwak Malays would be deprived of the means of receiving any supplies or succour from the interior, cutting off as it did the most facile route to Sambas, whose chiefs, as I have said, were accused of aiding the insurrection by many underhand means.

Now took place the only battle, if such it can be called, of the whole war. The Sarāwak Malays, seeing the danger that menaced them, crossed the river with a small force and endeavoured to drive the besiegers from the stockade they were erecting. They advanced firing, and had nearly succeeded in their endeavour when intelligence was brought to Mr Brooke of the danger of his allies, among whom was the old Orang Kaya of Lundu. Against the warnings and remonstrances of Makota, Mr Brooke started with his own men, and crossing the intervening jungle to an open field, saw the enemy advancing close to the incomplete stockade. He instantly charged across the open with his dozen Europeans, his faithful Malay boatman Subu, and an ally, a brave Lanun, called Si Tundo. The effect was immediate: surprised by a manœuvre to which they were unaccustomed, as charging is not known in their warfare, the enemy turned and fled, throwing away arms and ammunition, and, jumping into their canoes, they escaped across the river. This bloodless victory raised the influence of Mr Brooke above that of all present: though no one was killed, the daring shown was equally admired, and from that time forward the better class of natives began to look upon Mr Brooke as their leader. They compared his prompt and energetic conduct with that of Makota and his followers, who arrived when all danger was over, and loudly sang their own praises. But from this time forward the bravest leaders of the grand army stuck to the English, and this increased the influence of Mr Brooke in all the subsequent operations.

From this time forward, however, the resistance of the enemy became weaker, and they gradually abandoned their upper forts on the left-hand bank, though they still held possession of the strong fort of Balidah on the same side.

During these skirmishes there arrived on the scene of action Sirib Jaffir and a party of warlike Dyaks from Lingga, and the chief soon informed Mr Brooke that the Siniāwan Malays were anxious to yield, stipulating only that their lives should be spared. These Sherifs, or Siribs, as the Malays pronounce the word, are for the most part descendants of Arab adventurers who formerly settled on the coast, and who, giving themselves out as of the lineage of the Prophet, were enabled to marry well, and acquire an influential position in the country. Their influence, however, was generally for evil. Their religious position led them to despise the infidel, and then greed induced them to encourage piracy and warlike expeditions in order to obtain slaves and plunder. On Mr Brooke's arrival in the country, there were two or three of these Arabs exercising a certain influence in every district; in some they governed almost unquestioned.

Sirib Jaffir was well acquainted with a Sirib Moksain, then with the Siniāwan Malays, and by this means was soon enabled to arrange an interview between his friend and Mr Brooke. All parties present had agreed to this step being taken, but the whole affair was nearly spoiled by the meditated treachery of a Borneon noble, who proposed that they should seize the envoy; but on this being repeated in Mr Brooke's hearing he sprang to his feet, and drawing his pistol, declared he would shoot dead any man who dared to seize, or even to propose to seize, an envoy who had trusted himself to his honour. The scoundrel slunk away, and no more was seen of him.

The place of meeting was arranged to be at Pangeran Illudin's fort, at Sekundis, and thither were collected Mr Brooke and his party, Sirib Jaffir and his warlike Dyaks, some Chinese, and a large crowd of armed Malays. I have often heard both Mr Brooke and Sirib Moksain describe this interview. The Sirib was a small man, perfectly polite in manner, and of good intelligence. He had travelled, and had seen something of Europeans at Singapore, and at the Dutch possessions of Sambas and Pontianak, and therefore knew that the word of a white man was to be relied on. He, however, felt somewhat nervous when, landing from his boat on a dark night, he found himself surrounded by armed enemies, among whom he could distinguish some of the most treacherous of the Borneons; but he showed no sign of fear, and soon found himself in the presence of the white stranger. He then explained his mission, which was that Mr Brooke should give his word that the lives of all should be respected. This Mr Brooke was not authorised to do. During the whole interview the bearing of the envoy was firm, and the only sign of uneasiness, as Mr Brooke remarked, was the quick glance of his eye from side to side.

Though this first interview was not completely successful it led to others, and at last the insurgents surrendered at discretion, gave up the fort of Balidah, and destroyed their stockades, under Mr Brooke's guarantee that they should not be plundered or ill-treated until the pleasure of the Rajah Muda Hassim was known. Then came Mr Brooke's great difficulty—the greater from his being imperfectly acquainted with the causes of the rising—to save the lives of the chiefs of the insurgents, which he was fully convinced Muda Hassim intended to take. For this purpose he went down to Kuching, and begged as a favour to himself that the Rajah should pardon them. To this the Malay prince would by no means consent. He urged that their lives were justly forfeited, and that they must pay the penalty of their rebellion. To this Mr Brooke could only reply that he had taken part in the war, and aided to bring it to a successful conclusion, under the full conviction that the Rajah would exercise clemency; but as he refused this favour to himself, he could not help doubting the sincerity of his friendship, and that therefore he bade him farewell. On this the Rajah yielded. Throughout the interview Mr Brooke was oppressed by the conviction that the leaders had justly forfeited their lives; whereas, had he known the truth—to what oppression the people had been subjected before they rose in arms—he would rather have demanded the punishment of Makota, who was the cause of all the evils that had occurred.

Thus ended the civil war. The Sarāwak Malays surrendered their arms, ammunition, and property, and the chiefs gave their wives and families as hostages. Siniāwan was now abandoned, the inhabitants were dispersed, the chiefs had fled, the army was disbanded; and the

Chinese, finding themselves alone at Siniāwan, destroyed the Malay houses, and built a village for themselves in the neighbourhood.

The civil war in Sarāwak is now looked upon by the Malays as one of the most noteworthy incidents in the history of their province; and as it was the indirect means of placing them under English rule, it is perhaps the most important event in their local history.

CHAPTER III.

WRETCHED STATE OF THE COUNTRY—BROOKE ASSUMES THE GOVERNMENT OF SARĀWAK.

1840–1843.

After these warlike operations there followed a period of comparative peace, during which Mr Brooke could look around and study the true position of the country. To those who are accustomed only to the regular march of affairs under European Governments, it is difficult to convey an idea of what passes in countries so far removed from civilisation as Sarāwak. Sarāwak was a dependency of the Sultan of Brunei, inhabited by Malays and Dyaks, and a few nobles from the capital. The governor was appointed by the Sultan, and Pangeran Makota had, before the commencement of the civil war, been nominated to the post. Under him were three Malay chiefs, —two who had charge of the Dyaks in the interior, called the Datus Patinggi and Bandar; the third, the Tumanggong, looked after the coasts, the mouths of rivers, the fisheries, and the villages on the sea-shore, and ruled those Dyak tribes established on the streams which ran into the sea between Cape Datu and the river Lundu.

The laws and usages of Borneo strictly defined the relative duties of these different authorities; but Pangeran Makota was no respecter of either laws or traditions, and immediately endeavoured to get all power into his own hands. Unfortunately for the inhabitants, a bright mineral was discovered, which was at first mistaken for silver, but which was in reality antimony, and would fetch a fair price in the British settlement of Singapore. Makota thereupon set every one to work for it on whom he could lay hands. Malays, Dyaks, and Chinese were pressed into his service, and shiploads were soon procured; but Makota did not choose to remember that workmen must eat to live, and the price he paid for the ore was so small that it was impossible for the people to obtain sufficient food. The natural consequence was that the mines were deserted, and force had to be employed to induce the people to return to their work.

The next manœuvre of Makota to obtain money was founded on a custom of the country, though doubtless a gross abuse. This is the *serra dagang* or forced trade. A Malay Rajah will send to a tribe specially under his control a certain amount of salt, iron, and cloths, which the Dyaks are expected to take at a fixed price—generally a very exorbitant one. When this is purchased the tribe is free to trade with whom it pleases. The Rajah's followers generally stick a spear in front of the Dyak chief's house, to show that the Rajah's trade is still going on; or if a narrow river leads to the village, a string is drawn across it, with the flowers of a palm attached to it, as a sign that no one is to pass. Any one trading with the tribe until these signs are removed is liable to a severe fine. In Sarāwak the Malay chiefs had almost similar privileges to the Rajahs, and exercised them with some severity, but seldom with enough to induce resistance.

On Makota's arrival, however, all this was changed. Not content with one *serra*, he would keep his followers thus engaged all the year round; and as he was followed by a number of the greatest scoundrels in Borneo, they worried the people on their own account quite as much as they did for their master. When the rice and other commodities of the Dyaks were exhausted, they seized on the best-looking girls and the most likely lads, and carried them off as slaves; and whenever the tribes murmured, they were threatened with an attack of the sea Dyaks, the piratical tribes of Seribas and Sakarang, always ready to come at any one's call, so that heads and plunder could be promised.

The poorer Malays fared as ill as the Dyaks: they were lent a few dollars' worth of goods at heavy interest—ten or twelve, even fifty per cent per month. This soon accumulated to a formidable sum, and as they could never pay, they and their families were seized as slave-debtors, and had to work for their masters, and were too often treated as real slaves and sold out of the country.

This oppression, pushed to its utmost limits, at length exhausted the patience of one Dyak tribe, the Sows, and they refused to give up the rent of their rice or to permit their women and children to be taken. This resistance was of evil omen, and had to be suppressed at once; so the sea Dyaks were called in, and 3000 wild savages, under the guidance of Sirib Sahib of Sadong, were let loose on the country. They surprised the villages of the Sow tribe, killed about thirty of the men, and carried off upwards of two hundred women and children. As one of the Sows feelingly said to Mr Brooke: "You might, sir, a few years ago, have sought in this river and not found a happier tribe than ours. Our children were collected, we had rice in plenty, and numerous fruit-trees; our hogs and fowls were in abundance: we could afford to give what was demanded of us, and yet live happily. Now we have nothing left. The Sadong people and the Sakarang Dyaks attacked us; they burned our houses, destroyed our property, cut down our fruit-trees, killed many of our people, and led away our wives and young children into slavery.

We could build other houses; we could plant fruit-trees and cultivate rice; but where can we find wives? Can we forget our young children?" Though premature, let me state here that Mr Brooke did ultimately force Sirib Sahib and his followers to return about two hundred women and children they had captured from the Sow Dyaks.

Such acts as these at length aroused all to resistance: the Sarāwak Malays and Dyaks combined to expel Makota from the country, but he found allies in the neighbouring provinces, with whose assistance he held his own. At length Muda Hassim was sent down from the capital to put an end to the civil war, but he could do little or nothing except add to the misery of the country. The opportune arrival and assistance of Mr Brooke saved him, and put an end to the insurrection. But pestilence and comparative famine then desolated the land, and reduced the remnant of the people to a fearful state of poverty. Half at least of the inhabitants had fled from the country—many of the Dyak tribes having crossed the frontier and entered Sambas, while the Malays were dispersed through all the surrounding districts. There was no trade, the antimony was scarcely worked, little rice was cultivated,—in fact the task of governing and regenerating the people appeared wellnigh hopeless.

Mr Brooke thought over these things, and often felt inclined to continue his voyage of geographical discovery, and leave Sarāwak to its fate. He reasoned that, with his very limited means, he could scarcely do much good, that he was surrounded by enemies, that Makota would naturally oppose all his measures and endeavour to ruin him, that he was only supported in a lukewarm fashion by Muda Hassim; and he had yet to discover what reliance he could place on the support of his late opponents, the Malays of Siniāwan.

However, he was not dismayed: he first of all determined to connect Sarāwak with Singapore, and therefore proceeded to the latter place in the Royalist and purchased a schooner, the Swift, to carry on trade between the two places. He loaded his two vessels with an assorted cargo, and returned to Kuching.

The first step taken by Mr Brooke showed how totally unfit he was to fill his new character as a trader. The Rajah Muda Hassim, on hearing of the arrival of the cargoes, promised to load the two vessels with antimony ore in exchange, and Mr Brooke permitted him to take all his goods without being assured that any antimony ore was ready, or even that any one was engaged in procuring it. Mr Brooke treated Muda Hassim as a gentleman, whose word was sacred, forgetting that though he might be of good faith at the time, there were others who were deeply interested in Mr Brooke's never obtaining payment for his goods. He soon found that where before there had been zeal, now there was lukewarmness: no one talked of the return cargo, nor was any mention made of the promise to hand over the government of the country to the Englishman.

Mr Brooke waited with patience, but finding months elapse, he determined to send away the Swift to Singapore with the few tons of antimony that he had been enabled to collect, and to start the Royalist for Borneo Proper to endeavour to relieve the crew of a shipwrecked English vessel that was detained there in confinement by the Sultan. He, with three followers, stayed behind in a sort of native house that Muda Hassim had built for him, and continued to urge on the Malay prince the necessity of paying for the goods which he had received.

Mr Brooke had also many other causes of anxiety. His two vessels were gone to sea, and he knew that they would have to run the gauntlet of the pirate fleets which were cruising outside. One of these squadrons which had visited

Kuching during the previous month consisted of eighteen vessels, well armed with guns, swivels, and musketry, and manned by at least 500 fighting men, and an equal number of rowers. The pirates were principally brave Lanuns, whose prowess is well known to all frequenters of the Eastern seas.

Other pirates were also off the coasts, less formidable to European vessels, but, if possible, more destructive to native trade, and these were the Sakarang and Seribas Dyaks, led by the warlike Malays of those districts. Before the sailing of the Royalist, above a hundred of their long swift *bangkongs*, or native war-boats, had ascended the river—nominally to pay their respects to the Rajah, but really to obtain permission to attack an inland Dyak tribe living on the borders of Sambas. Although Muda Hassim knew perfectly well that these men would respect neither the lives nor the property of his own Dyaks, he gave them leave to proceed up the river to attack the tribe in question. All the land Dyaks and Chinese were in fearful alarm, knowing what was in store for them, as all heads are good that come in the way of the Dyak on the war-path. Fortunately Mr Brooke was warned, and by energetic remonstrances forced the Rajah to give the order for their immediate return, which they did sulkily, and with an evident intention to try their strength with the white power that had interfered with their raid. But the signs of complete preparation induced them to waive their intention: they soon fell down the river, and disappeared for the time. This action of Mr Brooke raised his fame among the people, and they now began to put their trust in him, and the belief soon grew general that he could and would befriend them.

The conduct of Muda Hassim on this occasion was another proof of the incapacity of the Malay princes to rule that or any other country; and was, in fact, but an instance of what is constantly done by other Rajahs in those countries in which they have influence. This particular raid, however, had been organised with a deeper motive than mere plunder. All the descendants of the Arab adventurers who infest the coasts of Borneo have an instinctive feeling that where the white man gains influence their power declines. Makota also felt that Mr Brooke was more considered and consulted by the Borneon prince than he was, and that should Mr Brooke be installed as governor of the province, farewell to his own power. He therefore entered into a plot with the Arab Sirib of Sadong, and it was agreed between them that the latter should collect a large force of piratical Dyaks, while the former obtained the necessary permission from Muda Hassim that the Dyaks might ascend the river to attack an interior tribe. The bribe offered was a large share of the slaves taken, and a proportionate amount of the valuable plunder. As the Rajahs were interested in the business, all having to receive shares, there was no one found near the ruler to counsel him not to commit so great an infamy.

Makota thought that if once the Dyak force were in the interior, Mr Brooke's authority would decline among the Sarāwak Dyaks and Malays, and also among the industrious Chinese, as it would be a proof of his want of both power and influence to protect them. He also reasoned that Mr Brooke, disgusted by the breach of faith shown by Muda Hassim, would sail away and leave the Malays to themselves, when his own influence would be again supreme. But the event had a contrary effect. Mr Brooke was kept in a state of constant uneasiness by the prolonged absence of his vessels, particularly as he knew that large fleets of pirates were cruising outside. At length the Swift and the Royalist arrived, but the latter did not bring the shipwrecked crew, as the Sultan had refused to deliver them up; but immediately after, the H.E.I. Company's steamer Diana arrived, and finding that the Royalist had not succeeded in obtaining the captives, steamed off to Brunei. The Sultan was thoroughly frightened when he saw the "fire-ship" advance into the midst of his town, and hastily surrendered his captives. The visit of the Diana greatly strengthened Mr Brooke's hands. Makota, however, continued his evil ways: he plundered the Dyaks to their little children and slaves, and, under threats of fine and imprisonment, he prevented the natives from visiting Mr Brooke, for fear that they should complain of their ill-treatment. He also continued his intrigues with the Sultan of Sambas, in order to retard the pacification of the country, fearing that Muda Hassim would then fulfil his promise to Mr Brooke, and return himself to the capital. At length a robbery and attempted murder, by some of Makota's followers, brought things to a crisis. Mr Brooke determined that a settlement should be made, one way or the other; and therefore he armed his vessels, landed his crew, and marched to the palace, where he explained to the Malay prince the treachery and crimes of Makota. Mr Brooke soon found that he might count upon sufficient support. The Siniāwan Malays sent and offered their assistance, while none but a few of his immediate slaves rallied round Makota. As Mr Brooke says, "After this demonstration, affairs moved cheerily to a conclusion. The Rajah was active in settling; the agreement was drawn out, sealed, and signed; guns fired, flags waved; and, on the 24th September 1841, I became the governor of Sarāwak with the fullest powers." One thing, however, Muda Hassim completely forgot to do, and that was, to pay for the cargoes.

The acceptance of the government of Sarāwak was indeed a venturesome step. We are so accustomed to hear of the riches of the isles of the Eastern seas, that many imagine we have but to step in to reap a rich reward.

But the fact is, that the riches spoken of have to be developed or created; that, though the soil of Sarāwak will bear most tropical productions, it did not then yield anything in a commercial sense—and even now, almost forty years after, it produces very little but the natural resources of the forest, minerals of course excepted. Capital and an intelligent government are necessary to develop a country; and these necessities Sarāwak has not had to the extent one might have expected.

When Mr Brooke took over the government of Sarāwak, the Dyaks did not cultivate sufficient for their own sustenance; the Malays were dispersed, and picked up a precarious living by fishing, and by the produce of a few small gardens; two or three hundred Chinese were washing for gold and working antimony on a small scale. From what sources, then, were revenues to be raised? Mr Brooke's faith in the future was all he had to support him in these trials.

Mr Brooke's three chief objects in assuming the government of the country were: to obtain the women and children of the Siniāwan Malays, whom Rajah Muda Hassim still held as hostages; to collect together again the old Malay population; and, thirdly, to conciliate the Dyaks. In the first he fairly succeeded, though he could not obtain all the women, as the Rajah's brothers having honoured them with their notice, they could not be permitted to leave their harems. However, all but ten were returned to their families.

Makota, after the conclusion of the civil war, had made every effort to collect the Siniāwan Malays together; but he was too little trusted to succeed. He might force some families to establish themselves in the town, but in a few days they or others disappeared, so that his efforts had but little success. Mr Brooke, however, gave out that the Malays were free to come or go as they pleased—to live at Kuching or elsewhere, as it best suited their convenience —and that no one had a right to interfere with them. This and the recovery of the hostages had a great effect; and gradually the population collected, and began to form the nucleus of the present town.

The greatest difficulty, however, was with the Dyaks: they had been so long used to oppression, that they looked with suspicion upon any one who talked to them as a friend. Mr Brooke, however, assembled their chiefs, and it was agreed that each Dyak family should pay a *pasu* (¾ bushel) of rice as a yearly revenue, and that then they should be free to trade with whom they pleased; and that no man had a right to demand anything further of them. Should any one attempt such a course, the chiefs were instantly to bring the case before the governor.

From the 24th September 1841, when Mr Brooke was appointed Chief of Sarāwak, to April 1843, he was employed in consolidating the government. He remained during the whole of this time in Borneo, to watch over his rising settlement, to protect the Dyaks, curb the licentious followers of the Rajahs, and oppose the intrigues of Makota, whose chief end was to disturb the confidence that was beginning to show itself.

The most difficult task, however, was the administration of justice. Almost every day the Court was opened in a long room in Mr Brooke's residence. He sat as chief, with the brothers of Muda Hassim beside him; to the right and left were rows of chairs, on which were seated the three Malay chiefs of Kuching, and any other respectable native who chose to take part in the proceedings; in front sat on mats the persons whose case was to be adjudged; and behind them a miscellaneous crowd of Malays, Chinese, and occasionally a few Dyaks, or some wondering stranger from the coast.

My own experience entirely confirms Mr Brooke's, when

he states that he received the greatest assistance from the knowledge, acumen, and sense of justice shown by the Siniāwan Datus. Unless some personal interest of theirs was involved—unless a relative was engaged in the litigation—these three chiefs might be completely depended on; and those who succeeded them were equally, if not more, worthy of trust. And it now became an established rule, that if a Datu was in any way interested in a case he stayed away, and left it to be settled by the others.

As long as the cases only involved the interests of the Dyaks, the Chinese, or the Siniāwan Malays, all went smoothly—the Rajahs cared for none of these; but directly their own followers were accused and brought up for justice, their susceptibilities were aroused. The Borneons who had accompanied Muda Hassim from the capital were bad specimens of a very bad class: they robbed, cheated, even murdered, with comparative impunity, and were rarely, if ever, punished by their chiefs. To be a follower of a Borneon Rajah was to secure immunity from every punishment. But Mr Brooke was decided, come what may, that, whenever a case came before the Court, he would have even-handed justice administered; and this was soon so well understood, that the Rajah's followers, rather than be brought up before the stern white man, would hasten to compromise a case, and thus give an instalment of justice.

The education of a Malay Rajah is such that he cannot be made to comprehend that others are entitled to any consideration at his hands. He thinks, like the French lady, that the Almighty will consider twice before punishing persons of such distinction. Makota himself, clever as he was, could not constrain his avaricious greed. He often said, "I know that the principles on which you govern are good; I have seen the success of Singapore under your rule; but I was brought up to plunder the

Dyaks, and it makes me laugh to think that I have fleeced a tribe down to its very cooking-pots."

Mr Brooke's courage and patience were, however, equal to the calls made upon them, though a great difficulty from which he suffered was the want of power. By one or two stern examples he had put down head-hunting and resistance within the territory of Sarāwak, yet he could not completely prevent his own Dyaks from the attack of his neighbours. Sirib Sahib of Sadong was the most mischievous: he incited the piratical Dyaks to make inland incursions, and news was continually reaching Kuching of the death of now three, now eight of a tribe. Sirib Sahib also often sent parties over to collect the taxes that were really due to Mr Brooke, and occasionally attempted, with the aid of Makota, to raise disturbances in Sarāwak itself. As yet Mr Brooke was not strong enough to suppress these disorders: he did something, however, and the little he could do inspired confidence. Tribe after tribe came over the frontiers, for, as one of them said, "They had heard, the whole world had heard, that the son of Europe was the friend of the Dyak."

There is one thing that I may note here, and on which I shall hereafter have often to dwell, and that is, that whatever was done in Sarāwak at this time was done by Mr Brooke alone. He had no aid or assistance: his followers were, a coloured interpreter from Malacca — a useful but not a very trustworthy man; a servant who could neither read nor write; a shipwrecked Irishman, formerly a clerk, brave as a lion, but not otherwise of much use; and a doctor—a first-rate companion, but so little interested in the country that he never even learned the language.

When Mr Brooke left on his last visit to Singapore, Muda Hassim made a promise to build for him a suitable house. As usual, however, he did not keep his word, so that on Mr Brooke's return to Kuching, he found that it had not even been commenced. However, better late than never: his Highness at length set to work, and the promise was fulfilled.

As it will give a good idea how Mr Brooke lived at that time, I will let him give the account in his own words.

"I may now mention our house, or, as I fondly style it, our palace. It is an edifice 54 feet square, mounted on numerous posts of the Nibong palm, with nine windows in each front. The roof is of Nipa leaves, and the floor and partitions are all of plank. Furnished with couches, tables, chairs, books, &c., the whole is as comfortable as man could wish for in this out-of-the-way country; and we have besides a bathing-house, cook-house, and servants' apartments detached. The view from the house to the eastward comprises a reach of the river, and to the westward looks towards the blue mountains of Matang; the north fronts the river, and the south the jungle. Our abode, however, though spacious, cool, and comfortable, can only be considered a temporary residence, for the best of all reasons, that in the course of a year it will tumble down, from the weight of the superstructure being placed on weak posts.

"The time here passes monotonously, but not unpleasantly. Writing, reading, chart-making, employ my time between meals. My companions are equally engaged,—Mackenzie with copying logs, learning navigation, and stuffing specimens of natural history" (he was afterwards killed by Chinese pirates); "Crymble" (the brave Irishman) "is teaching our young Bugis and Dyak boys their letters for an hour every morning, copying my vocabularies of languages, ruling charts, and the like; whilst my servant Peter learns reading and writing daily, with very poor success, however. Our meals are about nine in the morning and four in the afternoon, with a cup of tea at eight. The evening is employed in walking, never less than a mile and a half, measured distance; and after tea, reading and a cigar. Wine and grog we have none, and all appear better for it, or at least I can say so much for myself. Our bedtime is about eleven."

At this time the coast of Sarāwak was completely blockaded by large fleets of Lanun and Balagñini pirates who with perfect impunity cruised along the coast, frequented the Natuna and Tambilan groups of islands, carrying off captives by hundreds, and taking every native trading vessel that came in their way. The only intercourse with the neighbouring districts was by small boats, which kept inshore, and generally made their voyages under cover of the night.

The head-hunting Dyaks were also out on their forays, which made it dangerous even for small boats, and for the fishing villages, which were generally concealed up small creeks.

In 1842 Mr Brooke had his first specimen of the obstinacy and deceit of the Chinese. In recompense for some services, Rajah Muda Hassim had granted to a company of Chinese permission to work gold and antimony on the right-hand branch of the river, in Upper Sarāwak. This company was placed under the strictest discipline by its own self-elected chiefs, and every member was expected implicitly to obey the behests of the governing council. Affairs proceeded quietly enough at first, but at length some other Chinese expressed a wish to establish another company on the left-hand branch of the river. This roused the jealousy of the older company, and they announced their intention to oppose any concession to their countrymen. A long conference ensued, when it was found that the Malay document, simply allowing them to work, had been converted in the Chinese translation into a deed of gift of the interior. It is not necessary to enter into the details of this affair, as it ultimately came to nothing. Mr Brooke firmly adhered to his determination to grant to a new company on the left-hand branch the same privileges which had been granted to those on the right, but it was only by a display of force that the latter could be brought to give their consent. Here was evidently an *imperium in imperio,* from which Mr Brooke foresaw future trouble; but his own power increased so rapidly that he had no serious difficulty with the Chinese for fifteen years, until the union of a number of unfortunate circumstances weakened his prestige, and brought about the bloody insurrection of 1857.

I may at once say that the company on the left-hand branch never came to much, and, after a sickly existence, vanished from the scene.

About this time Mr Brooke calculated that the revenue of the country was about £5000 a-year. How he arrived at this estimate I do not understand, as the whole income of the country consisted of a few hundred bushels of rice, a little profit from opium, and the net proceeds of the antimony. I can readily imagine that he was incorrectly informed by his treasurers, who were such poor accountants that, on examining their books, I found that all expenditure was put under the head of revenue. His agents, too, were very careless, and on one occasion an error was discovered in which a bill of £1000 drawn by him was put to his debit. Mr Brooke knew nothing of accounts, and those he employed knew very little more. In the course of a Chancery suit, it became necessary to give detailed explanations of what had been the revenue and expenditure of Sarāwak before 1848. Mr Brooke naturally employed his treasurer to get up the figures, but the Court of Chancery sent back the document: it was so confused that no one could understand it. One of the chief officers, Mr Crookshank, could have mastered the figures, but he was otherwise engaged, so later on Mr Brooke in despair asked me to undertake it. Never was there such confusion seen. The poor treasurer could give no explanations. Dollars valued at 4s. 2d. and reals worth 3s. were treated as equivalent coins, and added together. It required weeks to unravel all these mysteries, and in the end it was possible to arrive at only an approximate result.

Thus Mr Brooke never really knew what was the true state of his affairs. What he did know was, that every now and then he was informed that there was a balance against him, and he drew bills on his private fortune, until it began gradually to vanish to nothing.

The revenues of Sarāwak never met its expenses until after the great influx of Chinese in 1850, which enabled Mr Brooke to establish various imposts in the indirect form of farms: then things went on prosperously, and have continued so until the present time, opium being the most productive of all the taxes.

Mr Brooke soon became convinced that as long as Muda Hassim, with the other Borneon Rajahs and their unruly followers, remained in Sarāwak, it would be quite impossible to develop the trade, or to inspire true confidence in the people. No trading *prahus* would enter with cargoes from the neighbouring districts, as it would have been impossible to prevent the Rajahs from going or sending on board to obtain as presents or on credit the most valuable articles. Besides, the natives would not believe that Mr Brooke might not some day tire of his task and sail away. So nothing was done: no trading vessels were built, no cultivation undertaken, and a feeling of restlessness prevailed all over the country.

Mr Brooke therefore made up his mind to visit the capital of Brunei, and see the Sultan himself. Muda

Hassim was pleased, directed two of his brothers to accompany the governor, and prepared letters for his royal nephew and chief. After many tedious formalities and an affectionate farewell, Mr Brooke got away from Kuching on July 14, 1842, and immediately sailed for the capital. The parting between Muda Hassim and his two brothers was very touching, and is thus feelingly described by Mr Brooke: "The Rajah addressed a few words to his brother, requesting him to tell the Sultan that his heart was always with him; that he could never separate from him, whether far or near; and he was, and always had been, true to his son. Bedrudin then rose, and, approaching the Rajah, seated himself close to him, bending his head to the ground over the Rajah's hand, which he had grasped. The Rajah hastily withdrew his hand, and, clasping round, embraced, kissing his neck. Both were greatly agitated, and both wept; and I could have wept for company, for it was no display of State ceremony, but genuine feeling. It is seldom, very seldom, they show their feelings, and the effect was the more touching from being unexpected. It is a part of our nature—our better nature—to feel when we see others feel. Pangeran Marsale followed. Both brothers parted from Muda Mahomed in the same way, and they certainly rose in my opinion from this token of affection towards each other. My adieux followed. We all rose; the Rajah accompanied us to the wharf, and as we embarked I could see the tears slowly steal from his eyes. I could not help taking his hand and bidding him to be of good cheer: he smiled in a friendly manner, pressed my hand, and I stepped into my boat."

Mr Brooke had now an opportunity to view the north-western coast of this great island. It is not very interesting seen at a distance from the shore—generally lowland, with a few bluff points, and lofty mountains in the interior. The true beauty of the coast begins to the north of the capital. In a few days the Royalist reached the mouth of the river Brunei, and cast anchor off the low, sandy island of Muara, famous for its wild pigs and its good fishing.

The Brunei river is really a pretty one. As you enter the small bay into which flow so many rivers, to the south you see a row of pretty hills. As you advance you pass several islands,—to the right, one whose summit is cut flat, with a heavy embankment and embrasures fit for the heaviest guns — an old Spanish construction, it is said. From this island a mole has been thrown across, so as to completely shut up the true entrance of the river; but as the water was forced to seek a passage, it has found one on the eastern side, where the river deepens to four fathoms. It is a difficult entrance. You then leave Pulo Cherimin, or Looking-glass Island, on the left, and advance up the stream between two rows of lovely hills, from 500 to 800 feet in height, gay with every variety of colour. Generally the verdure in the East is sombre; here it is not. You have the dark green of the forest, the bright green of cultivation, the waving cocoa-nut, the graceful areca palm. Cottages scattered here and there, a fisherman's house poised on the banks, and occasionally patches of red soil, add brightness to the scene. And the waters of the Brunei river are generally gay. You meet the Singapore trader heavily laden with sago; the unwieldy but picturesque Malay *prahu;* the fast fishermen's boats pulling as no one but Borneons can pull, to be the first arrival at the market; and then the prawn-catchers gracefully poising themselves on the prows of their little canoes, and throwing the cast-net with such marvellous skill that, though often eight yards in diameter, it falls in a perfect circle on the water. If you lazily pull up in a small boat, you will see the alligators swimming, or rather floating, near the surface of the water, a line of their back and their two ravenous eyes alone visible. After a few miles' advance you turn sharply to the right, and then catch sight of the first houses of this "Venice of hovels," as Mr Brooke called it. It is at high water a pretty place —a sort of basin into which several rivers and streams flow. It is almost surrounded by hills; and on the mud-banks formed in this little lake, but seen only at low water, are built the houses, on slight piles, so that the inhabitants can pull round or under or between long rows of houses that at the distance look picturesque. It is, however, but a wretched place; the dwellings are mean, being mostly composed of the stems of the Nibong palm, and thatched and walled with different kinds of mats made from the leaves of the Nipa palm. A few are roughly planked, but it is a tumble-down, wretched place. I knew it well, having lived there so many years that I seemed to recognise every face in it; and every nook and corner in its neighbourhood were familiar to me.

Mr Brooke had visited Brunei for three reasons. The first was, to assure a reconciliation between the Sultan and Muda Hassim, in order that the latter might be enabled to return to the capital; to secure the release of twenty-five shipwrecked Lascars of a recent wreck, the Melbourne; and thirdly, to obtain from the Sultan a confirmation of his appointment as Governor or Rajah of Sarāwak. After some preliminary ceremonies Mr Brooke went up to the capital, and was much struck and pleased with the lively scenery of the river. The town he saw in all its filth. He was received and lodged at the palace in a small room behind the audience-hall, and soon became familiar with the people around him. I must describe a few of them, as they are all important personages in this biography. The first in point of dignity was the Sultan. Mr Brooke, in his Journal, has well described him; and as he knew him so much more intimately than I did, I shall rather trust to his account than to my own recollections. At the time of this visit (1842) the Sultan was a man of about fifty years of age, short and puffy in person, with a countenance that very obviously showed the weakness of his mind, which, as indexed by his face, appeared a perplexed map of confusion, without astuteness, without dignity, and without good sense. He was ignorant, mean, and avaricious, fond of low society and of stupid jokes; and, when I knew him, he was suffering so much from a cancer in the mouth, that it was disagreeable to approach him. He was, however, full of pride, and had a high opinion of his own dignity.

Pangeran Mumein, the next in position, and the present Sultan, was more fitted to be a trader than a ruler; but he was not cruel, and I found him a likeable man.

Pangeran Usup, however, was the ruling spirit in Borneo; very active and intelligent, and though nominally a great friend of Muda Hassim's, was in his heart that prince's most determined opponent. At this time, however, Mr Brooke did not see through his ambition.

The objects of Mr Brooke's visit were soon settled: the most flattering invitations to return to Brunei were written to Muda Hassim; the imprisoned Lascars were released; and that most important document, the confirmation of Mr Brooke's appointment as Governor of Sarāwak, was signed, sealed, and delivered.

When these affairs were arranged, Mr Brooke took leave of the Court, and set sail for Sarāwak, where he was received with such demonstrations of welcome as were most gratifying to him. The Borneon Pangerans were delighted with his success, as they longed to return to the charms of the capital. The scene that took place at the reception of the despatches was highly curious and characteristic.

On the evening of the 18th August the Sultan's letters

were produced in all the state which could possibly be attained. On their arrival they were received and brought up amid large wax torches, and the person who was to read them was stationed on a raised platform. Standing on the step below him was Muda Hassim, with a sabre in his hand; in front of the Rajah was his brother, Pangeran Jaffir, with a tremendous *kampilan*, or Lanun sword, drawn; and around were the other brothers and Mr Brooke, all standing, the rest of the company being seated. The letters were then read, the one appointing Mr Brooke to hold the government of Sarāwak last. After this the Rajah descended from the steps of the platform and said aloud, "If any one present disowns or contests the Sultan's appointment, let him now declare it." All were silent. He next turned to the native chiefs of Sarāwak and asked them,—they were obedient to the will of the Sultan. Then the question was asked of the other Pangerans, "Is there any Rajah that contests the question? Pangeran Makota, what do you say?" Makota expressed his willingness to obey. One or two other obnoxious Pangerans who had always opposed themselves to Mr Brooke were each in turn challenged, and forced to promise obedience. The Rajah then waved his sword, and with a loud voice exclaimed, "Whoever dares to disobey the Sultan's mandate now received, I will split his skull!" At the same moment some ten of his brothers jumped from the veranda, and drawing their long knives began to flourish and dance about, thrusting close to Makota, striking the pillar above his head, and pointing their weapons at his breast. This amusement, the violence of motion, the freedom from restraint, this explosion of a long-pent-up animosity, roused all their passions; and had Makota, through an excess of fear or an excess of bravery, started up, he would have been slain, and other blood would have been spilt. But he was quiet, with his face pale and subdued, and as shortly as decency would permit after the riot had subsided, took his leave. Had he been slain on this occasion, many hundreds, nay, thousands, of innocent lives might have been saved.

An interval of tranquillity followed, and Mr Brooke could indulge in serious reading. No. XC. of the Oxford Tracts fell into his hands, and he set to work to answer it, as he considered it a very Jesuitical performance. I have the original MS., but I must confess that I have not read it. Though he himself at this time complained that he had read every book in his library, I do not think he went so far as the English planter in Java, who, finding himself banished into the interior for twenty years, turned to the only book he had, the 'Encyclopædia Britannica,' and beginning at the letter A, read it through to the end, and with such profit, too, that he was enabled to take an important post in Singapore, and to creditably fulfil its duties.

Things appeared now so quiet, the country being tranquil, and having some desire to refresh himself with a little civilised society, Mr Brooke thought that he might safely indulge in a trip to Singapore, from which he had been absent about two years; but before he started he intimated to Pangeran Makota that he must leave the country, as it was certain that as long as that clever, but always in the end unsuccessful, intriguer remained, the people would be uneasy. Makota began therefore preparing his *prahu*, and collecting his people, but was so slow in his movements that Mr Brooke had left for Singapore (February 8, 1843) before his arrangements were complete.

The idea of visiting Singapore was indeed a happy one, as he was destined there to become acquainted with Captain Keppel, whose energy and dash, and quick appreciation of the earnest purpose of Mr Brooke, had so great an effect on the future of Sarāwak.

CHAPTER IV.

FIRST EXPEDITIONS AGAINST THE SERIBAS AND SAKARANG PIRATES—CAPTAIN KEPPEL.

1843–1844.

MR BROOKE arrived in Singapore in February 1843, and after a short stay proceeded to Penang, and found there H.M.S. Dido, Captain the Hon. Henry Keppel, the senior officer in the Straits Settlements. They soon became acquainted, and that acquaintance ripened into friendship, as they at once recognised each other's good qualities. Mr Brooke could readily admire the dash, the vigour, the eagerness for action, which was displayed by Captain Keppel; while the latter, above all littleness and petty jealousy, qualities too often shown on these occasions, was eager to aid in the good work of pacifying the coast of Borneo, and saw at a glance, with the instincts of a gentleman, that he had no adventurer, no schemer to deal with, but a true-hearted man, whose whole soul was concentrated on one object—which was, to raise in the scale of civilisation and comfort the natives of his adopted country.

It was about this time that Mr Brooke began to dwell on a subject which was, in my opinion, the great error of his life, and one of the causes of the comparative want of success of Sarāwak. In his eager desire to establish a permanent government on the coast of Borneo, he began to doubt his own power to command success, and to urge on the British Government the necessity of taking over the country. Already he hinted at other alternatives—a powerful company, or even a foreign Government, if our own would have nothing to do with it. There was at this time some cause for uneasiness, as although he had established perfect tranquillity in Sarāwak, yet the coast was infested with the Lanun and Balagñini pirates, and the strong marauding tribes of Seribas and Sakarang, under their Arab and Malay chiefs, were continually cutting off the traders who ventured to leave the ports for Singapore and Java. In England his chief trust was first in his agent, Henry Wise, who in reality cared nothing for Mr Brooke or for Borneo, but who thought only how he could best serve his own ends by working the question. He was a clever, active, plausible man, who hung about the public offices requesting and obtaining audiences of Ministers, and acquiring thus a little reflected importance. The second friend was Mr Templer—familiarly termed Jack Templer. I never knew him; but, judging by his actions and his correspondence, I should say he was clever, but so injudicious in his conduct and advice that he was the cause of much of the estrangement between the Rajah and the British Government.

In May 1843 the Dido set sail from Singapore with Mr Brooke as a guest on board, and, being a good sailer, soon arrived on the coast of Borneo. After cruising for pirates during a few days, Captain Keppel sent away his heavy boats on the same errand. The expedition was commanded by Lieut. Wilmot Horton, and Mr Brooke accompanied the party, as he alone was versed in the language, and had a sound knowledge of the kind of vessels used by the pirates. The very day they left the ship they came on a fleet of Balagñini vessels; but though distant shots were exchanged, the marauders got away easily on account of the superior swiftness of their vessels, propelled as they are both by sail and by thirty to fifty oars. As they sailed in the direction of the Natunas, the English boats steered also on the same course, and anchored for the night under one of these islands. At daylight, while reconnoitring in a cutter, Lieut. Horton and Mr Brooke came upon six war *prahus*, which advanced upon them, beating their gongs, shouting, and making every warlike demonstration. Lieut. Horton returned to his force, and there drew up his three boats in line of battle: the enemy, confident in their numbers and in the size of their vessels, advanced with all the speed that their oars could give them; but as they approached near, Mr Brooke felt confident that there was some mistake, as the build was not that of pirate vessels. A white flag was therefore hoisted, and he shouted to them for a parley, but the only reply was a discharge of musketry. Lieut. Horton kept his men well in hand, and not a shot was returned until the *prahus* were within 50 yards, when he poured in a most destructive fire of grape and canister, and the marines bowled over every one who attempted to show near the enemy's guns. In a few minutes the affair was over; the largest *prahu* surrendered, and the other five fled, keeping up a running fire on the chasing cutters.

The captured *prahu* turned out to be one belonging to the Rajah of Rhio, an island dependent on the Dutch, that had been sent with the others to raise tribute in the Natuna group. They were greatly astonished to find that they had been engaged with the boats of an English ship-of-war,—declared that it was all a mistake, that they were at that moment searching for a fleet of pirates that had plundered one of the islands, that the sun being in their eyes they had not distinguished the flag, and urged every other excuse possible. Lieut. Horton, thinking perhaps that they had been sufficiently punished by the loss of twenty-five killed and wounded out of a crew of thirty-six, accepted their excuses, and sent the surgeon on board the *prahu* to attend to the wounded, and gave up his prize, as well as the others which the cutters had captured, deserted on the beach. These rascals had in fact mistaken the English force for boats from some wreck, or as some pretended, Dutch boats from Sambas; and as the Rajah of Rhio, in these days, sent the greatest scoundrels in his dominions to collect tribute, they would have had no hesitation in plundering or murdering any unfortunate survivors from a wreck who could not defend themselves.

After this adventure, the boats proceeded to Sarāwak, and found the Dido anchored at the entrance of the Tabas, called the Muara Tabas,[1] or mouth of the Tabas. Captain Keppel could also laughingly tell of how he had been deceived by the artful trick of three pirate boats, and how they had got away scot-free from under the guns of the Dido.

This was an eventful day for Mr Brooke. It was the first time that the natives could understand that true Englishmen are rarely abandoned by the force that above all others is the visible sign of the power of England. The Dido was a beautiful 18-gun corvette, in splendid order; and when she first anchored in the Sarāwak river below the town, her tapering masts overtopping the tallest jungle-trees, her crowds of white-dressed sailors running up the rigging, her 32-pounders booming forth a salute, were enough to strike the coolest native with awe, and to drive wild with excitement the warmest well-wishers of Mr Brooke's success. It was a beautiful sight, the whole river covered with boats and *prahus* of every description, draped out with gay flags, crowded with half the population in gala costume. Every old musket was brought into requisition, guns were fired, muskets were discharged, gongs were beaten, and an everlasting dinning was kept up. Presently Mr Brooke landed with

[1] Generally spelled "Morotabas."

all the honour of a governor, under another salute of the heavy guns, which awoke the echoes in many a surrounding mountain, and no doubt made many a Dyak heart jump, as the roar of these guns must have appeared to him worthy of a combat of the gods.

Captain Keppel visited Muda Hassim in state, and no ceremony was left undone which might please the royal prince, and cement his growing friendship for the English. He soon returned the visit, and was evidently astonished with what he saw, as previously nothing had entered the Sarāwak river but the Royalist and a few small trading vessels from Singapore. Here was order, cleanliness, and power, and this he felt.

Captain Keppel, in his 'Voyage of the Dido,' has given us a very good account of the house in which Mr Brooke lived in 1843, and of which I have already given Mr Brooke's description, though the engraving in that work representing a Sarāwak house was really that of Pangeran Makota. Captain Keppel says that Mr Brooke's residence, although equally rude in structure with the abodes of the natives, was not without its English comforts of sofas, chairs, and bedsteads. It was larger than any other house in the place, but like them was built on Nibong piles, and to enter it, it was necessary to make use of a ladder. The house consisted of but one floor: a large room in the centre, neatly ornamented with every description of firearms, in admirable order and ready for use, served as an audience-hall and mess-room, and the various apartments around it as bedrooms, most of them comfortably furnished, with matted floors, easy-chairs, pictures, and books, with much more taste and attention to comfort than bachelors usually display. But the fact is, you could never enter any place where Mr Brooke had passed a few days without being struck by the artistic arrangement of everything: his good taste was shown even in trifles, though comfort was never sacrificed to show. The house was surrounded by palisades and a ditch, forming an enclosure, in which were to be found sheep, goats, pigeons, cats, poultry, geese, ducks, monkeys, dogs, and occasionally a cow or two.

Then, as later, the great hour of meeting was sunset, when, after the preliminary cold bath to brace the nerves relaxed by the heat of the day, all the party met to dine. When Captain Keppel was at Kuching, all the officers of the Dido were welcome, and many a merry evening was passed at Mr Brooke's house. I have often heard Mr Brooke speak of that glorious time: then the future was all hope—no disappointments had soured the mind, and the cheerfulness of the host was sympathetic. I have never met any one who in his playful mood was more charming at a dinner-party: he told a story well, he was animated in discussion, fertile in resource, and when beaten in argument would shift his ground with great dexterity, and keep up the discussion, to the entertainment and admiration of us all. An appreciative observer once wrote: "The Rajah certainly has a most uncommon gift of fluency of language. Every subject he touches derives an additional interest from his mode of discussing it, and his ideas are so original that to hear him speak is like opening out a new world before one. His views about Sarāwak are so grand that it is with real pain one thinks how very little has been done to aid him in his noble efforts." Captain Keppel was also a capital story-teller, so that between the two, with occasional assistance from others, the time never lagged, and it was often well on in the small hours before the party broke up.

Already the custom of the natives coming in during and after dinner was observed. The house was open to all—rich or poor, Malay, Dyak, or Chinese, any were welcome. Often a very poor man would creep in, take up his position in the most obscure corner, and there remain silent but attentive to all that passed. There he would wait till every other native had left, neither addressing Mr Brooke nor being addressed by him, but when the coast was clear the Governor would call him to his side and gently worm his story from him. Generally it was some tale of oppression, some request for aid. None of these stories were forgotten: in the morning careful but cautious inquiries were made as to their truth, and rarely was it found that the suppliant had attempted to deceive wilfully. Redress or aid soon followed; and the custom was kept up, and should have been kept up to this day, but the presence of ladies and the advancement of civilised ways of governing have made those who have at various times acted for Mr Brooke during his absence and since his death neglect a custom, not without its inconveniences, but productive of much good, not only to the poorer natives, but to the governors, as a hint may be then thrown out of conspiracy or of danger which a Malay or Dyak would never dare give but under the shelter of night, and when almost alone with the white man.

But gaiety did not absorb the time. Captain Keppel had come with the firm determination to attack both the Seribas and Sakarang Dyaks, and put an end to their piracies; but sudden orders for China forced him to content himself with giving one blow, and Seribas was selected to receive it. Captain Keppel had heard much of their piracies before he reached the coast of Borneo, but here he was indeed surprised at the extent to which it was carried, and at the horrors that were almost daily enacted, as it was rare that during the fine season one or two squadrons were not foraging for heads, slaves, and plunder.

To understand the operations now about to be described the map should be examined, and it will be readily perceived that in the deep bight lying between Datu and Serikei Points, are the rivers Batang Lupar and Seribas.

These were the districts occupied by the two marauding tribes, who had even pushed their way down to the banks of the next great river, the Rejang, and occupied the Kanāwit and Katibas branches. The Seribas is a fine-looking river, and is navigable for a considerable distance; but the tides are rapid, and near the town of Boling there commences a "bore" that is exceedingly dangerous. The first of the flood-tide comes up with great force, and when it meets a narrowing river and sandy shallows it rises like a foaming wall from bank to bank, and upsets whatever boat or *prahu* it may meet in its impetuous passage, unless these are in deep water or sheltered in regular anchorages; and the crews, when such accidents occur, are generally drowned.

The Seribas and Sakarang Dyaks are of the same tribe and speak the same language, and in 1843 they were under the influence of Malay chiefs and Arab adventurers, who had first taught them piracy and encouraged them to frequent the seas for plunder and slaves. At first the Dyaks were contented with the heads taken during the expedition as their share, but now they had become the real masters, and only shared the plunder when Malays accompanied the expedition with guns and musketry.

It was soon known that a force was preparing to attack the pirates in their stronghold, and every effort was made by them to perfect their defences. Their forts were strengthened, booms were placed across the river, and lofty trees were partially cut through, so that when the supporting ratan ropes were severed they would fall athwart the stream and stop the progress of the attacking force.

Muda Hassim having officially called upon Captain Keppel to put a stop to the piracies of the inhabitants of the Seribas and Sakarang, preparations were made for the expedition. Mr Brooke determined to accompany the boats of the Dido with as large a force of natives as he could collect. At first all hung back, and his native chiefs entreated him not to go. The Seribas were so great a terror to them that they thought them invincible, but, seeing Mr Brooke determined, they gave in and commenced fitting out their war-boats.

While these preparations were going on, Mr Brooke was fortunate enough to get rid of Pangeran Makota. Though ordered to leave in January, he was found in May still getting ready, but the sight of the Dido was enough for him. He instantly packed up and disappeared with his people, going off to join Sirib Sahib of Sadong, the greatest instigator of piracy on the coast.

It is strange, but true as it is strange, that those who have attacked the policy pursued of putting down the piratical communities of Seribas and Sakarang have rarely found fault with Captain Keppel's proceedings, but have reserved their censure for those of Captain Farquhar. I will follow their example, and reserve my account of the conduct of these tribes until I have to treat of the affairs of the year 1849, when the "great battle" of Batang Marau was fought.

The force which left Sarāwak during the first week of July 1843 was a large one. It consisted of the pinnace, cutters, and gig of the Dido; a native-built boat called the Jolly Bachelor, manned by thirty Didos; and a vessel laden with stores and ammunition.

The native force consisted of about 300 Malays and 400 Dyaks, all thirsting for revenge, but considering the expedition as almost desperate. As the mosquito fleet ascended the Seribas, it passed, on the right bank, first the Rembas branch, and then the Paku, up each of which there were fortified towns, which it was their intention to destroy later. Padi, the furthest inland town, was now the object of attack. The boats brought up at Boling the first night—a deep bend where the dreaded "bore" has but little effect, though the tide runs with great force. During the night no enemy came to disturb them; but the distant beating of gongs, and the occasional discharge of ordnance, served two purposes: it kept up their own courage, and might intimidate the enemy. Next morning the force advanced with the flood-tide: the beating of gongs, the yells of thousands of Dyaks in the woods, heralded the approach to the scene of action. Mr Brooke was with Captain Keppel at the head of the force; and as they were swept up the river by a strong flood-tide, it was indeed an exciting scene. Presently they came upon a cleared hill, with a fortified house on the summit, upon which dozens of men were performing a most awe-inspiring war-dance. As they neared, however, hundreds of warriors sprang up from the long grass, and rushed down upon what seemed an easy prey; but after giving them a couple of shots, the boat swept past them. An instant after, the report of a heavy gun showed that the pinnace was near. On they went, and in a few moments found themselves opposite a strong boom, with a very narrow opening in it. For this, Captain Keppel had steered his gig; and the speed at which they went enabled them to squeeze through, and to find themselves in front of three formidable-looking forts, which instantly opened fire on the little gig. Fortunately the guns were laid for the boom, and the shots went over their heads; but hundreds of warriors rushed down to the banks, and hurled their heavy javelins at the daring strangers. The pinnace was now thrown up against the barrier, and all the other boats were stopped; but the natives, with their sharp-cutting *parangs* or choppers, soon severed the fastenings, and the boats got through and opened fire on the enemy with both grape, canister, and musketry. While the pinnace used her 12-pounders, the others dashed on, and the blue-jackets, under their gallant officers, sprang ashore, and rushed at the nearest fort. This method of warfare was so new, so unexpected, that the garrison, without for a moment considering their superior numbers, or the strong position they held, fled panic-stricken into the jungle, and the fort in a few minutes was in the hands of the Didos. After this stockade had fallen there was little further resistance, and the native force dashed on for the town, and after having plundered it, burnt it, as well as all the surrounding villages. Next day parties ascended both branches of the river to complete the work; but finding that the left-hand branch was the most encumbered with obstructions, a strong party was sent up this. The Seribas Malays and Dyaks determined to make a last effort, and to endeavour to destroy this detached party. As night closed, an attack was made on the boats from all sides, particularly from a sloping bank, where the pirates could wade to within a few yards of the invaders; but the pinnace's heavy gun, and the steady musketry-fire, proved so destructive, that they were forced to beat a retreat. There was no sleep for any one that night: the rain poured down, but the marines had to stand sentry through it all; and the big gun had constantly to be used to stop the work of obstructing the river. A simple signal-rocket having been fired, however, the enemy retired before what they did not understand, and left the English at peace.

Next morning preparations were made to continue the advance, when a white flag was seen to be hoisted by the enemy, and in a few minutes a truce was established. The chiefs came down; and after Mr Brooke had fully explained to them the causes which had led to the attack, called upon them to give up piracy, and take to honest trading. This they promised to do, but they pointed out that, unless the same punishment was inflicted on the two other towns, they could not prevent pirate squadrons from leaving the river, while they were assured that the compact would not be neglected by the English. During the previous skirmishes, there had arrived reinforcements in the shape of about 1000 Balow warriors—brave Dyaks, from the Lingga branch of the Batang Lupar—who, though of the same tribe originally as the pirates, had never joined in piratical acts. These men had suffered so much from the Seribas that they did not miss this occasion to retaliate, and the country was laid waste for miles around—a most just and necessary measure, and the only way to put an end to piracy when it is encouraged and fostered by a whole people. Let them feel in their own homes the misery they so often inflict on others. The Padi forces were estimated at 500 Malays and 6000 Dyaks, brave against native enemies, but incapable of opposing any disciplined force.

The English and native forces now dropped down the river, and soon found themselves at Boling, where the store-vessel was anchored. She had received and returned a few shots, but no serious movement had been made against her.

Then followed attacks on the pirate settlements of Paku and Rembas. There was little resistance, as the defenders, both Malay and Dyak, had been cowed by the defeat and destruction of their allies at Padi. It is worth while, however, mentioning that at Rembas there was found a vessel 92 feet in length, and of 14 feet beam, capable of carrying several guns, and a crew of 150 men. The inhabitants of these places soon hoisted flags of truce, and the chiefs agreed to proceed to Sarāwak to arrange the terms of future peace. The chastisement they received was enough to discourage them for the moment—and even for a few years they ceased sending fleets to sea; but the spirit of piracy was too ingrained in them to be driven out by one lesson.

Sakarang was the next point of attack; but when Captain Keppel reached the Dido he found imperative orders for China, so that expedition had to be put off. On their arrival at Kuching, they were treated as conquering heroes; and, to the native mind, their achievements merited that honour—for in the space of a fortnight they had done what, for fifty years, the whole power of the Sultan had failed to do,—they had bruised the head of piracy as represented by the Seribas, and the other black sheep tremblingly waited their turn.

Twelve months elapsed before Captain Keppel could return a second time to the coast, and during that year many interesting events occurred. No sooner had the Dido left than the Samarang, commanded by Sir Edward Belcher, arrived. This visit was strictly an official one, and was as useless as such visits usually are. What can the most acute naval officer in the world understand of a country during a few days' or weeks' visit? He can describe more or less accurately its outward appearance; but to understand its internal politics is not possible in the time. And yet on such comparatively valueless reports the British Government relies in a majority of cases. Mr Brooke suffered more than any other pioneer of civilisation from this system.

Mr Brooke explained everything fully to Sir Edward, showed him the country, and pointed out to him the favourable and the unfavourable side of the question. After a short sojourn at Sarāwak, Sir Edward invited Mr Brooke to accompany him to Brunei, where he wished to inspect the coal; but in descending the river the Samarang grounded on a rocky bank, and when the ebb came, fell over on her side and filled. This, in some respects, was a fortunate circumstance for Mr Brooke: it drew the attention of the Admiral to Borneo; and Mr Brooke was enabled to aid so materially in the recovery of the vessel as to merit the official thanks of that chief. The recovery of the vessel was a triumph of naval skill, which reflected the greatest credit on the captain, officers, and all concerned.

Among the young officers on board was a midshipman named Brereton, a bright intelligent lad, who afterwards joined the Rajah in his arduous task of governing the sea Dyaks.

In August Mr Brooke proceeded to Brunei in the Samarang, that vessel having been recovered in the short period of eleven days. A squadron—sent to her aid, but arriving too late—was with them, consisting of H.M.SS. Harlequin, Wanderer, Vixen (steamer), and Diana, with the Royalist, and a merchant vessel, the Ariel. Never had such a squadron appeared in Sarāwak waters.

I cannot help noticing here what false notions are given of countries by the imaginative pens of young travellers. One of the officers, in describing the kind way in which Mr Brooke housed the shipwrecked, speaks of the nightly visits of wild hogs, porcupines, wild cats, guanas, and various other animals, not to speak of swarms of mosquitoes, scorpions, lizards, and centipedes, that failed not to disturb them with their onslaughts during the whole night. All this is pure imagination. A wild hog might come near the house, and so might a guana. During thirteen years in Borneo, I saw two guanas near a house, once a wild hog, never a porcupine or wild cat. Scorpions and centipedes were almost equally rare, except among old wood; while the lizards are of the smallest and most harmless description. So much for travellers' tales.

During this visit to the Sultan, his Highness confirmed his grant of Sarāwak to Mr Brooke, and, in addition to the previous stipulations, gave him the power of naming his heir in the government. As this concession was possibly due to the presence in the river of a British squadron, Mr Brooke, with great delicacy, considered that H.M.'s Government was now entitled to the refusal of the country. Mr Brooke soon returned to Sarāwak and passed three months of quiet there, necessary to him, as he had been living almost in public for the last five months. No man enjoyed those periods of rest more than he. After having carried on active operations, after having passed through a round of entertainments—receiving every day a dozen officers at dinner—he rejoiced in repose, when he could enjoy his books, his writings, and meet his people, and carry on the government in a quiet, unpretending manner. Besides, after so much suspense, the almost inevitable fever and ague followed, and he thought it better to make a short voyage to Singapore, where he hoped to meet Sir William Parker, the Admiral, who then commanded the station, and explain to him the position of affairs in Borneo. But on his arrival there he found that the Admiral had left for Penang, and he received at the same time the news of the death of his mother. This was a great affliction, for not only had he the greatest love for her, but she was one of the few who thoroughly understood her shy, sensitive boy, and who urged him on in his career of usefulness. To the last hour of his life he ever spoke of her with that tender affection which was one of the most winning points in his character.

Mr Brooke followed the Admiral to Penang, and arrived about the time that an expedition was preparing to punish the piratical towns on the coast of Sumatra. These towns were nominally under the sway of the Sultan of Achin, but in reality were independent; in fact, the coast of Sumatra was, like the coast of Borneo, a series of nests of pirates. Mr Brooke, finding that his services would be highly appreciated, offered to accompany the expedition, as a knowledge of Malay and an acquaintance with the court life of Asiatics were unknown accomplishments in the fleet.[1]

H.M.SS. Harlequin and Wanderer, with the steamer Diana, were told off to punish the pirates of Batu and Murdu; but first of all they set sail for Achin, to meet the nominal suzerain of these districts. Achin is a decaying empire, but the late war with the Dutch shows that courage is not wanting. As the Sultan could do nothing to punish the pirates, the squadron sailed for Batu, and all satisfaction being refused the town was burned. At Murdu greater preparations were made for defence, and various stockades were erected to repel the invaders. The last crime committed by the people of Murdu was to seize a merchant vessel, pillage the cargo, and murder a portion of the crew. It was intended to land a party above the stockades, but a very strong tide carried some of the boats past the landing-place, among others the gig in which Mr Brooke was a passenger. It was swept right under a stockade full of Malays, who instantly opened fire on the English. There was nothing to be done but spring on shore and dash for the defence, and try and shoot down every one who showed himself. The Malays resisted with vigour, and Mr Brooke soon found himself among the wounded: a shot struck him inside the right arm, and as he approached the stockade, a spear was jobbed over, which caught him on the eyebrow, and cut through to the bone. The rush of blood from the wound blinded him and made him look a ghastly object, and gave the idea of a much more serious injury. The blood streamed over his face and clothes, and almost closed his eyes. The action was kept up for about five hours, the

[1] Miss Jacob, desirous to give her hero every accomplishment, speaks of the facility with which Mr Brooke picked up native languages, including the Lanun. In reality, however, Mr Brooke never spoke but one native language—the Malay—but that, though acquired slowly and laboriously, he spoke well; and in court Malay he had no equal among the Europeans.

resistance being obstinate; but ultimately discipline prevailed over irregular valour, the town was taken and destroyed, the enemy losing heavily, while the English had two killed and twelve severely wounded. Among the latter was Lieutenant Chads, who, dashing ahead, found himself face to face with a Malay chief, who boldly advanced to the combat. They both struck together: the *kris*, with the edge of a razor, nearly cut through the left arm raised instinctively to guard the blow, and inflicted a severe wound on the body; while the cutlass did not penetrate the silk jacket. Before, however, the Malay could finish his work, a ball from a marine's musket laid him low. It is a curious circumstance that Lieutenant Chads was considered one of the best single-stick players in the fleet, and yet, in actual combat, he forgot his sword exercise, and cut instead of parrying. Single-stick exercise is generally of little use, as players seldom strike home.

The battle of Murdu took place on the 12th of February 1844, and Mr Brooke's gallantry was so appreciated by the blue-jackets that they asked permission to give him a parting cheer when he left the Wanderer; and this they did right heartily, manning the rigging as he went over the side. On their return to Penang, Mr Brooke met the Admiral and his friend Keppel, and was delighted to hear that the Dido was to be sent again to the coast of Borneo. Sir William Parker had highly approved the conduct of Keppel and his dashing achievements on that coast, and wished him to complete the work. Sir William was most kind to Mr Brooke, and fully appreciated the enlightened views which he laid before him for the development of British influence and commerce in the Eastern Archipelago: but these views are unappreciated by the public still, and particularly by statesmen and politicians, though the governors of Singapore have commenced a new system which is likely to awaken from its torpor the fruitful peninsula of Malacca. The British Government usually leaves to its agents the initiation of a new policy, and its agents are in general afraid to undertake the responsibility.

Mr Brooke remained in Singapore awaiting the arrival of Keppel and the Dido till the end of May, but finding when she came in that she was carrying treasure to China, he left in the Harlequin for his home. Captain Hastings was in command. They found a large pirate force on the coast. At the time, Mr Brooke was disappointed that nothing was done to destroy this fleet; but even had the boats of the Harlequin been a match for the enemy, such an engagement would not have had the moral effect which arises from the destruction of the strongholds of the pirates, who, after months of preparation, reckoned confidently on repulsing any attack.

Mr Brooke, after six months' absence from Sarāwak, arrived to find all prospering: there had been no serious crime committed, no attack had been made on his people, trade was increasing, and now he could write—"I like couches, and flowers, and easy-chairs, and newspapers, and clear streams, and sunny walks." All these he had in the new house which he had built on the left bank of the Sarāwak river, on a rising knoll between two running streams, with the broad river flowing below. It was a pretty spot. A four-roomed, lofty house, surrounded by broad verandas; in front his well-stocked library, a splendid hall or dining-room, with a couple of bedrooms behind them. When I knew it, a special wing had been added for Mr Brooke's own use, and the rest was given up to his followers. Around the house was the thick foliage of fruit-trees, with lawns and paths bordered by jasmine plants and the *Sundal Malam*, that only gives out its fragrant perfume during the night. Pigeon-houses, kitchens, and servants' rooms were partly hidden by trees, and here and there were planted and tended with uncommon care some rose plants, Mr Brooke's favourite flower. "All breathes of peace and repose, and the very mid-day heat adds to the stillness around me. I love to allow my imagination to wander, and my senses to enjoy such a scene, for it is attended with a pleasing consciousness that the quiet and the peace are my own doing."

While in the full enjoyment of this repose, he was rudely awakened by the news that the piratical Dyaks had attacked and killed a couple of his people. Instantly, with his usual energy, he called out his warriors, manned his war-boats, and started off in pursuit, and was fortunate enough to inflict some loss on his savage enemy. "Oh for Keppel!" was his constant cry, and he could not but spur the willing horse.

At length, on the evening of the 29th July, the Dido anchored in the Morotabas entrance of the Sarāwak river, and found there the Phlegethon steamer, that had been sent on ahead to await her arrival. At daylight on the 30th Keppel arrived at Kuching, to receive a welcome such as so appreciative an officer deserved.

As everything had been prepared for his arrival, the expedition was ready to start on 5th August, and this was a more formidable one than the last, when fear of the pirates caused many to hesitate and hang back. Even Pangeran Bedrudin had insisted upon accompanying his friend, and Muda Hassim had given his consent, for all had confidence in the "Red-haired Devil,"—for by that euphonious name was Henry Keppel called by friend and enemy.

The Batang Lupar river, into which the Sakarang flows, was to be the scene of the new expedition. Its entrance is marked by two hills, one on either bank, with the island of Trisan in the centre. It is a broad, noble-looking stream, being from three to four miles wide for the first six leagues; but the land is low, and there is nothing striking in the scenery. The forests are dense, and filled with fruit-trees, affording nourishment to herds of pigs, which boldly swim the stream when in search of pastures new. We once fell in with a drove there, which gave us an exciting chase; when, in endeavouring to decapitate one with a cavalry sword, I nearly performed the same office for a too eager follower. Here also is found the Mias Papan or the gigantic Orang Utan,[1] which is, in my opinion, a very different creature from the Mias Rambi about which Mr Wallace writes. About twenty miles up the river on the left-hand bank the Lingga joins the great stream. This is inhabited by the numerous tribe of Balow Dyaks, brave, but untainted with piracy. There was also a village of Malays under the command of an Arab, Sirib Jaffir, the chief who initiated the peace negotiations with the rebels at Siniāwan in 1840. The river still maintains a great breadth, but it is more encumbered with shoals, and soon the spot is reached where the bore commences, a terrible enemy to strangers or unskilful mariners. Passing the hill of Tisan, the rapid stream carries you on to the town of Patusan, and then, after twenty miles' further ascent, to where the river divides into three—the Undup, the Sakarang, and the Mani, still called the Batang Lupar. Two Arab pirate chiefs commanded her. Sirib Sahib, formerly of Sadong, had retired to this river, and had strengthened the forts at Patusan, until he believed them impregnable; and Sirib Mulla had a position further in the interior. Makota, or the Serpent, had established himself as Sahib's chief adviser; and, secure in the difficulties of the river, the strength of their fortifications, and their numerical superiority, these men awaited the attack of the English.

I have often been up the Batang Lupar river, and shall never forget the fearful velocity of its current, when at

[1] In Malay, "Man of the forest."

new or full moon the bore is at its height. And this was the place that, with the steamer Phlegethon, the boats of the Dido, and the native squadron, Keppel determined to attack. The unknown was the greatest element of danger, as Borneon pilots were not accustomed to handle steamers. The expedition moved up the river, and when off the mouth of the Lingga, directions were sent to Sirib Jaffir not in any way to support or countenance the pirates. It then moved on and anchored a short distance below Patusan, Sirib Sahib's stronghold. Next morning, with a slashing stream, the force moved up the river, and soon came in sight of the enemy's forts, that opened fire immediately on the coming foe. The Phlegethon's anchor was let go, the boats formed in line, and a rush was made for the nearest defences. The forts continued their fire until the blue-jackets and marines were at the embrasures, and then the garrison broke and fled, not understanding this peculiar mode of fighting. In a few minutes the enemy were driven out of all their stockades, and the whole town was in the possession of the combined forces, as this time the Sarāwak men were but little behind their white allies. The town proved extensive, and nearly all newly built, as it had been made the rendezvous of the neighbouring pirates, when Sirib Sahib, no longer feeling himself safe at Sadong, left that place for a stronger. At Patusan there is a small river, the Grāhu, up which Makota had established his village. He was not forgotten: in fact, three days were spent in destroying every vestige of dwelling, embarking sixty brass guns, and throwing a large number of iron ones into the river. Thus the rendezvous of the worst pirates established near Sarāwak was destroyed with the loss of but one killed and a few wounded.

There were still three other communities to punish,—the Sakarang, the Undup, and the Batang Lupar. The last two attacks were but repetitions of previous ones: trees cut and dropped across the river to arrest progress, firing from the banks, dreadful yells, followed as usual by defeat, destruction of forts and villages, and the dispersion of the pirates. In the attack on the Undup, the first lieutenant (Wade) of the Dido was killed, from rashly rushing ahead of his men. Keppel in vain warned him, but as the two were 100 yards ahead of the others, it would seem that the commander did not practise the caution that he preached.

During the attack on the Sakarang, the Dyaks showed much skill in defending the approaches to their capital, and one of their stratagems, which caused the attacking party considerable loss, is worth recording.

After proceeding a certain distance up the river, the flotilla either anchored or the boats were fastened to the banks of the stream, and all hands were piped to breakfast. While Captain Keppel and Mr Brooke were thus engaged, Patinggi Ali, the most daring of the Sarāwak chiefs, asked permission to go ahead with the fast spy-boat to reconnoitre. He was permitted to do so, but strict directions were given that at the least sign of the enemy he was to return and report. Unfortunately, a Mr Stewart, a young merchant, who had volunteered to accompany the expedition, found means to conceal himself under the mats in the spy-boat, and passed on with the rest. It is supposed, and it is highly probable, that being unacquainted with native warfare, he urged on the brave but usually cautious Patinggi Ali to proceed further from his supports than prudence warranted. The spy-boat proceeded onward until the river narrowed, and the confined stream came down with a swift current which slackened the speed of the advancing party. No enemy was seen, no yell heard : this ought to have warned them that some ambuscade was prepared. The moment the Sarāwak party was well in the narrows, loud shouts and yells arose on all sides, and at the same instant bamboo rafts were thrown across the stream to cut off retreat, and six large war *bangkongs* came sweeping round the point, and closed on the doomed party. The spy-boat had no chance,—the crew could only sell their lives as dearly as possible. As their boat sank under them, Mr Stewart and Patinggi Ali endeavoured to board the nearest of the enemy's boats, but they were soon overpowered and their heads secured. On the first shot being fired, and the diabolical yell heard, Keppel pushed off in his gig, and soon came upon a mass of confusion,—rafts, boats, war *bang-kongs*, all pell-mell together; the Sarāwak Malays in reduced numbers still defending themselves, and fighting single combats on the rafts; enemies and friends mixed so together that it was impossible for the English to fire. At last a raft caught a snag in the river, which made an opening through which Keppel pushed his gig, and giving the helm to Mr Brooke, he opened a rapid fire on the enemy. Seeing only half-a-dozen whites on their side of the confused mass, the Dyaks rushed down to the banks to secure their prize; but the steady fire kept up at so short a range disconcerted them, and soon another English boat came up, and the rockets dispersed the enemy who crowded the banks. The entire force was soon on the ground, and the Dyaks fled in all directions, after incurring very heavy loss. The attacking party had also suffered: the killed amounted to twenty-nine, and the wounded to about sixty. No further serious resistance was offered to the advance; and shortly after, the flames and smoke issuing from near the banks told all the country round that the Sakarang capital had been reached by the invincible invaders, and that the "Red-haired Devil" had inflicted a severe lesson on its piratical inhabitants. The deaths of Lieut. Wade, Mr Stewart, and Patinggi Ali, were greatly regretted, as each in his own manner had nobly distinguished himself.

The expedition now returned to Sarāwak, meeting on its way the boats of H.M.S. Samarang, which had come to help, but was too late to join in the attack. No sooner, however, had the force reached Kuching, when news arrived that Sirib Sahib had fled to the inner waters of the Lingga branch of the Batang Lupar, and that, with the assistance of his brother, Sirib Jaffir was rallying his forces. With his usual energy, Captain Keppel immediately returned to Lingga, and the boats of the Dido and Samarang proceeded in search of the enemy. This energy was too much for the late fugitives from Patusan: they fled in all directions, Sirib Sahib taking refuge over the mountains in the Dutch tributary states.

To complete the work, Jaffir was deposed from the government of Lingga, and new authorities named in his place. Thus these three Arab adventurers disappeared from the scene, and almost ceased hereafter to exercise any influence. Makota was also taken prisoner, but was allowed to go free: far better would it have been, however, for the inhabitants of north-western Borneo, had Mr Brooke suffered Muda Hassim to punish his crimes with death.

It is important to notice that when these proceedings were brought before the High Court of Admiralty in England, the Sakarangs were declared to be pirates.

The Dido and Samarang soon left—the former for England; while the latter, after visiting Singapore, returned with the Phlegethon to remove Muda Hassim and his family to Brunei, and to search for a white woman said to be held captive at Ambong, a pretty bay to the north of Brunei.

The families of Muda Hassim and his brothers were with the greatest precaution removed from the houses to the steamer. Screens were put up, each woman well wrapped up, and covered with cloths, so that no indecent eye should be able to view her charms. To render the operation more secure, midnight was chosen, when the dull light of the torches showed little more than that bundles of dirty linen were being carried below. As a rule the female relatives of the royal family are not more worth looking at than the ugly men. Sir Edward Belcher thought that by the light of the torches he could detect some fair women among them, but the fairest are of a light yellow,—a sickly yellow, obtained from living in darkened rooms and being never exposed to the sun; a sickly yellow, like the leaves of plants which have never been exposed to the light of day. The poor women, who with their slaves might be numbered by dozens, were crowded into the small cabins below, and were suffocated by the heat, and one died during the voyage from the effect of the vitiated atmosphere, and all were covered with rash. Yet Muda Hassim would not permit them to come on deck, even under the protection of screens. It was an absurd jealousy, as from his own experience he might have known that none of his precautions would render his women faithful: on the contrary, enough was seen on board the Phlegethon to show that the harem system was an absurdity.

I have seen many of the young women belonging to the royal family, and my own experience confirms that of Mr Brooke. They are generally broad-faced and ugly, though their manners are pleasing and gentle. A favourite pastime among them is to go on picnics, and when they passed any house in covered boats, the women would pull down the mats to have a good look at the stranger; and after I had built the Consulate-General at Brunei, there was quite a movement among the wives and daughters of the Rajahs to inspect the imposing building, and see the wonderful mirrors and other furniture supposed to exist there. On one occasion, notice was sent me that some of the Sultan's relatives were coming to inspect the house, and a request was made that I would send all the men-servants away. I offered to go myself as well, but that was not permitted. They came, about eighteen in number, the elderly taking little care to conceal their features, and about a dozen with silk *sarongs* over their heads, showing only a bright black eye. Presently one of them sprang into an American rocking-chair, which immediately turned over. As I ran to pick up the young girl, the rest dropped their head-covering, as if by accident, and came forward to help. I then had a good look at them all, and certainly there was not a pretty one among them. They were from fourteen to twenty years of age, almost all flat-faced, but with bright black eyes, and long black hair. The impression was certainly not favourable, and I had a fair chance of judging, as after I had seen them once, little trouble was taken to conceal their faces during the rest of the visit.

The Phlegethon started for Brunei, and the Samarang followed: on their arrival at the mouth of the river, they found the whole place in arms. The English were coming to seize the country, and all the batteries were manned. However, it was soon known that Muda Hassim and his family had arrived, and the hostile party were obliged to succumb and to receive him. His family were disembarked from the Phlegethon, to the relief of the officers, and Muda Hassim was established beside his nephew, the Sultan, as his chief adviser; while his late minister, Pangeran Usup, was allowed to occupy an inferior position.

Mr Brooke and Sir Edward both expressed their surprise that any hostilities should be dreaded from the English after all the late amicable intercourse; but the piratical party, ignorant of the strength of the foreigner, were longing for a rupture. Muda Hassim, however, with the consent of the Sultan, sent large parties down the river to destroy the batteries, and peace appeared to be established in the capital. War squadrons then left for Ambong Bay to search for the lady said to be in captivity, but it was soon proved that no such person had ever been held captive there. Mr Brooke was delighted with the country,—in fact, there is no part of Borneo equal to that neighbourhood for beauty. Taking leave of Sir Edward, who sailed for Manilla, Mr Brooke returned in the Phlegethon to Sarāwak.

It was during this visit to Brunei that Mr Brooke obtained from the Sultan the offer to cede the island of Labuan to the British Crown; and at the same time coal was found near the north-eastern point of the island. As I shall have much to say hereafter on the subject of Labuan, I will defer any description of that island until the period of establishing our Government there. Coal was also found at several places near the capital, and the quality was pronounced good.

CHAPTER V.

EVENTS IN BRUNEI AND ON THE NORTH-WEST COAST—MURDER OF MUDA HASSIM AND HIS FAMILY—CAPTURE OF BRUNEI.

1844–1847.

MR BROOKE returned to Sarāwak in November 1844, and began now to really govern the country. He had previous to this time been hampered by the presence of the Malay Rajahs, but the removal of Muda Hassim and his insubordinate followers swept at one stroke the chief difficulties from his path. He now remained the only chief to whom all looked up; he had his Europeans and his three Sarāwak Datus to aid him, and quiet being restored in the interior, trade commenced to prosper. The Dyaks living in the surrounding countries began soon to send emissaries to Sarāwak to discover whether it was true that the "son of Europe was the friend of the Dyak;" and on their reports, hundreds of families began to move into Mr Brooke's territory. In two months over two hundred had passed the Sarāwak frontiers, to the intense disgust of the Malay rulers, from whose tyrannical governments they were flying.

On the coast, also, peace was almost established. The severe lessons given by the Dido had for the moment awed the piratical tribes, and now there were only the Lanuns and Balagñinis who infested the neighbouring seas. And even these had observed that ships of war were now more numerous on the coast, and therefore they themselves visited it less, as these pirates' trade is to plunder, not to fight.

While enjoying this precious tranquillity, Mr Brooke was startled by the arrival of his agent, Mr Wise, and of Captain Bethune in H.M.S. Driver, which anchored in the Sarāwak river on the 17th of February 1845, and brought a despatch from Lord Aberdeen appointing Mr Brooke confidential agent in Borneo to her Majesty, and directing him to proceed to Brunei[1] to convey a letter on the subject of the suppression of piracy; and Captain Bethune was authorised to select a spot on the coast where a British settlement could be formed. Mr Brooke spoke of Captain Bethune as intelligent and liberal, and he proved himself to be so.

Little time was spent in Sarāwak, as the Driver started on her mission the 21st of February, and on the 24th anchored off the Brunei river, to find all quiet at the capital. The Queen's letter was received with all honour, the presents were accepted with pleasure, and every one pretended to be delighted with the determination of her Majesty's Government to put down piracy. I say pretended, as the nobles felt that if piracy were effectually put down, slaves would be more difficult to procure, and they were too blind to see how their riches would increase by the natural development of their country. It was quite clear, however, that Muda Hassim could do nothing to put down piracy, unless backed up by our English force.

On March 12th Mr Brooke and Captain Bethune arrived at Labuan to examine its capabilities, and from thence returned to Sarāwak and Singapore in order to meet the new Admiral, Sir Thomas Cochrane. I do not dwell on these journeys, as they were so similar in most respects, but

[1] It may be mentioned here that Brunei is the "Borneo Proper" of old maps and voyagers.

reserve details for more important expeditions. Another visit to Brunei convinced both Mr Brooke and Captain Bethune of the dangers to which Muda Hassim and his party were exposed from the intrigues of Pangeran Usup, supported as the latter was by the piratical communities of the north, headed by the dreaded Sirib Usman, an Arab adventurer, who had gained considerable influence at Maludu Bay and in the neighbouring countries,—an influence which was much increased by a marriage with one of the daughters of the Sultan of Sulu. The Sultan of Brunei also, who was a pirate at heart, secretly favoured the enemies of Muda Hassim: in fact, though this was not known at the time, the Malay prince was only supported by the traders, or the men of the district of Burong Pingé. Mr Brooke saw the danger to his friends, and returned to Singapore again to lay the whole case before the Admiral. Sir Thomas Cochrane now determined to act, and collecting a squadron, the largest that had ever appeared on the coast of Borneo, sailed for Sarāwak, where after a short stay he proceeded to Brunei. Sir Thomas felt that there could be no peace on the coast until Pangeran Usup was driven out of the capital, and until the piratical stronghold at Maludu was destroyed. Fortunately, Pangeran Usup held two British subjects in slavery, and had refused to give them up. Upon this the Admiral, who had ascended the river to the capital with H.M.S. Vixen and two other steamers, called upon the Brunei Government to punish the man who not only had committed this crime, but who was in constant correspondence with the pirates, and was encouraging them in their pursuits. The Sultan, awed by the fine present, pretended to be willing, but said that Pangeran Usup was too strong for him, and begged the Admiral to take his punishment into his own hands. Usup was sent for, but refused to come: he assumed a defiant attitude, loaded his guns and said he would resist force by force,—in fact, he did not believe that the English would act. There were three war-steamers in the river, besides the boats of the squadrons—enough force to capture the whole island of Borneo; but still this boasting noble would not yield, though during the night he removed his treasures and his women to a place of security, and gave out that he intended to surprise one of the steamers. Precautions were, however, taken on board the ship of war.

Next morning a shot was fired over Pangeran Usup's house, which he returned by firing at the steamers. In a few minutes such an iron shower had fallen on the place as never before had been known in Borneo. Pangeran Usup and his followers, however, had discreetly disappeared after firing their first volley. Having thus energetically supported Muda Hassim's authority in Brunei, Sir Thomas set sail for Maludu, on the north coast of Borneo, to look into Sirib Usman's proceedings. Mr Brooke, though on board H.M.S. Agincourt, the flagship, was not actually present at the attack on Maludu, Sir Thomas very properly thinking that a civilian should not expose himself unnecessarily in such expeditions. But the account given of the affair shows how severe a struggle it was. I have been to the spot several times, and am rather surprised at the comparatively small loss. The attacking force, under the command of Captain Talbot, consisted of 24 boats and 550 men—blue-jackets and marines. They ascended the river to a spot where a strong boom prevented further progress. A treacherous attempt was made to take prisoners some of the leaders of the force, but this failing, the action began by the enemy opening fire upon the boats. This was returned; but for fifty minutes the boom resisted all attempts, and during this time the English were exposed to a very heavy fire indeed. Fortunately Malays and Sulus fire badly, or we should have suffered worse. Once the boom was cut and broken, the boats pulled on, and in a few minutes the forts were in our power. We had but about twenty killed and wounded, but the enemy had suffered severely. Sirib Usman was carried away severely hurt, and soon after died: many of his most warlike chiefs were likewise killed, and all their riches were lost. The place was full of evidences of piracy —the spoils of many a captured European vessel. In this action one of the officers engaged was Charles Johnson, the present Rajah of Sarāwak, who here, for the second time,[1] became practically acquainted with the pirates of Borneo.

Among the little incidents which marked this affair, I may notice that a woman was found, twenty-four hours after the action, lying in a canoe with her arm fractured, unable to move, and dying for a drink of water. Playing about her was a little child, who was vainly seeking its accustomed nourishment. She was taken on board one of the vessels, her arm amputated, and Mr Brooke, on finding that she was a slave, offered to take her to Sarāwak, to which she assented. I often saw her there coming up to visit her benefactor, with her little child; and although she quickly found a husband, she occasionally showed herself in order that Mr Brooke might not forget his accustomed largess.

The destruction of Maludu was the greatest blow that had been struck at piracy, and did infinite honour to Sir Thomas Cochrane's judgment. The effect was great; for although it afterwards became necessary to punish other pirate communities, no stronghold was again formed, and the good done was permanent.

After the action a visit was paid to Balambangan, an island lying off the north of Borneo, and formerly held by

[1] Mr Johnson was present at the attack on Patusan with the Dido's force in 1844.

the English. The first time we were driven out by the Sulus, the next it was abandoned as useless.

The Admiral now departed with his squadron, and Mr Brooke and Captain Bethune left in the Cruiser to touch at Brunei, and from thence to Sarāwak and Singapore. At Brunei they had the welcome intelligence of the energetic action of Bedrudin, Muda Hassim's favourite brother, in repelling an attack of Usup, and in driving him out of the country, forcing him to leave the capital and take refuge in Kimanis, a district about thirty-five miles to the north of Brunei. When it was known that Usup had taken refuge there, a *chop* [1] or order was sent by the Sultan to the headman to seize and put to death the rebellious noble. I have listened to the chief's account of what occurred, and it strangely illustrates the ways of the natives. The Orang Kaya [2] of Kimanis was a very respectable man, who was exceedingly puzzled what to do. Usup had been kind to him, but if he disobeyed the order he would very likely suffer death himself on the first occasion that he visited the capital. Usup had naturally great suspicion that the Orang Kaya had received the order for his execution : he therefore took every precaution. His followers had dispersed, and there remained with him but one man, and that was his younger brother. When Usup slept, his brother sat by him with a drawn *kris*, or sword, in his hand ; when he bathed, similar precautions were taken : Usup in his turn performed the same office for his brother. Two or three weeks passed in this way, the Orang Kaya doing the honours of his house as if nothing had occurred, but at the same time keeping near him two or three strong men who were to take advantage of the first favourable opportunity. "It was here," said the Orang Kaya, pointing out to us the

[1] A mandate with the seal of the Sultan or ruler.

[2] Orang Kaya means chief, or literally "rich man."

steep steps that led to the stream, "that I at last seized them. Usup was bathing; his brother sat at the head of the steps, *kris* in hand; my followers and I sat near chewing our *siri;* when Usup called from below that he should like a little of that leaf. The brother bent forward to give it, and in doing so let his *kris* rest flat on the floor. At this moment I gave the signal, and my followers sprang upon him and secured him and the arms. Usup would have fled, but it was useless: my men rushed down the steps and secured him. They were taken into an inner room, and there strangled with all the respect due to their relationship to the Sultan. Usup bitterly reproached me with my ingratitude and treachery, but I could not but obey the orders of my sovereign. There is his grave," added he, pointing to a neighbouring low hill, where a stone marked the spot of Usup's last resting-place.[1]

Pangeran Usup was an able man, but unfortunately could not read the signs of the times: he thought the English were but birds of passage, and knew nothing of our power. Had he frankly accepted our advent, his ability might have been of singular service in a country where ability is rare. Mr Brooke returned to Sarāwak, and Captain Bethune, after another tour among the Dyaks, departed for England in order to lay his report on the north-west coast of Borneo before the British Government.

A little before this time an American frigate had visited Brunei, and had endeavoured to secure a cession of the coal, offering in return to protect the Government. This induced Mr Brooke to compare the energy of the Americans with our own slow proceedings. But a little reflection would have shown him that the captain of the frigate wanted to get the cession, and that the subsequent protec-

[1] There were various versions of this story, though differing only in minor details. I think the one I now give is perhaps the most correct.

tion would have been nominal, as the United States had only one vessel in those seas, and nothing but a force constantly in or very near the capital could have saved Muda Hassim's Government.

The rest of this year was passed in peace. An attempt made by some Arab adventurers to lead a Sakarang fleet to sea was defeated by the bravery of the Bolings, who surprised them; and the pirates, thinking that these Dyaks were but the advance-guard of the Sarāwak force, fled, leaving eighteen war *prahus* in the hands of the brave warriors of Lingga.

In Sarāwak everything was peaceful—trade was improving; the people, instead of being half starved, now began to import rice, which showed what quiet could do for their prosperity. Thus closed the eventful year of 1845, during which much had been done to secure Mr Brooke's position.

Before proceeding further, let us glance at those who now aided Mr Brooke in his attempt to govern Sarāwak. His chief foreign supporters were Messrs Ruppell, Williamson, and Crookshank. Mr Ruppell was a good-hearted and a good-natured man, but did not understand administrative work: though bred to trade, he showed a similar incapacity in account-keeping. Mr Williamson, originally an interpreter, had the faults of those who are born and educated in the East: he was able, and understood the natives, but was not sufficiently frank in his dealings with the poorer Malays. He was distrusted by Mr Brooke during the last years of his life, as it was evident that he was too much influenced by the female relatives of the Malay chiefs. It is necessary to tell the exact truth about his officers, or otherwise Mr Brooke's career would be but half understood. Mr Crookshank, who had been bred to the sea, was one of those few men who are not thereby rendered unfit for shore pursuits. Of all those who entered Mr Brooke's service, Mr Crookshank was the one who best understood the natives: no other ever arrived at that perfect knowledge of the language, that almost instinctive insight into Malay character, that patience to follow a difficult case, or to unravel a web of native intrigue. He was certainly a most useful and competent officer, and did what none other did,—he remained a faithful follower till long service entitled him to a pension. Perhaps I should add that he was one of the few sportsmen who succeeded in providing the larder with venison; and his hunts of the wild boar were as famous as was his good dog Kejang.

Two melancholy events ushered in the new year: the first was the death of Mr Williamson, who, falling from his boat, was drowned. The circumstances were indeed sad. I have mentioned that he was too much under the influence of the relatives of the Malay chiefs,—so much so indeed, that Mr Brooke had reason to suspect that justice was not administered to the poor during his repeated absences; that, in fact, they were left without protection, and that some of the old abuses of Malay government were creeping in. Inquiry having proved this to be the case, Mr Brooke had been forced to leave the administration in the hands of Mr Ruppell, and to direct Mr Williamson to act as his assistant: he treated the latter also with marked coldness, and never invited him to join the pleasant dinners at Government House. At length the Malay chiefs interfered: they pointed out to Mr Brooke that Mr Williamson's errors had probably arisen from thoughtlessness; and the Datu Patinggi also felt inwardly that if the English subordinate had erred, he himself had been the great culprit, as it was through his instigation that his relatives had exercised the wrong influence. At length Mr Brooke relented, as he felt that the severity he had already shown would probably deter Mr Williamson in future, and be an example to be avoided by others. The next morning he sent him an invitation to dinner, and in the evening he came. Everybody noticed how cheerful Mr Williamson was, now that the cloud between him and his chief had passed away. He stayed later than usual, and after a most affectionate farewell to all, went down to the river's side and stepped into his canoe. Instead of sitting down, as was his custom, he stood up, and his servant began to paddle him across. Unluckily the canoe bumped against some driftwood, Mr Williamson lost his balance, and was pitched head-foremost into the river. He did not rise; and though the neighbourhood was roused by the shouts of his servant-boy, and people hurried thither in canoes, it was not possible to do anything, as the tide swept the body away from the spot where he fell. Thus Mr Brooke lost an experienced and useful follower, who with many faults had yet many qualities, which made his death appear a public loss. I have described what occurred thus fully, as in later days, when passion was at its height, Mr Brooke was accused of having murdered this young officer.

I have already noticed the removal of Muda Hassim, his brothers, and their followers to Brunei. Sir Edward Belcher having kindly undertaken this task, and established this branch of the royal family in the capital, it was believed that they would be advantageous to British interests. I have also noticed the opposition with which they met, and the spirited conduct of Pangeran Bedrudin in driving Pangeran Usup, their most prominent enemy, from the neighbourhood of the capital. The death of this redoubted chief, however, did not clear the city of the enemies of the legitimate branch of the royal family.

The Sultan of Brunei, like many men of weak intellect, delighted in the society of persons of inferior rank, and had collected around him a set of scoundrels inferior to none in villany. He had also near him his own illegitimate children, or those who passed for his children, and these had gained a decided influence over him. The presence, therefore, of Muda Hassim and his brothers was irksome to all those who saw that they were aiming at supreme power, and that their own ambitious projects were likely to be foiled. When, therefore, Muda Hassim forced the Sultan to recognise him as heir to the throne, although a strictly legal act, conspiracies arose, and it was determined to get rid of this branch of the royal family. Mr Brooke at the time thought that the determination to massacre Muda Hassim and his brothers had its origin in the engagements entered into by them with the English to aid in putting down piracy, and to protect legitimate commerce. No doubt this added fuel to the fire; but I believe that, whether they had entered into engagements or not, they would have been destroyed, as the Sultan would have no equal near his throne, and the comparatively respectable character of Muda Hassim and his family was utterly opposed to his own conduct. The advocates of friendship with the pirate communities no doubt supported the scheme, Haji Saman being at the head of this party.

Numerous meetings were held in the palace, and it was some time before the Sultan would give his consent to the massacre of these his nearest relatives; but after Muda Hassim had obtained from him the declaration that he was the legitimate heir to the crown, he no longer hesitated, and preparations were made to surprise all the brothers. So secretly was this scheme conducted that no details of the plan reached them. They did, however, receive some warnings; but feeling confident after the destruction of Maludu and the deaths of Sirib Usman and Pangeran Usup, and thinking that they had adequate support in the friendship of Mr Brooke and the aid of the English navy, they took no efficient precautions.

One night, when the brothers were scattered, the signal was given: bands of armed men left the palace, and pulling silently in the darkness, arrived unobserved near the houses of the different brothers. They attacked simultaneously. The young princes had but few followers with them. Bedrudin fought gallantly: he defended the entrance of his house for some time, but with three or four followers he could do little against a murderous band of forty or fifty. Finding that he with his *kris* held his own, and that they could not force an entrance into the house, one of the assailants fired. The shot took effect in Bedrudin's left wrist, and as that arm fell he received a severe wound in the right shoulder and several wounds in the body. His few followers were either killed or fled. He managed, however, to gain the inner apartments, where he found his sister, a favourite concubine, and Japar, a slave lad. The latter he commanded to reach down a barrel of powder, and spread the contents on a mat. He then called the women to sit near him, and turning to the lad said: "You will take this signet-ring[1] to my friend, Mr Brooke, tell him what has occurred, let him inform the Queen of England that I was faithful to my engagements, and add," he said, "that my last thoughts were of my true friend, Mr Brooke." He then ordered the lad to save himself. Japar opened the lattice-like flooring, slipped down a post into the water, and swimming to a small canoe was enabled to paddle quietly away, while the murderers, suspicious, were cautiously making their entrance into the house. Japar had not proceeded many yards when a loud explosion told him that the gallant prince had set fire to the powder, rather than fall into the hands of his enemies.

Muda Hassim was attacked at the same time, and probably would have escaped had he pulled directly to the Burong Pingé *kampong*, the inhabitants of which would

[1] One which Mr Brooke had given him.

have protected him. After a gallant defence, he too, wounded and overpowered by numbers, was forced to destroy himself. Of the fourteen brothers, but two or three escaped: one, Muda Mahomed, whom I knew afterwards, was desperately wounded; another became insane; and this unfortunate family ceased to exist as a power. Although Mr Brooke endeavoured to do something for the survivors, they have almost disappeared as a political element in Brunei.

Mr Brooke has often said before me that the destruction of this family was a misfortune to their country. Perhaps it was; but I who lived several years in the capital heard many things which accounted for the unpopularity of these princes. Malay court etiquette, when carried to extreme, is etiquette run mad. With all the apparent servility of the Malays, they are a democratic people, and during late years had become more so. One of the customs of Brunei was, that when a non-noble passed before a house inhabited by a royal personage, he was obliged to fold his umbrella and expose himself either to the hot rays of the sun or to the rain. The custom had fallen into desuetude, but these princes determined to revive it. The principal street of Brunei is the main river. Whenever a non-noble was seen passing before Muda Hassim's palace with his umbrella up, officers were ordered to pursue and bring his canoe to the landing-place, and he himself was to be brought before the Rajahs to be fined. This gave rise to much abuse. The insolent followers of the princes, secure from all punishment, beat and otherwise ill-treated the most respectable members of the commercial class, and thus alienated from the cause the most devoted partisans of Muda Hassim. I give this as but one instance; but similar efforts to revive an obsolete etiquette, and many acts of great oppression whilst raising revenue, practised by irresponsible agents, loosened the bonds of respect which once united the cause of the people with that of the family of Muda Hassim.

With these faults of education, however, it was not possible to treat with indifference the claims to respect to which Bedrudin was entitled. Brave, courteous to a degree even to be remarked by the most punctilious of a punctilious service, ready to be influenced by a foreign civilisation, and wanting near him but a wise and appreciative European to render him a superior ruler, I have ever regretted the death of Bedrudin at a moment when English influence was about to be established on the north-west coast of Borneo.

I can well imagine the effect which was produced on Mr Brooke by the news of the death of his friends. His excitable nature was roused almost to madness: it made him ill to feel that he was without power to fall upon their murderers and exterminate them. He blamed the British Government for not having followed his advice, and stationed a vessel in the Brunei river until the power of his friends was fully established; but he could not but afterwards confess that such an act of treachery on the part of the Sultan, after the recent triumphs of Bedrudin, was not to be foreseen. In fact, all appeared in their favour; but we had yet to learn how utterly without forethought are the Malays,—how, in their ignorance of the power of foreigners, they brave their enmity, when it could be avoided by the dictates of the most ordinary prudence.

Directly the news of the massacre of Muda Hassim reached Sarāwak, Mr Brooke applied to the Government of the Straits Settlements at Malacca for a steamer to enable him to keep the coast quiet pending the decision of the Admiral. At that time Lieut.-Colonel Butterworth was the Governor, and he readily complied with the requisition. He sent the E.I.C. steamer Phlegethon to Borneo, and directed Commander Scott to place himself at the service of H.M.'s Agent. With the aid of this steamer Mr Brooke was enabled to keep the piratical rivers quiet, and to prevent the contagion from Brunei from spreading. Already secret emissaries had been sent from the capital to endeavour to incite disturbances in Mr Brooke's province, and, if an opportunity offered, to get rid of that agent of new ideas, the disturber of the good old order of things, and the possible future avenger of the death of his friends.

Mr Brooke chafed at his inability to move with his own force, and the necessity of waiting for the decision of Sir Thomas Cochrane; but he afterwards felt how much better it was to act with the power of England at his back.

Directly the Admiral heard of the death of those who had so earnestly urged an alliance with England in order to extirpate piracy, he determined to look into the matter, and if necessary to act. He felt, as the responsible agent of England in these seas, that we had no right to interfere in the internal arrangements of the country, or to punish crimes that might arise from local jealousies, or from causes with which England had no concern. He had admired Bedrudin, had felt how superior he was to those around him, but at the same time he knew that the Sultan who had committed these crimes was the lawful ruler of the country, and that England had no right to call him to account for them, unless at the same time he broke his engagements with his new ally. It is seldom that a naval chief has either the time or the inclination to master the politics of a native state, but it is to the honour of Sir Thomas Cochrane that he did attempt to understand the position of parties in Borneo Proper, and, understanding them, had the courage to act.

As soon as circumstances permitted, he set sail from Singapore with a powerful fleet, and in a few days reached Sarāwak. Anxious to promote the peace of the coast, he cruised quietly along, and left the squadron, to visit the Rejang, the finest river on the north-west coast; steamed above a hundred miles up it in the Phlegethon, and warned the different chiefs against giving any aid to the piratical tribes, or listening to the evil advice of the emissaries from Brunei. The sight of this powerful squadron had a great effect for the moment, but the natives used to look upon us as birds of passage, so that the effect was not permanent.

The fleet anchored at length off the island of Muara, at the entrance of the Brunei river, and preparations were made for a visit to the capital. Mr Brooke had received full information that the Sultan had summoned all his supporters, that the inhabitants of the capital had been forced to work at extensive fortifications, that guns were mounted on many batteries, and that the armed bands comprised more than 5000 men. While the fleet was preparing for an expedition, a boat was observed pulling towards the flag-ship, and the yellow umbrellas denoted that some man of rank was approaching. It soon drew alongside, and two persons dressed in great state came on board bearing a letter from the Sultan, addressed to Mr Brooke. It requested him to pay no attention to rumours; but should he desire to visit his Highness, to come in two small boats only, as no other force would be allowed to pass the batteries. Mr Brooke asked the messengers a few questions, and instantly suspected that these were low men who had been sent to act the part of nobles; and it proved to be so.

The Admiral determined to take no notice of the restriction as to numbers, but to ascend the river as he had done the year before, with a similar force to support him. Two steamers and two sailing vessels were told off, and all the principal boats of the fleet were prepared. On passing the bar of the Brunei river the Hazard grounded, and the Admiral was forced to proceed with the other three vessels. The Phlegethon led the way. For some time nothing of an opposing force was seen; but no sooner did the squadron arrive at the reach below the great bend than they saw several formidable batteries before them, crowded with armed men. These batteries were posted with great judgment, and were mounted with very formidable artillery. When the advance party was within about 1000 yards, the enemy opened fire, fortunately aiming too high, so that the whole volley passed over the English force and ploughed up the water in their rear The fire was quickly returned; guns, rocket-tubes, and musketry were brought into requisition; the blue-jackets and marines dashed ashore, and in a few minutes the enemy abandoned their guns and fled into the jungle. The squadron then advanced and steamed up the main street. In front of them, where the mosque now stands, Haji Saman, the head of the piratical party, had erected a battery that he considered invincible. As soon as he saw the English force before him he fired, and the heavy shot from his guns did some damage to the advancing ships; and even his grape pierced the thin iron plates of the Phlegethon, which would have sunk had she not been built in compartments. But resistance was useless; the English guns soon drove the Malays from the batteries, and the landing-parties took possession of the guns. These batteries were formidably armed. Some of the cannon were 30 and 40 pounders, mostly of old Spanish manufacture, and were elaborately ornamented. Thirty-nine pieces (nineteen of which were brass) fell into the hands of the British force.

The marines were landed and occupied the hill behind the palace to prevent annoyance during the night; but the Borneons were too cowed by the large force to attempt doing anything further.

The city was found completely deserted, and thousands of houses full of property lay at the mercy of the conquerors. But private property was strictly respected. The Sultan fled into the interior, and tried to fortify himself, but he was followed up by armed parties under the command of Captain Mundy of the Iris, and he had again to fly into the swamps, leaving his property to be destroyed. Some of the less guilty chiefs, finding that the town was respected, began to return, and in a few days Mr Brooke was enabled to establish a sort of provisional government. The Admiral went so far as to say that, although the ultimate decision remained with the English Government, yet, on proper submission, the Sultan would be permitted to return to his capital.

The capture of Brunei by the English squadron was a great event in the annals of the country. When, some years later, I travelled in the interior, the oppressed aborigines, after cautiously glancing round to see that no Borneon was present, would express to me their delight at the defeat of their own Government, and describe in vivid colours the flight and terror of the nobles. "You should have kept the country," was the invariable finish to their stories.

The provisional government which was established consisted of Pangeran Mumein, who, though not of really royal blood, was at the time considered the best man that could be found—a reason which some years subsequently induced all parties to make him Sultan,—and of Pangeran Muda Mahomed, brother to Muda Hassim—a man originally of little wit, and that little much lessened by the terror of that night of murder, and the desperate wounds which he himself had received. Affairs being thus settled, the squadron left Brunei, a single ship remaining to preserve order, and proceeded along the north-west coast to look after the piratical communities of Pandassan and Tampasuk. These were punished for recent marauding acts. Maludu was visited, but found deserted. The Admiral then proceeded to China, Captain

Mundy of the Iris being left to look after the coast and punish Haji Saman, the promoter of piracy, who, when driven out of Brunei, had retired to Membakat, where he fortified himself. He was quickly routed out of this, and Mr Brooke proceeded to Brunei, where he found the Sultan ready to make every submission. Our agent insisted on the punishment of the murderers of Muda Hassim, but as the true murderer was the Sultan himself, only inferior agents could be punished. The Sultan was very humble, renewed all his engagements, even to a re-gift of Sarāwak, and granted the right of working coal to Mr Brooke. This last he accepted to prevent any other nation obtaining exclusive rights, and held it at the disposal of her Majesty's Government.

Having thus finished the political part of the business, Mr Brooke collected the survivors of Muda Hassim's family and removed them to Sarāwak, where for some years they remained at the sole charge of Mr Brooke. There were about thirty or forty women and children, and a few men, none of whom were calculated personally to excite any interest.

While these events were taking place, the English Government had determined to take possession of Labuan, an island opposite the mouth of the Brunei river. There had been a change of Ministry, and Lord Palmerston acted, as usual, promptly. It was during this year that my attention was first turned to Borneo. I had read Keppel's book with interest, and my father being acquainted with Mr Wise, Mr Brooke's agent in England, we were kept informed of all that was passing on the momentous Borneon question. My father had constant interviews on the subject of Labuan with Lord Palmerston, both before and after he returned to the Foreign Office, and that enlightened statesman was glad to be able to obtain information from one who had really studied and taken great interest in the question.

When the orders arrived to take possession of Labuan, Sir Thomas Cochrane directed Captain Mundy to proceed with a small squadron to Brunei and obtain a cession of the island, and then to take possession in the name of her Majesty. This he did, while Mr Brooke proceeded to Singapore to meet the Admiral, with whom he was desirous to communicate before his return to England.

Mr Brooke remained absent from Sarāwak four months, only returning there in May 1847, having passed the interval with the Admiral at Penang, where the delicious climate of the Government house, 2500 feet above the level of the sea, aided in restoring his health, somewhat shattered by exposure.

Having received instructions from her Majesty's Government to negotiate a regular commercial treaty with the Sultan of Borneo Proper, he left Singapore in the East India Company's steamer Nemesis, Captain Wallage, and after touching at Sarāwak to find all quiet and prosperous, proceeded to Labuan, and from thence to Brunei, accompanied by Commander Grey of the Wolf, and Lieutenant Gordon of the Royalist, and a guard of marines. He found the capital quiet, and had little difficulty in inducing the Sultan to sign the treaty, by which Labuan became definitively a British possession, and future commercial and other arrangements settled.

With the treaty in his possession, Labuan under the English flag, and secure of the Sultan's conduct after the lesson he had received from Sir Thomas Cochrane, Mr Brooke began to think of home. He made up his mind to pass a few days in Sarāwak, and then to start for England overland. But before he left he was destined to have another brush with his old enemies the pirates. As the Nemesis steamed towards the mouth of the river she was hailed by a Borneon *prahu*, and Mr Brooke was able to make out that there was a squadron of Balagñini outside that had chased the Borneon to the entrance of the river. Whenever the steamer cleared the point, the look-out man at the mast-head shouted that he could see some *prahus* chasing another; in a few minutes he could make out eleven in full pursuit of a fishing-boat that was being urged with vigour towards Labuan. As soon, however, as the smoke of the steamer became visible, the pirates gave up the chase and pulled into shallow water, and tried with might and main to get down the coast to some small river in which they might seek temporary shelter. But the Nemesis, though forced to make a detour to avoid the shoals, gained rapidly on the Balagñini, who, seeing flight impossible, as their vessels made slow way in a chopping sea, drew up in battle array in a small bay. Their sterns were placed towards the shore, their bows to the sea, and to keep them well in line they were connected by a line of cable stretched across their bows. The Nemesis approached them, and Commander Grey, the senior officer, to prevent any idea of mistake, directed a boat to be lowered, and an officer volunteered to approach them and discover who they were. It proved a useless precaution. Before the boat could shove off from the side, the pirates opened fire on the steamer, by which one man was killed. The fire, of course, was immediately returned, and round-shot and grape were vomited forth from two 32-pounders on this formidable-looking enemy. The fire was kept up by the pirates with great spirit. The practice on board of the Nemesis was not good, even allowing for the roll of the sea, as the action lasted three hours and a half before it was thought advisable for the boats to make a dash, and even then it was too soon. When the pirates saw that the boats were about to attack them, three put to sea and steered to the northward, a fresh breeze having sprung up. The Nemesis gave chase. No sooner did the Balagñini observe that the steamer was off, than the crews of three *prahus* which they had abandoned ran down and manned them, and boldly engaged the boats. Had they fired decently they would have got the best of it, but the steady fire from our side kept them off. In the meantime Captain Wallage of the Nemesis, becoming aware of the danger of the boats, gave up the chase and returned. This soon changed the face of affairs; some of the *prahus* were again driven on shore, while three more were enabled to escape. The action, while it lasted, was severe. The English lost two killed and many wounded, while the enemy had above fifty killed, whose bodies were seen, besides those who may have suffered in the other *prahus:* [1] there were also fifty or sixty men who were forced to take refuge on shore.

The firing was heard at the capital, which, in a straight line, is not very distant from the coast, perhaps eight miles, while not a sound was noticed on board the English war-vessels at Labuan, though not more than fifteen miles distant, otherwise not a *prahu* would have escaped. As it was, of the six who got clear of the Nemesis, three foundered from the injuries received during the engagement, so that but few returned home to tell the tale.

The Nemesis, after this encounter, communicated with the capital, Mr Brooke writing to the Sultan to request him to take care of the captives taken by the pirates, many of whom had been enabled to escape ashore. They were taken care of, and subsequently forwarded to Singapore. They were mostly either Chinese or Dutch subjects.

The firing being, as I have said, heard in the capital, much alarm was felt, and the authorities collected together their armed men to be ready for any eventuality. Very

[1] These *prahus* carry from 40 to 50 men each,—besides captives, who row in double tiers. The vessels are lateen-rigged, and are armed with two bow-guns, several swivels, muskets, spears, swords, &c.

soon the villagers came flocking in to report what they had seen from the neighbouring hills. In the meantime the pirates who had landed were left without any means of escape. Six of their *prahus* had disappeared, and the English had seized the five remaining. Being without provisions they marched inland, and coming upon some small cottages after dark, they surprised the inhabitants in them and wantonly killed those who had no time to escape. When this news spread, the whole country was in arms, and by the morning hundreds of men arrived from the capital and surrounded the houses where the pirates were staying. They were summoned to surrender, and being without food or chance of escape, they agreed to do so provided their captors would take them into the Sultan's presence. This was done. His Highness first offered to permit the escaped captives to kill with their own hands those who had treated them so barbarously, but they declined the task. Upon this the Sultan gave the signal, and the young Rajahs rushed upon the pirates and hacked them to pieces. One of the young nobles who was engaged in this affair told me that these Balagñini behaved with the utmost courage, not one having begged for his life, but all declaring that their death would be thoroughly avenged by the Sultan of Sulu.

The result of this action was most important: for many years after the Balagñini gave the north-west coast a wide berth, not caring to meet again one of the English Rajah's fire-ships. This engagement proved also how dangerous it is to send small boats to attack large pirate *prahus*, whose crews must fight or be taken. The inhabitants of all the north-west coast greatly rejoiced at the result of the fight. For many years the Borneon seas had been rendered insecure by them, and trade had languished, as every one feared to expose himself to a hopeless captivity. It took the Balagñini about fifteen years to forget the lesson.

Mr Brooke left Sarāwak under the care of Mr Arthur Crookshank, to whom I have before referred, feeling perfectly secure that nothing would go wrong in his province. His reputation was then at its height on the coast, and no one dared to oppose him. Dr Treacher and Mr Ruppell, with their subordinates, were associated with Mr Crookshank in the task of holding Sarāwak during its Rajah's absence.

I have no details of the voyage to England. Mr Brooke embarked with his friend Mr Hugh Low. He left Singapore in July, but on arriving at Point de Galle found the Calcutta steamer gone, and that he must wait in Ceylon for a month. Not to have to pass this time in such a place as Galle, he determined to visit Colombo, and then the up-country, where he fell in with an old friend, Mr Jolly, with whom he spent his time very pleasantly. He used often to recur to this visit, and express his admiration of the beautiful culture to be seen in the coffee plantations. In August Mr Brooke was enabled to start again, and finally reached Southampton on the 1st October 1847, after an absence of nearly nine years. He was met at Southampton by Captains Keppel and Mundy, many members of his family, and by some of his old friends. What a happiness it must have been to his affectionate heart to find assembled there all whom he loved best!

This visit to England was one of the most interesting episodes of his life. He could not complain that his countrymen had overlooked his merits: on the contrary, no one was ever better received, nor has any one's qualities been more generally appreciated. I fear, however, that but few materials are at my service to give a clear account of all that took place, as his letters during that time are brief, and personally I saw but little of what passed, though I had the pleasure of being introduced to him at Mivart's.

CHAPTER VI.

MR BROOKE VISITS ENGLAND.

1847–1849.

Mr Brooke's visit to England was in every way satisfactory. He found himself surrounded by an affectionate family and by a crowd of enthusiastic admirers, was treated with friendly consideration by the Ministers, and he acquired the respect and friendship of many who ever after remained steadfast to him. Among the most influential of these were Lords Ellesmere and Grey, the latter of whom has lately shown how fully he understood and admired Mr Brooke's noble qualities.

Mr Brooke could not but feel that his countrymen thoroughly appreciated his services. London presented him with the freedom of its ancient city; the principal clubs and City companies paid him every attention, and made him an honorary member; and Oxford honoured herself in honouring him with her distinctions. The undergraduates received his name with enthusiasm, for he was pre-eminently the man to create that feeling among young men.

I find by his notes that he was dining with Lords John Russell, Grey, Lansdowne, Leicester, and Auckland, making a round of country visits to Lords Lansdowne, Ellesmere, Haddington, and to Sir Patrick Stewart, yet giving as much time as he could spare to his family and friends.

Whilst in London he lived at Mivart's Hotel, and there surrounded himself by his most intimate friends, and continued his Borneon habit of staying up half the night in merry converse. One day, at breakfast, a waiter brought in a letter, which Mr Brooke asked permission to open immediately. He then said, "This is a very curious one." It was from some lady, who, enamoured of his deeds, proposed herself in marriage. The letter continued, that if Mr Brooke had no intention of marrying, he was to destroy a note which was enclosed, and which contained her name, her address, and all particulars as to her family and fortune. The guests laughingly said, "Have an intention to marry, and open the note:" but Mr Brooke immediately rose from the table, saying, "I have no intention of marrying," and put the letter and the enclosure into the fire. If the lady be still living, it will be a comfort to her to know what became of her communication.

My father used greatly to enjoy these meetings, and many a night has he sat up with his friend talking to two and three in the morning.

I must notice one incident particularly, as it may tend to explain many an attack subsequently directed against Mr Brooke. In 1846, Mr Wise called to express his thanks for the tone of the articles my father had written about Borneo in the 'Morning Chronicle,' 'Foreign Quarterly Review,' &c. From this there arose a great intimacy, and Mr Wise urged me to turn my attention from Persia, where I expected an appointment, to Borneo. We noticed how enthusiastic Mr Wise always appeared to be when he referred to Mr Brooke; but as my father used to say, he was too enthusiastic. This lasted about a year, when one day, after a very long conference with Mr Wise, my father said to me, "Our friend has been letting out a great many things about Mr Brooke: he accuses him of all sorts of crimes, but says these must be kept a profound secret.

Perhaps you had better give up the idea of going to Borneo." When, however, Mr Brooke arrived, Mr Wise was the first to get him to attend a large dinner at his house, and then made a most fulsome speech when proposing his health, citing him as a paragon of goodness, philanthropy, and humanity. When my father told me this, I instantly set Wise down as a humbug; and further intercourse with Mr Brooke made us understand that some secret cause of enmity existed on the part of his agent. I afterwards discovered that his change of tone commenced from the date when he found that he could not induce Mr Brooke to intrust him with full powers to form a great company which was to acquire Sarāwak for a very large sum, of which Mr Wise was to have half. He had stomached the expression that "a friend was worth a dozen agents," which by carelessness he was allowed to see in one of Mr Brooke's letters to his mother, but he could not stomach the loss of the great fortune which he had anticipated. This particular account is necessary to understand what followed.

Another circumstance must not be forgotten. Captain Mundy having decided to bring out an account of the voyage of the Iris, had received permission from the Rajah to incorporate into his book that portion of his journals which had not already been published by Captain Keppel. My father undertook to prepare this work for the press; but as his partial blindness prevented him reading the proofs, that share of the work fell to me, and I regret to add that my want of familiarity with the names of Dyak tribes caused several transpositions, which were subsequently used to attack the Rajah out of his own mouth. The Rajah tried to correct the proofs, but his engagements were too numerous. However, he did try, as I find by the following extracts from notes to my father: "*January* 13, 1848.—I send this corrected, but it is nonsense, written on my first acquaintance with Borneon affairs." "I cannot write any more journal." "The names are frightful—much of the information valueless, from subsequent inquiry." It is important to note these remarks, as even to this day his detractors have endeavoured to attack his fame on account of some contradictions in his journals.

Not only did the Ministers show their appreciation of what Mr Brooke had done in Borneo, but her Majesty testified her approbation by inviting him to Court—so that among the numerous visits which Mr Brooke paid, one of the most interesting was that to Windsor Castle. He gave a description of what occurred in a short letter to a niece, and although it has been already published, I will reprint it. It is as follows :—

"WINDSOR CASTLE, *25th October* 1847.

"I know, my dear Mary, that all you young ladies will be dying of curiosity to hear all about my visit here, and I hasten therefore to tell you that I am sitting in a room, a very comfortable room, with a good fire, a neat bedstead, and every other comfort and luxury which a gentleman could desire. I am sitting in this said room writing to my dear niece, and I much regret to say that I have not met with one single adventure, nor have I seen one precious face since being in this celebrated Castle, excepting Prince Albert's *valet de chambre*, who is a very well-spoken, well-dressed, civil gentleman—at which circumstance I am rather astonished, as I had always entertained an idea, a very vague and indistinct one, that all subordinate persons in all palaces were addicted to insolence and vainglory. Thus, my dear girl, you have the wonderful and entire history of all the events which have befallen me since I arrived; and as the time draws towards half-past seven, I must lay down my pen and dress for dinner. By the by, I asked the civil and well-spoken valet (whom you must know is a German) what dress it was strictly proper to appear in, and he very discreetly informed me that a black or blue coat, white waistcoat, white cravat, tight pantaloons, with black stockings, was the right thing—your shoes without buckles, and neither hat nor gloves. Heaven help me! how little I dreamed once that I should ever think of dress more! how little I thought in my wildest imagination that I should be here, her Majesty's guest! So let us say with all our hearts, God bless the Queen! I will go on to-morrow morning, if I have time.

"26*th.*—Three minutes before eight, the groom of the chambers ushered me from my apartment in the York Tower, conducted me along a splendid gallery, resplendent with lights, and pictures, and statues, decorated with golden ornaments, the richest carpets, and bouquets of fresh flowers, and ushered me into a drawing-room as fine as mortal eye could wish to see. Directly afterwards Lady Westmoreland and Lady Peel, with Lord Westmoreland and Sir Robert, entered, with the lord-in-waiting (Lord Morley), equerries, and grooms: then came the Duke and Duchess of Bedford, &c., &c., and last the doors were thrown wide open, and the Queen and Prince Albert and the Duchess of Kent were ushered in, attended by the Court ladies. I had to kiss hands on my presentation: her Majesty said very sweetly that she was happy to make my acquaintance. I bowed to the ground. The Queen took the arm of the Prince, and led the way to the dinner-room. I handed out Lady Emily Seymour, the music played, the lights glittered, reflected from golden ornaments, the hall was splendid, the sideboard resplendent. What shall I say of the dinner, the plate, the cookery, all befitting royalty, and neither coldness nor stiffness? The Queen was seated between Prince Albert on one side and the Duke of Bedford on the other, and—can you conceive it, Mary ?—she ate and drank just like any other mortal, but all in the most ladylike manner possible. After dinner we all stood, and her Majesty conversed first with the Duke of Bedford, then with Sir Robert Peel, and lastly with myself. She said all that was kind, talked with me for nearly ten minutes, and then we returned to the drawing-room. Prince Albert likewise honoured me with a long conversation, the Duke of Bedford talked to me, Sir Robert shook me by the hand, and said I was no stranger to him. I was presented to Lady Peel, *et hoc genus omne,* which being translated means, "all them there nobs." Such are my adventures, my dear Mary; and in return I beg of you not to let any person read or see this letter, excepting our own families of Lackington, Cheltenham, and Hillingdon. Farewell, dear Mary. It is now nearly time to go to chapel, after which I breakfast with the equerries, and start for London; and I may conclude by saying that, highly honoured as I have been, delighted and pleased, yet I shall be glad when it is over."

The Queen asked Mr Brooke, how was it that he found it so easy to manage so many thousands of wild Borneons? His reply was characteristic. He said, "I find it easier to govern 30,000 Malays and Dyaks, than to manage a dozen of your Majesty's subjects. At which reply the Queen laughed, and said, "I can easily imagine that." Mr Brooke ever spoke of that visit to Windsor Castle with pleasure, and was pleased in after-years to learn that though not again invited to Court, it was from no want of sympathy on the part of her Majesty, but rather from the desire which Prince Albert expressed, not to appear to take even an indirect part in the discussions which waxed warm before Mr Brooke again returned home.

Among the many schemes to which Mr Brooke's arrival in England gave rise, was that of establishing a mission in Borneo. A great meeting was held in Hanover Square Rooms, where friends were collected and speeches made. Among the speakers was the Rev. F. T. McDougall, who had been selected to be the chief of the mission. The tone of his speech made Mr Brooke sigh, but it was hoped that his actions would be more sensible than his words.

Many of Mr Brooke's friends and admirers thought that the Government would have taken the opportunity of his visit to England to give him some special mark of its approbation; but whatever hints he may have received of coming honours, he left England as he arrived, plain Mr Brooke.

It was during this visit to England in 1847 that the late President of the Royal Academy painted that life-like portrait of the Rajah which shows what he was in his bright days, when hope still kindled his eye, and disease and sorrow had not bowed his frame. The Rajah fully appreciated, and often spoke of, this handsome gift of Sir Francis Grant.[1]

During his stay in England, Mr Brooke was appointed Governor of the new settlement of Labuan, and several officers were selected to serve under him. As these were numerous, it was determined that they should proceed to their destination in a ship of war, *viâ* the Cape of Good Hope, instead of taking the overland route—a most unwise determination, as during this long voyage were sown the seeds of strife and heartburning, which sprouted and grew, and later on ripened into very unpleasant fruit.

Besides being Governor of Labuan, Mr Brooke was

[1] I bought this picture at a sale of the effects of the late Rev. Charles Johnson, Sir J. Brooke's brother-in-law.

Commissioner and Consul-General in Borneo, and through this appointment I became associated with his work. I had read Keppel's book, had taken a warm interest in everything appertaining to Borneo, and had written innumerable articles on the subject; and, as I have said, when Mr Wise suggested to me that I should seek an appointment in Borneo, I quickly caught at the idea, and determined to carry it out. My father introduced me to Mr Brooke at Mivart's, and I was then struck with that winning manner which in those days made every one who approached him his friend.

Mr Brooke applied to Lord Palmerston to name me as his secretary, but as there were some difficulties in that straightforward course, I received my appointment in a peculiar manner. Mr Hawes, Under Secretary of State for the Colonies, wrote to Governor Brooke that I was appointed his private secretary, and Lord Palmerston allowed me in a roundabout way £200 a-year. I received this announcement on the 24th January, with orders to be ready to start from Portsmouth on the 1st February 1848.

From that time to the day of his death I was in constant communication, either personally or by letter, with the subject of this biography.

As I have said, our party being a large one, the Government determined to send us out in the Meander frigate, Captain the Hon. Henry Keppel. The vessel was specially prepared for action against the pirates, being furnished with extra boats drawing but little water, so as to be able to carry a large force up the shallow rivers.

How full of hope we all were on that, to me, memorable day when we sailed from Portsmouth under our renowned Captain, February 1, 1848! We had on board the following passengers: Mr Brooke, Governor of Labuan, and Commissioner and Consul-General in Borneo; Mr Napier,

Lieutenant-Governor; Mr Scott, Surveyor, now Sir John Scott; Captain Hoskins, Harbour-Master; Mr Hugh Low, Secretary to the Government, now H.M.'s Resident at Perak; Mr Spenser St John, Secretary to Mr Brooke as Commissioner, now H.M.'s Minister Resident at Peru. There were also several other passengers, as Mrs Napier, Miss Napier, Mr Gwynne, Captain Peyten, &c.

There is no greater error in the world than turning vessels of war into passenger-ships, particularly when ladies are concerned. Every spot is occupied beforehand, so that the unfortunate passengers soon discover that they are *de trop* wherever they endeavour to find a resting-place. And the comfort of the officers and the discipline of the ship suffer from having a miscellaneous crowd of idlers on board.

Though every desire was shown both by the Captain and officers to render the passengers comfortable, it had but poor success. Mr Scott, Mr Hoskins, myself, and a Lieut. Muller, were stuffed into one small cabin with only two beds, and I had to resign myself to swinging in a hammock during the voyage, and dressing as I could in the cabin of an officer, a good-natured fellow of the name of Jeans. Mr Brooke was naturally better off, and was provided with a very spacious cabin constructed on the main deck; while Mr Napier and the ladies had all the cabins on the starboard side to themselves.

The beginning of our voyage was a rough one. Scarce had we left the harbour when a heavy wind and a head sea drove the unfortunate passengers to seek shelter where they could. The luggage was banged about, the sea broke in the glass ports and deluged us, and all was discomfort until, after calling in at Plymouth and Cork, we found ourselves at Funchal.

We did enjoy Madeira. Mr Brooke called on and afterwards dined with the Queen Dowager, who was residing on the island for the sake of her health; and balls and picnics and parties marked our stay. We were sorry to leave.

I now began to have opportunities to study Mr Brooke's character, and I notice frequent references in my journal to the growing enthusiasm which our intercourse created, though I was not blind to the inconvenience of the presence of so good-natured a man on board of a ship of war. Mr Brooke, as I have said, had a large cabin, and this was the rendezvous of as unruly a set of young officers as it has been my fortune to meet. He had a nephew on board, Charles Johnson, a staid sub-lieutenant, who endeavoured to preserve order, but it was of little avail. The noisy ones were in the ascendant, led by a laughing, bright-faced lad, who, when he was a midshipman in the Agincourt in 1845-47, had become acquainted with Mr Brooke, and whose fondness for cherry-brandy was only equalled by his love of fun. No place in the cabin was respected: six or seven would throw themselves on the bed, careless whether Mr Brooke was there or not, and skylark over his body as if he were one of themselves. In fact he was as full of play as any of them.

The grave Secretary seated at the writing-table could but look on with astonishment at the liberties taken with his chief, for whom he felt then almost veneration, so highly did he esteem the work he had been performing in the East. But these young imps thought of nothing but fun: they ate his biscuits, drank his cherry-brandy, laughed, sang, and skylarked, till work was generally useless, and nothing was done.

One can readily imagine how all this was injurious to discipline. There were some twenty in the midshipmen's berth, and nearly all considered themselves at liberty to use Mr Brooke's cabin as a sort of club. The consequences were soon felt: the senior officers thought themselves slighted in favour of their juniors, whose natural impatience of control was heightened by the injudicious encouragement they received; and I, who lived in the gun-room, soon began to fear that this coolness augured ill for our future proceedings in Borneo.

To add to our troubles, disagreements soon arose among us about the lady passengers; and what was intended to cement the friendship of all those who embarked on board the Meander, had the effect of producing disagreements from which we all suffered afterwards. Mr Brooke passed the principal portion of his time in reading, reclining in his bed or on an easy-chair; for although capable of great exertion, he was very fond of lounging. Occasionally he wrote, but much work was impossible, unless the bolt was drawn, and intruders thus shut out.

From Madeira to Rio Janeiro the voyage was pleasant. The trade-winds blew steadily, and we forged ahead at a good pace, sometimes touching 13 knots an hour. Nothing worthy of notice occurred at this capital of Brazil, and we soon passed on, making the shortest cut to Anjer, in Java, by trying the great-circle sailing. We did not actually go far enough south, but quite far enough for comfort, as we had heavy winds, rough seas, and bitter cold.

Captain Keppel had a passion for making rapid passages, and carried on sail until the ship was half buried in the sea: the consequences were loss of spars and canvas to the ship, and loss of comfort for all. We grumbled—officers, men, and passengers—but it was of no use; drive ahead we must, and drive ahead we did.

It was a great pleasure to be invited to one of Captain Keppel's dinners. When in good spirits, Mr Brooke kept the table alive with his talk, and Keppel was an excellent *raconteur*. On one of these occasions the band kept on playing after dinner; and on striking up a polka, Mr Brooke jumped up from his seat, seized hold of an officer, and danced about the cabin to our great amusement: the band changed into a waltz, and Mr Brooke immediately continued his dance with one of the ladies, and, after every species of fun, sat down to piquet, and laughed and talked the whole evening, to our great satisfaction.

As we approached towards the end of our journey, every one seemed to gather fresh spirits; and on the quarter-deck in the evenings games were inaugurated, and high-cocko-lorum was greatly appreciated. Mr Brooke, active as a kitten, made the most wonderful jumps, only approached by one or two on board. It was an amusing sight—first a little naval cadet, then the tall Lieutenant-Governor of Labuan, then a fat mid, till the line was most irregular in height, and a heavy-weight would send the whole floundering on deck together.

One day we had suffered from calm and light winds; the sea was unruffled as a mill-pond; the mids were whistling for a wind; Mr Brooke and I were leaning on the taffrail and talking of the future. The ship was still, the sun had set; we gradually sank into silence, from which I was drawn by a sort of ejaculation from my companion, "What a beautiful scene!"—and beautiful indeed it was. The moon, but a few days old, appeared brighter than usual, and shed its soft rays on every side, tinging the waters, and illuminating the ship's wake. The sky was of the deepest blue, and the stars appeared to stand out from it, while occasionally the light fleecy clouds coursing above added charm to the scene, and gave signs of the coming breeze.

What an interesting moment it is on board to watch the springing up of a breeze after a long calm! We stand anxiously at the stern, and watch with intense anxiety the approaching wind, as it raises the waves and curls them in the distance. The course of the breeze is easily distinguished by the darker appearance of the water. This time, as the breeze caught us, we flew before it, and presently found ourselves in a calm: again the wind came up, and sent us merrily on our way.

Mr Brooke was so great an admirer of these scenes at sea that he would often get up and keep the middle watch with a friend, walking the deck from 12 P.M. to 4 A.M., or at least a good portion of it. When in London, Mr Brooke was describing to a very ingenuous American lady the pleasures of a long voyage. "For example," he said, "you can take a friend's watch." "What! Do you gentlemen on board steal each other's watches?" was the startled reply.

I seldom troubled the deck myself during the middle watch, but as we approached land I so longed to sight it that I got up at half-past two one night to catch a glimpse of Java, and asked Mr Brooke to accompany me. When we came on deck we found the ship almost still, although the wind blew freshly; but we had hove to, for fear of running on shore. The rain fell heavily, and towards the north a deep loom proclaimed the vicinity of land. The scene, however, looked cheerless: the rolling of the vessel, the dripping rigging, the sloppy decks, the splashing of the rain, all combined to render it desolate, but I felt an inward satisfaction in knowing that we were near land. Mr Brooke soon went below, while I stayed to enjoy a mid's simple supper, served on the capstan, where five, possessing only a couple of plates between us, managed to stow away a goodly amount of salmon and biscuit, watered by the everlasting cherry-brandy and a little grog.

As the voyage drew to a close, Mr Brooke began to draw up minutes for the guidance of the different officers; and as I did most of the copying, the bolt of the cabin-door was more frequently drawn than before, much to the disgust of our mids, who must have looked on me as a dreadful bore, not being able to work amid their noises.

I find very few references to this voyage in the Rajah's correspondence, except as to the Meander being an unhappy ship. I am afraid that the passengers, not excepting Mr Brooke himself, were the original cause of the troubles on board.

At length we sighted the hill of Singapore, and a boat came off with all the wonderful intelligence of the French Revolution and the fall of Louis Philippe. How interested Mr Brooke was,—how eager for news! The post-office boat brought off our correspondence, and we spent half the night in reading it. We had had no news whatever from England since we left Cork on the 15th February, and now we were arrived at May 20. How different from the present time, when the telegraph pursues you everywhere, to tantalise you with but fragments of news!

We were all pleased to reach Singapore, particularly Mr Brooke, who had become tired of the eternal squabbling that marked the last few weeks on board. The day after our arrival in Singapore Roads Mr Brooke landed in state, and was received at the wharf by Colonel Butterworth, Governor of the Straits Settlements, and a guard of honour. Next day a ball was given to the new arrivals, which was really well worth seeing. I have spoken of the discords which broke out on board the Meander,—these separated the Lieutenant-Governor from the Navy; an incident which occurred at this ball separated him for good from the Governor of the Straits, and produced evil fruit for our new settlement. I will not say that the Lieutenant-Governor was altogether in the wrong, except that in a public position an officer can be too sensitive to apparent slights.

After about three weeks' stay in Singapore, Mr Brooke

sent on Messrs Scott and Hoskins to prepare the site of the town which was to be built at Labuan, while the rest of the party remained in Singapore. Here was the first error. The officers sent on were unaccustomed to a tropical climate, while those who remained knew exactly what should be done. This arose from too kind a feeling towards the ladies of our party, and an unwillingness to separate them; but the presence of one conversant with Malay was so apparently necessary, that we were surprised at the time that our chief could be talked into the arrangement, which proved almost disastrous.

We passed above three months in Singapore. There was much to be done in preparing framework houses for Labuan, but the work was more suited to a carpenter than to a couple of governors. Singapore is a very favourite place,—a settlement where the sagacity of its founder, Sir Stamford Raffles, has been borne out by its prosperity, and where the merchants, fairly prosperous, are hospitable to every one who will be at the pains to seek their society. No one who has any employment need be dull in Singapore.

During our stay, news reached us that her Majesty had been pleased to confer upon Mr Brooke the Order of the Bath,—that he was now a K.C.B.—Sir James Brooke for the future. The installation took place in Singapore. How we regretted that the news arrived before we sailed! as we should have liked to have had the ceremony performed amid the wild beauties of the Borneon forest.

CHAPTER VII.

RETURN TO SARĀWAK.

1848–1849.

EVERY arrangement being completed, the Meander sailed for Sarāwak on the 28th of August, leaving the Lieutenant-Governor to follow after the arrival of the mail. We were a small party, consisting of Sir James Brooke, Mr Grant, and myself. Mr Grant had been a midshipman on board the Meander, but had left the service to become Sir James's private secretary. He was the one with the laughing eyes, who was the leader of the noisy fun in Mr Brooke's cabin.

We soon reached the coast of Borneo, to be welcomed, as Sir James had been during his first voyage, by thunder, lightning, and rain; but a different welcome greeted us the day after our arrival at the entrance of the Sarāwak river.

I must give, in words written at the moment, an account of the reception which the Rajah of Sarāwak received from his faithful subjects. As his followers ever after called him "the Rajah," I shall use this term as synonymous with Sir James.

Sept. 4.—About ten o'clock in the morning native war-boats commenced issuing from the Morotabas entrance of the Sarāwak river, and sailed towards the frigate. These were manned by the Sarāwak people, come to welcome back their Rajah to the country of his adoption. They were in long light *prahus*, with tapering masts and "butterfly" sails, ornamented with flags and streamers, and with all the crews dressed in gala costume. The chiefs came on board, and greeted their Rajah with heartfelt gladness, while from their own boats we were deafened by a continual beating of gongs. Some of our visitors were rather fine men, but on the whole their outward appearance was very homely: their jackets, however, were tastefully ornamented with gold embroidery, and when clustered together they looked picturesque. About one o'clock we left the Meander under a royal salute and manning of yards,—a truly royal treatment. Manning the yards has a singular effect—the whole of the spars covered with men in their clean white dresses, standing apparently hand in hand, and all of the same height. When the last echoes of the salute had died away, the blue-jackets gave three hearty cheers, and then swarmed like bees down the rigging. The war *prahus* kept up a constant firing of guns, much to our and their own amusement. The pull up the river was a long one; but the appearance of the country was worth the *ennui* of six hours in a boat. Near us, the scene was ever the same; but in the distance the fine outlines of ranges of mountains afforded a striking contrast to the low jungle around. Occasionally we passed fishing huts and boats, and once a small Chinese junk that fired a royal salute of three guns as we passed by.

As we drew near the town, the shades of evening fell,—not, however, before a most brilliant sunset had delighted us. The *prahus*, sailing up at irregular intervals behind, the long snake-like Dyak boats, all kept up their firing and beating of gongs and drums, forming to me, a stranger, a most wild and picturesque scene. As we rounded the last point and came in view of the capital, we were surprised to find the whole town illuminated. Every house had rows and clusters of lamps, whose light, reflected in the placid waters, rendered the effect doubly brilliant.

We were welcomed at the Rajah's house by all the European inhabitants, and soon felt quite at home. The house was quickly filled by the native chiefs, while every available window and door presented crowds of eager gazers. There was a genuine feeling of delight at the Rajah's return, and I could perceive a glow of satisfaction beaming in his face, gradually settling down into a look of quiet happiness, and his observation proved it—"I feel more happy here than anywhere else in the world; this is my home."

Sir James Brooke's house was, as I have before remarked, oblong in form, and consisted of four rooms—a large dining-room, a library, and two bedrooms—while at one end was built a wing, in which he had his private apartments. Another wing was added afterwards for his nephew, and little cottages were built for his suite. Altogether it was a very charming abode. The interior arrangements were at a later period modified and improved, a noble library added, which was my chief delight; in fact, in no other home have I spent more happy hours.

When we arrived in Sarāwak in September 1848, Kuching, the capital, was but a small town of perhaps 6000 or 7000 inhabitants, with a few Kling[1] and Chinese shopkeepers. There was but little trade,—the unsettled state of the coast prevented *prahus* from arriving; the staple article—antimony ore—was unsaleable, on account of the disturbances in Europe; and the revenues, *minus* the unpaid royalty on antimony, scarcely exceeded £1500 a-year. It was a dull prospect, but we were all full of hope.

To show the style of intercourse which the Rajah kept up with his people, I will give an account of a visit he

[1] Natives of the Malabar coast of India.

paid, four days after our arrival, to Mina, one of the wives of the principal Malay chief, the Datu Patinggi. I will give it in words written at the time, as they have the merit of freshness.

Sept. 8.—We started up the river at one o'clock in a light canoe: every house and every landing-place were crowded with people to gaze on the Rajah, as the boat was rapidly paddled up the stream. There was every preparation made for receiving us in state—flags, streamers, and gongs as usual, with men assembled near the guns holding lighted matches; and as we approached the landing-place, the salutes began, and were continued at frequent intervals during the whole of our visit. From the wharf to the house white cloth was spread for us to walk on; and as we entered we were greeted with a shower of yellow rice for luck's sake. Passing through an outer room, we entered the hall of audience—a large square apartment, without an ornament on its bare plank walls save a solitary old pistol. The floor was covered with mats, and on either side was ranged a file of chairs, facing each other, for Mina's European guests. At the upper end was a seat, with a piece of gold brocade thrown over it, for the Rajah, while cloth-of-gold was spread under his feet. At his right sat Mina, on the prettiest of mats. She had received us with a good deal of elegance. After a few words of welcome, she rose, and, with her maidens, began to shower on us yellow rice, touched our foreheads with a golden ring, and then sprinkled our hair with gold-dust. These greetings being over, I had leisure to look round the room and examine the assembled crowd. Behind us sat the women and girls, a few of the younger tolerably good-looking, and, for Malays, perhaps handsome; but in general they were very plain. The other portions of the room were crowded with men and boys—some of the latter naked, others half dressed, while a few were decently clothed. Among the Malays, however, it is difficult to judge the rank of a man by his dress. Mina sat, as I before observed, on the Rajah's right, and entered readily into conversation. One of her observations is well worth recording. The Rajah paid her a compliment on her neat house, when she answered, "Ah, sir, were it not for you, I should not have had this house; it is yours. We never had such a place as this until you came to live among us." Her voice is occasionally very sweet; she is pleasant in her manner, and tolerable in her appearance — plainly dressed in a long black robe, with large gold buttons down the front, and rows on either sleeve. After partaking of a luncheon of sweetmeats, we retired with the same honours that had greeted our entrance.

We paid many other visits to the wives of the different chiefs,—among others, to the second wife of the Datu, and we were much amused by his plaintive complaints, that now Mina had a new house, the other wife, Inda, insisted on one for herself. A model of a large house was brought forward, and the wife and her fine-looking daughter, Fatīma, playfully insisted that the Rajah should order the Datu to comply with their desire. He answered, amid much laughter, "If you tease him enough, he will do it;" and he did it shortly afterwards.

A few days before the latter visit, Captain Brooke of the 88th Regiment, Sir James Brooke's nephew and aide-de-camp, arrived,—a pleasing, considerate companion, with whom, for fourteen years, I lived on the most intimate and affectionate terms.

The object of Sir James Brooke in visiting Sarāwak before proceeding to Labuan was not only to see his people, but to organise a league of the different trading rivers against the pirates of Seribas and Sakarang, who, during the Rajah's absence in England, had recommenced their marauding expeditions. A great meeting was held on the 21st September, at which the flag of Sarāwak was hoisted amid general cheering and a salute of twenty-one guns: then the European inhabitants of Sarāwak presented Sir James with a handsome sword, and he, in return, presented the native chiefs with swords; the Meander's band played, complimentary speeches were made, and the ceremony was over.

The important object of the meeting then came off. The Rajah explained to the assembled chiefs that the flag of Sarāwak might be used by any community allied with Sarāwak for the repression of piracy, but at the same time he warned the chiefs of the neighbouring provinces who were present that if they held any intercourse with the pirates, they would be considered equally guilty. The object chiefly aimed at was to stop as far as possible the supply of salt, iron, and firearms, and thus facilitate the submission of the pirate communities. Neither the Seribas nor Sakarang made salt, and were therefore dependent on their neighbours; but many chiefs, influenced by the large profits of the trade, clandestinely furnished the pirates with this indispensable article.

The question of the Sarāwak flag gave rise subsequently to much discussion. It was argued that in introducing this flag, Sir James Brooke not only desired to create a distinctive mark of country for his own people, but that he was influenced by the ambition to combine the whole coast under one Government. This is no doubt true: it was impossible for good government to exist in Sarāwak without its neighbours desiring the same advantage, and the feeling was almost unanimous that the Rajah should take possession of the adjoining districts. Those who opposed it were certain chiefs, who feared both loss of power and loss of income.

I should here notice that a few months before our arrival at Kuching, the Society for the Propagation of the Gospel had established a Mission in Sarāwak, at the head of which was placed the Rev. F. T. McDougall. His remarkable medical talent was of the greatest service to us all, and no one in Sarāwak can ever forget his unwearied attention to them when struck down, as the officers often were, by the weakening fevers of the country. I have publicly differed with him as to the management of the Mission, but it is unnecessary to revive a forgotten controversy. Mr McDougall deservedly exercised more influence over the European inhabitants than any one before or since: it was a happy party, until, many years later, the demon of discord entered there.

Three days after the ceremony I have described, we left Kuching, and joining the frigate Meander at the mouth of the river, immediately sailed for Labuan, where we arrived in a few days.

Labuan, the island chosen for an English settlement, was 11 miles long by 7 miles broad at the base of its angular shape. It was covered with forest, but had an open swampy plain, which formed one side of its fine harbour on the southern shore. Coal had been discovered, and was already rudely worked by a Mr Miles.

When we reached Victoria harbour we found that Mr Scott had erected a series of comfortable-looking mat houses on the dry sandy beach between the swamp and the sea. No more unfortunate place could have been chosen; the site could not fail to be fever-stricken. I do not blame Mr Scott for this, as not only was he unacquainted with tropical countries, but he acted under orders. The proper site for the first houses was, however, on the ridge of hills, where Mr Low subsequently built his permanent residence.

As soon as possible after our arrival, Sir James Brooke was sworn in as Governor, and then we started for Brunei in the Meander, to ratify the treaty with the Sultan. What a day of disasters! The south-west monsoon was blowing fiercely. Under easy sail the Meander was forging ahead so fast as to drag under water her launch which she was towing astern, and seriously to endanger her little steam-tender. Men and gear were seen floating in the distance as we shortened sail.

Standing by the side of Sir James Brooke, at the stern port of the captain's cabin, I observed this curious scene. The wind was blowing hard, driving masses of cloud across the sky—the rain descended in torrents, and there was a heavy sea. Now and then appearing above the waves we could see the heads of the swimmers in the distance, the launch filled with water, the steam-tender cut adrift to save her from being dashed to pieces against the ship's side, the two cutters and the jolly-boat in search of the men and gear. Fortunately the men were saved, and a glass of grog put them to rights: one of them who could not swim had already had four such upsets on this coast. While occupied in getting in the launch, the signalman reported that H.M.S. Royalist, which had left three days before, was in sight without a single mast standing. When she passed us she appeared a perfect wreck. A squall had taken her aback off Baram Point, and had swept everything away. Fore, main, and mizzen masts, had all fallen along the deck without seriously injuring a man, though several of the crew were sent sprawling under the weight of the rigging.

These disasters induced Captain Keppel to return to Labuan without accomplishing the object of our mission, and on the 14th October the Meander left us for Singapore.

We now began to suffer from the effect of the ill-chosen site of our houses. One after another the officers were struck down with fever. In the midst of this distressing period, Rear-Admiral Sir Francis Collier made his appearance in the Auckland. He was a short, stout, shrewd man. I sat next to him at dinner, and he amused me greatly by his remarks. He was a clever, prosy, coarse, though often gentlemanlike man, a Tory of the old school. "Did I think," said he, "I had a drop of Whig blood in my veins, I would have the barber to come and bleed me until it was all out of my body." He had once been summoned before a bench of magistrates, and having uttered a great oath, the chairman fined him five shillings for swearing. He looked grimly at the bench of three magistrates, drew a gold coin from his pocket, threw it on the table, and said, "D—— your eyes all round, and that will make up the sovereign."

Either alarmed by the sickness on shore, or for some other reason, the Admiral took a violent dislike to the new colony, and ever after tried to run it down. He remained but a few hours, embarked some invalids, and was then off. His report was no doubt deemed highly valuable and interesting. I cannot much wonder at his alarm, for in the midst of dinner I was called out to attend to one of our chief officers, Dr Treacher, and found him sinking fast. I ran back to tell the Rajah, who instantly came, and, seeing the danger, forced a glassful of brandy-and-water down his throat, and thus revived him. No wonder that the poor Admiral turned pale at the idea of the hotbed of fever in which he found himself.

Most of our party having recovered from their fevers, an expedition was organised to enable us to proceed to Brunei to ratify the treaty. We had the little steam-tender of the Meander, called the Ranee; a sailing gunboat, the Jolly Bachelor, belonging to Sir James himself, but manned by a crew from the frigate; and a couple of other boats. This was to me a most interesting expedition. Our party, consisting of the Rajah, Govt.-Secretary Low, the Rajah's nephew Captain Brooke, his private secretary Grant, an energetic midshipman named Suttie, a few others, and myself, were all inclined to enjoy the change from Labuan and the novelty of the visit. I have before described Brunei and its lovely river. We were as usual well stared at, and our little steamer enjoying the monopoly of favour, was well saluted; but at length we landed and took possession of a large shed-like house near the shabby palace and still shabbier mosque.

The ceremony attending the ratification of the treaty was very simple. We visited the Sultan in his hall, of which the only ornament was a rough sketch of a ship done in charcoal by some lad. Facing us as we entered, his Highness was discovered seated on a china bedstead as his throne, with Mumein, his prime minister, at his feet, while a few nobles were squatted around. At our entry his Highness rose, and advancing seized Sir James Brooke's hand and led him to the bedstead, and seated him on his left. A few complimentary speeches were made, the treaty was brought in, salutes fired, and the ceremony was over. One of the causes of his Highness's lively satisfaction was, that the Rajah had brought with him £1000 to be divided among the members of the Government; and his eagerness to obtain the money was only equalled by his childish curiosity to know if there was not something else coming.

Sir James, however, aided by Mr Low, entered actively into negotiations with the authorities in order to insure the welfare of the new colony: he endeavoured to calm their apprehensions and to excite their interest, and before we left a much better feeling prevailed. We visited many sights of interest in the neighbourhood, particularly the coal-formations on the Kianggi stream. On the 31st of October we returned to Labuan, and next day there commenced the most distressing month that I have ever passed through. The weather was very unpropitious: the south-west monsoon blew with all its force; the sea rose to an unprecedented height and swept over the beach, filling the lower parts of the houses with water, and damaging the provisions stored there; the rain fell in such quantities as to turn the swamp into a huge pool of fetid water, and the consequences were obvious. The Governor, the Lieutenant-Governor, the Doctor, Mr and Mrs Low, Captain Hoskins, Mr Grant, Captain Brooke, the marines,—all were down with fever at the same time. The only civilians who escaped were Mr Scott and myself. Mr Scott, as surveyor and engineer, was the most actively employed of the whole party, and was always well. I, being half an idler, turned hospital nurse, and for a fortnight at a time slept on a mat beside one patient or another. The water rushing under our houses and through the store-rooms alarmed the rats, and dozens of them invaded our bedrooms. So bold were they, that while watching by the side of my patients I could prod with my sword the boldest that ran near, but very quietly, for fear of awakening my sleeping friends. A panic seized the servants, and most of them refused to aid under the plea of sickness, so that much unusual work fell to my share. The Governor's fever so reduced him that I began to entertain grave apprehensions, when on the 27th the Auckland arrived, to be followed next day by the Meander.

Captain Keppel saw at a glance what was to be done. His friend was evidently in a low state. He must leave the fever-stricken island, and go for a cruise with him. To combine business with the pursuit of health, it was determined to start for the Sulu Archipelago, and to call at various rivers on the way to open relations with the chiefs. Sir James Brooke, Captain Brooke, Mr Grant, Dr Treacher, and myself embarked therefore on the 3d December, and once more found ourselves among the companions of our outward voyage. We visited Kimanis, whence the chief came off to see the Rajah. We were, however, prevented from carrying out our design of visiting other rivers by the arrival on board of the master of a British merchant-vessel, the Minerva, that had been wrecked on the island of Balambangan, at the extreme north of Borneo. He had not been able to save anything, as he had been frightened from his ship by the crowds of armed natives who rushed on board of her for the purpose of plunder.

In the hope of saving something, Keppel set sail immediately for Balambangan; but when we arrived there, we found the wreck burned to the water's edge. Nothing could be saved, nor could we find any natives, or a trace of the missing crew. We therefore stood over to Maludu Bay, and I took a message to the principal chief to invite him to visit Sir James Brooke aboard, as from him we hoped to obtain tidings of the missing crew, who, however, eventually turned up in Labuan.

It is not my intention to write a full description of our voyage, but only to touch on it when anything happened in which Sir James was concerned. We visited the lovely island of Cagayan Sulu, with its pretty lake, apparently an extinct crater, filled with the clearest of fresh water.

I may notice an incident which occurred during this voyage to show what imagination will do. Sir James had taken a great prejudice against quinine, saying it did not agree with him. "Can you take gentian?" asked Dr Clarke, our clever senior surgeon. "Yes." So a bottle marked "gentian" was sent up. I was still nurse, and when helping the Rajah I noticed that it was quinine I was giving. I directly remarked that it was so, when the Rajah said, "No, it is gentian." On going into the gun-room I told Dr Clarke what had passed. "If you want the Rajah to die, you will tell him it is quinine," was his reply. The stuff was taken as gentian, and the patient recovered. Eighteen years afterwards, when I heard the Rajah refusing quinine, I was tempted to tell him this story. He looked vexed, but he no longer made any objection to the quinine, and it again did him good.

We continued our voyage to Sulu, with the Sultan of which place Sir James was anxious to enter into friendly relations, in order to attract trade to Labuan. We struck land just off a large village surrounded by groves of cocoa-nut trees. The first appearance of the island is very striking,—two high-peaked hills, with alternate patches clear of grass-land, and noble forest, scattered clumps of trees, and a long belt of cocoa-nut palms separating the park-like scenery from the sea. We arrived off the capital, Sough, on the 27th. The Dutch had lately attacked this place, and had burned that portion of the town which was built on piles in the sea — the blackened remains being still quite visible at low water. Our visit was therefore regarded with very great suspicion. There were many other reasons for hostile feeling. The chief of Maludu, Serib Usman, who died from wounds received during the British attack on his town in 1846, was a relative of the Sultan; and numerous relations of the men killed during the Nemesis's action with the pirates in 1847 resided near Sough. We did not, therefore, expect a warm reception.

There was an Englishman living at the capital, one of those adventurous spirits to be found in almost every outlandish place. He had been an officer under the command of the famous Admiral Cochrane, when he so gallantly upheld his name while fighting to free the Spanish colonies from the mother country—a short broad-shouldered fellow named Wyndham. His had been a life of adventure; and if rumour did not belie him, all was fish that came to his net. Our arrival did not please him, as he was devoted to the Spanish cause; but blood is thicker than water, and he soon lent us his hearty assistance. Having through his aid arranged an audience with the Sultan, Sir James

Brooke, Captain Keppel, and a numerous party landed at Mr Wyndham's house, which was built out in the sea on piles. Our guide led us by a long shaky platform to the shore, where we found assembled crowds of armed natives. A messenger sent by the Sultan cleared the way, along a broad rough road with a high stockade on the left, and houses on the right. We passed on through ever-increasing crowds to a market-place, where the women were selling fish and vegetables. At last we came to a creek, over which a rough bridge was thrown, leading to the palace. This was also strongly stockaded, and here and there we observed mounted some heavy brass guns. Entering through a large gate, we found, on a small green, hundreds of men armed with muskets, spears, the heavy Lanun swords, and *krises.* Most carried shields, while a few were protected by chain-armour and ancient European helmets, probably taken in their wars against the Spaniards. The audience-hall was on the right and the mosque on the left. The crowd opening we mounted some steps, and crossing a broad veranda, densely packed with the Sultan's body-guard, we found ourselves in the presence of the monarch. The audience-hall was large, but perfectly bare of ornament, as all their brocades, silks, and hangings had been sent to the interior, on account of an absurd report that had been spread that we intended to bombard the place. In the centre of the hall was a round table, on the opposite side of which sat the Sultan and his chiefs; on our side a number of rude chairs, in which, after shaking hands, we took our seats. The Sultan then asked a few questions; among the first, singularly enough, "Was France quiet?"

The Sultan was very like the portrait given of him in Sir E. Belcher's voyage of the Samarang: he was dressed in white flowered silk, with a very broad gold belt round his waist, a handsome *kris,* and gold bracelets sparkling with jewels. Some of the chiefs were splendidly dressed in silks, gold brocades, handsome turbans, and head-dresses like tiaras of gold. The young men, as usual, were the most gaudily decorated, while the old were often in plain white jackets. The rest of the hall was filled with men evidently of a respectable class. Observing, after a short conversation, that Sir James looked hot and tired, the Sultan politely broke up the audience, and we returned by the path we came. Mr Wyndham dined with the Captain, and amused us by a variety of anecdotes relating to the Sulus and the pirates who frequented the port. Sir James stayed a few days longer to converse with the different chiefs, while we made some acquaintance with the country. I was particularly interested with two young chiefs, who, taking me by either hand, led me through the town to the race-course, where we found many young men putting their horses through their paces. The Sulus appear an energetic race.

Whilst the Meander was at Sough, strong parties were sent to a watering-place about a couple of miles from the town. I accompanied the second expedition, and to our surprise we found all the ground around the spring full of fish-bones, which wounded the feet of several of our men. While the water was being got on board the boats, some of us strolled through the groves of palms and fruit-trees, among which we observed many cottages. We noticed that the inhabitants regarded us with fierce and savage looks, and came to their doors with arms in their hands; but being well armed also, we took no notice. We afterwards found that the survivors of the action with the Nemesis in the previous year (1847) were quartered about this spring. No wonder they looked savagely at us. The fish-bones had been scattered by them,—a piece of wanton mischief, from which more of their countrymen suffered than ours did.

On the 3d of January 1849 we weighed anchor and stood along the coast, intending to visit the great island of Mindanau, and to touch at the Spanish settlement of Samboangan. We enjoyed our stay at this place, where we were received with the greatest hospitality. Sir James was particularly pleased with the Governor, an artillery officer, Colonel Cayetano de Pigneron. We found him very angry with the interference of the Dutch in Sulu, and the Governor-General of the Philippines had protested against it. Yet the Dutch had suffered enough from pirate-fleets, which obtained their supplies at Sulu, to warrant a severe retaliation.

On our return to Cagayan Sulu, we passed a coral-reef, on which were collected thousands of birds. A party landed to shoot, and singularly enough they found a grave on this lonely spot, out of sight of all other land.

Sir James Brooke, having shaken off his fever, enjoyed greatly his second visit to Cagayan Sulu. The week was passed in exploring the beauties of this very lovely isle, and in making friends with its chiefs. We had of course our usual adventures: the Meander was several times ashore, but she was always got off without any damage. At length on the 28th January we reached Labuan: the patients who had embarked were now in perfect health, and to our great delight we found that all those whom we had left ill were also convalescent. Every one had quitted the houses on the plain, and was installed on the hills.

We only stayed a fortnight at Labuan, as Sir James was anxious to commence operations for the complete suppression of piracy between Sarāwak and our colony. The Rajah had an immense correspondence, and used to write very frequently to my father. After upwards of twenty-eight years, I come across such words as these to warm my heart towards my old chief: "This place, as you may have heard, has been sickly, a nasty remittent fever having prostrated most of us. Your son has escaped, and, generally speaking, has been healthy. I cannot boast of having paid him much attention at our first outset" (that I had noticed), "but now I do so from selfish motives as well as better ones, for I am daily knowing him better, and better appreciating the goodness of his heart and of his abilities. His attendance on the sick did him the greatest credit, and I owe him a debt of gratitude for his tender care of me."

This was a year of expeditions. On our arrival at Sarāwak on the 16th February 1849, we found that the Seribas and Sakarang pirates had already commenced their ravages, and that the coast was so insecure for traders that few *prahus* were venturing out, unless in company and well armed, or large enough to go to sea, so as to be out of sight of land, beyond the observation of the Dyak pirates, who secretly kept inshore.

I will give one instance of the injury done by these pirates to the peaceful inhabitants of this coast. To keep as far as possible from the better-armed and brave pirates of Mindanau and the Sulu Archipelago (the dreaded Lanuns and Balagñini), the Malays of the north-west coast of Borneo usually built their towns and villages far up the rivers, generally from twenty to thirty miles from the mouths. But the land between their towns and the sea being the best farming ground, most of the inhabitants established themselves at regular seasons in small huts to plant their rice in the rich alluvial soil on the banks of the streams. This was the case at Sadong, a river about twenty miles to the east of Sarāwak.

The pirates of Seribas, of course, were well aware of this practice, and early in February they prepared a formidable expedition against Sadong. It consisted of 130 Malay and Dyak war-boats, and was commanded by a fierce old Malay, the Laksimana of Seribas. The estimated number of the men would be between 4000 and 5000—*warriors*, not *slaves,* as Mr Gladstone imagined. They started from their river about the 17th February, and reached the entrance of the Sadong at daybreak, and immediately commenced the ascent of the stream. As the foremost boats came abreast of a farmhouse, they stopped at the landing-place and allowed the rest to push on. Jumping on shore, they too often surprised the unfortunate Malays, and heads, captives, and plunder rewarded their nefarious activity. This continued until they reached a fortified house, where the men being prepared, opened fire on those Dyaks who landed, and thus gave the alarm to the people who lived higher up the river. The loss of some men and the necessity to push on rapidly induced the pirates to abandon the attack on the stronghold, but they obtained few heads after this, as the inhabitants, on hearing the firing, fled to the jungles. They, however, secured some plunder and a few captives. A well-known Seribas fighting-man named Dundong, arriving near a farmhouse, and observing a girl making for the jungle, gave chase. Being encumbered by his heavy spear, he stuck it in the ground, and darted after and soon came up to the trembling creature. Seizing her in his arms he hurried back, little suspecting that his movements had been watched by the girl's father, who, emerging from the bushes where he lay concealed, seized the spear and waited the pirate's return. As soon as he passed by, the father, springing from his place of concealment, thrust the spear through the ruffian's neck, and killed him on the spot; then, leaving the body, father and daughter escaped to the jungle.

In this raid the Seribas obtained about one hundred heads, a few captive women, and a small amount of plunder. Their loss was trifling, consisting of about half-a-dozen killed and wounded.

The news of this and other similar atrocities greatly excited Sir James Brooke: he grew restless, could not sit still or sleep, but was continually wandering about the house both night and day. At last he determined to cruise against the enemy with his native boats, and we started with about twenty war *prahus* and 700 men; but we were in a few days driven back by the recommencement of the north-east monsoon, when our native war-vessels could not live in the open sea. As what drove us to shelter would also drive the pirates home, our ineffectual expedition did no harm, but aided in disciplining the crews. We were all of us kept in constant practice, the Rajah encouraging his followers to exercise themselves either with the rifle, the pistol, or the sword, and showing us the example. He himself was a master of his weapons.

When the H.E.I.C.S. Nemesis arrived about the middle of March, it was determined to retaliate on the Seribas, and make them feel at home what they had made so many others feel. It would be wearisome to the reader to give an account of all our cruises against the pirates; but I intend to give in another chapter a full account of one expedition, to show the life Sir James Brooke led out in Borneo. We started on this occasion with four boats from the Nemesis and fifty-five war *prahus*. We visited many rivers, surprised and destroyed eight villages, drove back in dismay a large pirate fleet which was just issuing to sea, and made the pirates pause in their incursions.

During this time the inhabitants of the neighbouring districts, for above 100 miles along the coast, began to abandon their struggle with the pirates, and to fly to Sarāwak for shelter. A thousand arrived at a time. Though this raised the population of the capital from about 7000 to 13,000, it was at the expense of the wellbeing of the whole coast, and Sir James determined to strike one great blow at the pirates when the fine season should commence; but while the necessary force was being collected, he considered that he had sufficient time to proceed once more to Labuan, and to hold on his course as far as Sulu to negotiate a treaty with that power. We had received intelligence that the Dutch were preparing an expedition to make a final attack on Sough; and Sir James thought that England should now interfere to prevent the absorption of the few native states which remained independent.

The Nemesis took us to Labuan about the middle of May. We found it fairly healthy, although our good friend, Dr Treacher, had been obliged to take sick leave. We noticed also that there was an uneasy feeling among the officers, and that the Lieutenant-Governor, not acting with much tact, was incurring their systematic opposition. It appears the destiny of small communities to quarrel. Sir James Brooke felt these differences much: many of the officials were his personal friends, and had he only had to deal with men, he could have readily reconciled their differences; but the women were too much even for his conciliatory disposition.

We were glad to get away from their wretched squabbles, and find ourselves on board the Nemesis, steering away for Sulu, where we arrived without an adventure in four days.[1] We were received in the most friendly and intimate manner, and had no difficulty in negotiating a treaty. We then visited Samboangan, saw our old friend the Governor, and returned to Labuan to find its home politics more tranquil—traders arriving, but no one to buy. The Chinese merchants, having been frightened by the report of fever, scarcely ever visited the place.

As Sir James expected that the force for the great expedition against the pirates would be by this time collected at Sarāwak, we stayed but a few days at Labuan—long enough, however, for a fever I had contracted in Sambo-

[1] For a full account of this trip, *vide* 'Life in the Forests of the Far East' (vol. ii. chap. ix.), by Spencer St John, Esq. 1862.

angan to show itself in force. As I lay tossing about in my bed during the long wearisome nights, a gentle hand would smooth my pillow, present cooling drinks to my parched lips, and during the height of my delirium would watch for hours at my side. I cannot say whether he was always by the bedside; but whenever consciousness returned, or I awoke from sleep, I was sure to see the Rajah near, tender and true to all his friends.

On reaching Sarāwak we found two ships of war awaiting our arrival—H.M.'s brig Albatross, Commander (now Admiral) Farquhar; and the Royalist, Lieut.-Commander Everest. We waited for the return of the Nemesis before starting on our expedition against the pirates; in the meantime the native forces were organised.

As this expedition made a great stir at the time, and was much and sharply criticised both then and now, I will give the account of it at length, as I find it written in my diary. It will show the work in which Sir James Brooke was engaged—the fatigue, the exposure, which laid the groundwork of fever which prostrated many of us for years. It was not undertaken for pleasure or from ambition, but from a stern sense of duty,—for the protection of the trading and agricultural classes.

CHAPTER VIII.

THE SERIBAS AND SAKARANG PIRATES.

1849.

I HAVE written at the close of the last chapter that Sir James Brooke did not undertake expeditions against the Seribas and Sakarang either from pleasure or from ambition, but from a stern sense of duty—to protect the trading and agricultural classes of the coast. I had thought it unnecessary to go more particularly into this question, thinking that the occasional notices of these pirates which I have introduced into this biography would suffice; but after the covert and unjust attack made on Sir James Brooke's reputation by Mr Gladstone, I have thought it better to treat the subject of Dyak piracy in a separate chapter, and to examine subsequently the grounds upon which Mr Gladstone's attacks against the Rajah were based after I have given an account of the battle of Batang Marau. It may be a superfluous task, but it will bring into one connected narrative information that is now scattered over a great variety of works. It will necessarily involve many repetitions, but its perusal will be useful in order to understand the life of Sir James Brooke.

Whilst looking over my papers, I came upon a manuscript in which I had entered very fully into the whole question. It was written in 1850, during the great excitement on the subject, and contains a summary of the proceedings of these piratical tribes. After twenty-eight years I find nothing to change in the views I then expressed.

The Sarāwak, the Sadong, the Kaluka, and the Rejang, are the principal rivers which fall into the great bay between Datu and Sirik capes.

The Dyaks of Seribas, Sakarang, the Balows of Lingga, and the scattered Sibuyows, derive their origin from one great tribe. The Seribas inhabit the interior of the river of the same name, and the country near the sources of the Lipat, a branch of the Kaluka river. The Sakarangs live on the left-hand branch of the Batang Lupar, and on the Kanāwit, Katibas, and other tributaries of the Rejang. The whole of the divisions of the tribes have an inland communication, and when a piratical fleet is fitting out in one river, all who are desirous of joining in it cross overland to the place of rendezvous and assist in manning the *bangkongs* or war-boats.

The Balows partly resided on a small eminence that overlooks the Lingga branch of the Batang Lupar, and partly in villages in the interior; and the Sibuyows are scattered in the Lundu, the Quop, the Samarahan, and many of the neighbouring rivers. The Balows were the only Dyaks that were capable of offering any serious resistance to the Seribas and Sakarang; but although warlike and brave, they were never piratical, and always conducted themselves in a manner to gain the esteem and goodwill of their neighbours.

The Seribas and Sakarang Dyaks are essentially the same: the former, however, were the most addicted to piracy, the fiercest, and the most unrelenting. They commenced their piratical career about 130 years ago, while the Sakarangs did not follow their example till 60 years later.

The corrupting influence of the Arab Siribs, so justly complained of by Sir Stamford Raffles, and also the example of the Seribas, were the causes of the Sakarang tribe becoming piratical. Within the memory of many of the inhabitants who were living when I arrived in Borneo in 1848, the Sakarangs had been unaccustomed to the sea, and never ventured on its surface until these Siribs, for their own purposes, taught them how to construct and manage war-boats, and led them in their marauding expeditions.

The Siribs and Malays, who accompanied the Dyaks and supported them with their guns and firearms, at first continued to obtain the principal portion of the plunder, and the larger number of female captives, and left the refuse and the heads to their Dyak allies. Later on, however, the Dyaks came to feel their own strength too well to submit to the dictation of the Malays, and a more equal distribution followed.

In the Batang Lupar there were comparatively few Malays until the arrival of Sirib Sahib in 1844; but in Seribas they were very much more numerous, and were recruited from the neighbouring rivers by all the scoundrels whom gambling and thieving had corrupted, and rendered too idle and vicious to work for a living.

The population of the various districts inhabited by the unpiratical Dyaks is even now difficult to be stated. In 1848 we could, however, obtain an approximation by calculating the amount from the number of fighting men. From the best accounts, and by weighing the various statements, we reckoned that the Seribas had then about 6000 fighting men, and in the Sakarang there were about the same number; the mixed tribes inhabiting the Kanāwit might furnish about 4000 men, and the Katibas and Poé together, as many,—which would give about 20,000 fighting men, or 120,000 souls, scattered throughout that large extent of country. They themselves, in estimating their fighting population by the number of war-boats they could send out, and allowing for those that remained at home, would bring it to about the same number. In every inquiry I made of the natives as to the population of different villages, I always found them rate the inhabitants rather under than over the mark, and subsequently we were inclined to place the population much higher than this estimate.

Their nearest opponents, the Balow Dyaks, who inhabited the Lingga river, were much more numerous than had generally been represented; indeed, they could not have held their own, even with all the advantages of defence, had they been able to place fewer than 2000 men under arms.

The Seribas and Sakarangs, inhabiting the Kanāwit and Katibas branches of the great river Rejang, practically confine their cruises to the outlets of that mighty stream, on whose banks were situated some of the richest producing sago districts, as Mato and Bruit; or they extended their operations in a north-easterly direction towards the rich districts of Oya and Muka. These countries are inhabited by a very industrious but unwarlike race—the Milanows—who carried on an extensive trade with Singapore, and their large unwieldy *prahus*, laden with valuable return cargoes, fell an easy prey to the long, swift warboats of the head-hunters. If surprised, not one of the crew would survive, otherwise the vessels would be run ashore, and the men rush into the jungle to escape from their bloodthirsty pursuers. It is evident, from the remains of the deserted towns and villages that we saw in their districts, that the population was formerly much greater than we found it during our expeditions to protect their industrious people. We heard of almost monthly attacks on one or other of their villages, and few weeks passed without the Milanows having to add many to the list of their murdered relatives. The houses of the Kanāwit pirates we found crowded with the dried skulls of these unfortunates. The Milanows were too unwarlike and too disunited to retaliate on the marauders; and although, when driven to desperation, they defended their towns with success, the struggle in the long-run had become unequal, and a few years more would have reduced these countries to the condition of those bordering a hundred miles of the great Rejang,—to the condition, in fact, of a deserted wilderness.

The Sakarangs nominally acknowledged the sovereignty of the Sultan of Brunei, and were occasionally employed by his deputies in subduing refractory tribes; but they never paid tribute, and would not engage in any expedition that did not hold out a good prospect of heads and plunder. Oppression frequently drove the Dyaks of Sarāwak, Samarahan, and Sadong into opposition to their rulers; then the Borneon chiefs would threaten to let loose on them the Sakarangs, &c., and if the threat had no effect, they called in the ruthless horde, and allowed them to pillage and massacre the offending tribe.

While Sir James Brooke was in Sarāwak in 1841, above 100 Sakarang war-boats came up to Kuching, called in by Makota, and received permission to proceed up the river into the interior, under the pretence of attacking some Sambas tribes, but, in fact, to pillage and head-hunt for themselves and their employers. As I have already related, Sir James, by urgent remonstrances and a display of force, had them recalled, and thus saved his Dyaks from a great danger.

We often noticed, in the writings of those who attacked Sir James Brooke's policy, a pretence of believing that he had at first looked upon these hordes as inoffensive, till, for his own sinister purposes, he proclaimed them pirates; but, in truth, Sir James's opinion of them never varied. In 1840 he wrote: "To quiet this coast the Seribas should receive a severe lesson." "The Seribas are against all, and all are against them." "The Seribas and Sakarangs are not fair examples of Dyak life, as they are pirates as well as head-hunters, and do not hesitate to destroy all persons they meet with."

At another time he wrote: "The tribes of Seribas and Sakarang are powerful communities, and dreadful pirates, who ravage the coasts in large fleets, and rob and murder indiscriminately." "They" (the Sakarangs) "are the most savage of the tribes, the Seribas excepted, and delight in head-hunting and pillage, whether by sea or land." "By sea, the Sakarangs and Seribas reckon all they fall in with as fair prize, and acknowledge no friends."

When Sir James Brooke arrived off the coast of Borneo in 1839, piracy was at its greatest height, and fleets of Lanuns and Balagñini cruised among the smaller islands, while the swift-pulling war-boats of the Seribas and Sakarangs kept the whole coast in terror. The operations of their different squadrons could be traced with sufficient accuracy for us to know that they attacked the Milanow districts belonging to the Sultan of Borneo, cruised amid the lovely islands of the Natunas and Anambas, inhabited by Dutch subjects, ravaged the shores of the Dutch-protected states of Sambas and Pontianak, and one of their squadrons pushed on as far as Banjarmassin, on the southern coast. The Sultan of Sambas, whose subjects were most exposed to these attacks, did all he could to stop them, and even went so far as to hold communication with them to discover a method of clearing the seas of these pests; but they laughed at his proposals, and continued their work.

The Sultan of Pontianak lost his nephew in an attempt to drive them from the shores, and Dutch cruisers were taken by them.[1]

The Chinese of Sambas were also great sufferers, as many as 200 having lost their lives during a single attack; for

[1] See Timmink, chapter on Piracy; and 'Moniteur des Indes,' article on Piracy, compiled from Dutch official sources.

the pirates, when fighting against an enemy they understand, are bold and courageous, and in cunning are more than a match for any of the natives along the coast. In 1854 I had the opportunity of a long conversation with Monsieur Boudriot, a Dutch officer, who had held high rank on the coast of Borneo, and he fully confirmed all these accounts, and added, "They are the worst and most mischievous pirates that frequent these seas." Mr Earl, another impartial witness, who visited Sambas before Sir James Brooke arrived in Borneo, states that he found the rivers blocked up by booms; and no one would venture out even to fish, so much did they fear these dreaded Dyaks.[1]

Nothing was done to check these systematic piracies until Captain Keppel came over in the Dido, and boldly, and as wisely as boldly, carried out those operations which I have before described. The Seribas had laughed at the idea of retaliation; and no native power on the coast could have punished them. Borneo indeed owes a debt of gratitude to Sir Harry Keppel.

During the stay of the Dido in Sarāwak, Captain Keppel made a strict inquiry into the character of the Seribas and Sakarangs; and though perfectly convinced of their piratical character, did not desire to attack them in their own country without a previous reference to the nominal Government of the coast. He therefore officially entered into communication with the Sultan's representative and uncle, the Rajah Muda Hassim, who, in reply to his inquiries, addressed him the following letter:—

"This is to inform our friend that the Seribas and Sakarang people living in our neighbourhood are great pirates, who seize goods and murder people on the high seas. They possess above 300 war *prahus*, and extend their ravages as far as Banjarmassin: they are not under

[1] Earl's 'Eastern Seas.'

the control of the Government of Brunei. They plunder the vessels which trade between Singapore and the good inhabitants of this country. Our friend would be rendering us a great service if he would adopt measures that would put an end to these piratical outrages."

When Captain Keppel prepared to attack the Seribas with his boats, the attempt was considered a mad one by all the natives. They pointed out the strength and resources of the enemy, their strong and well-armed forts, the difficulty of ascending the rivers, and they considered that those who followed the English were going to almost certain destruction. Nevertheless, as Sir James Brooke accompanied the expedition, 300 of the best men joined him.

In seven days Captain Keppel assaulted their strongholds, carried them, captured most of their brass guns, burnt their forts and towns, and compelled them to sue for mercy. Thus by the brilliant dash of the leader of the Dido, these scourges of the coast were reduced to submission, and hopes were entertained that the submission would be permanent. For these services Captain Keppel very deservedly received the thanks of Admiral Sir William Parker and the Lords Commissioners of the Admiralty.

It is very remarkable how confirmed was the idea that the English were but birds of passage, and that the punishment of the Seribas should not have frightened the piratical chiefs; but no sooner had the Dido left the coast than Sirib Sahib endeavoured to intrigue with the Sarāwak chiefs to expel Muda Hassim and Mr Brooke from the country, offering his own assistance, and that of his allies, the Sakarangs. Having failed in this attempt, he left Sadong, and retired to Patusan, on the Batang Lupar, where he had built forts, which he fondly hoped would be able to resist any attack of the English forces.

Sir James Brooke, being anxious to prevent Sirib Sahib from thus acting, had given him notice that Captain Keppel intended to return to Borneo and complete his work by attacking the Sakarangs, and he recommended the Sirib to abandon all connection with the pirates; but, confident in his strength, the Arab chief despised these warnings, and, as I have said, left the seat of his Government at Sadong, and openly joined the pirates. He had with him, when he started from his headquarters, a fleet manned by at least 6000 Dyaks, and 600 Malays; and his followers represented to him that with this force he might consider himself invincible.

Before leaving Sadong, however, Sirib Sahib called a council of war, when his flatterers represented to him that with his power he might defy the English, and should they attack Patusan, he would disgracefully defeat them. They all promised to stand by him to the last. He found one man, however, who had the courage to tell him the truth. This was Datu Jembrang, a Lanun chief, with whose son I was very intimate, and he strongly advised the Sirib not to enter into a contest with the English. "These men around you, who recommend you to fight, will desert you at the first moment of danger. I remember the English well. I fought against them at Sambas, and you will never be able to resist them." Shortly afterwards this old chief died.

In August 1844 Captain Keppel again returned to Sarāwak; and as Sir James Brooke's detractors have expressed a fear that our naval forces were not acting legally in punishing pirates in the territories of a friendly power, it will be a satisfaction to them to know that Captain Keppel did not act until he was called upon to do so by the legitimate Government of the country. Muda Hassim wrote thus: "We beg to let our friend, Captain Keppel, know that the pirates of Sakarang, whom we mentioned last year, still continue their piracies by sea and land; and that many Malays, under Sirib Sahib (who is accustomed to send out or accompany the pirates, and to share in their spoils), have gone to Sakarang river with a resolve to defend themselves there rather than to accede to our wishes that they should abandon piracy.

"Last year Captain Keppel told the Sultan and myself that it would be pleasing to the Queen of England that we should repress piracy; and we signed an agreement at his request, in which we promised to do so: and we now would draw the attention of our friend to the piracies and evil actions of the Sakarang people, who have for many years past done much mischief to trade, and make it dangerous for vessels to sail along the coast; and this year many *prahus* who wanted to leave for Singapore have been afraid. We bring this information to our friend, as we are desirous to put an end to piracy, and to perform our engagements with the Queen of England."

Besides the general information connected with piracy contained in this letter, Muda Hassim drew Captain Keppel's attention to the fact that during the last few months no fewer than eight villages had been taken by the Sakarangs and burnt, the men killed, and the women and children carried away into captivity. He also referred to the trading vessels captured, which had produced so much alarm that the coast was in fact in a state of blockade.

By most natives Patusan was considered impregnable, but after having seen the Dido at work the previous year, the Sarāwak chiefs did not doubt of success. Sirib Sahib, however, had constructed a formidable fortress, defended by about 100 brass guns and 20 iron ones, and his followers were well provided with firearms and ammunition. In a quarter of an hour, however, the affair was decided: the pirates, astonished at the appearance of a steamer, and at the English boats, that, regardless of their fire, were pulled up almost to the muzzles of their guns, soon fled and left the town to their conquerors.

Captain Keppel continued his attack on the pirates further inland, with a loss of about 120 men on our side, principally natives; but the enemy suffered more heavily, for the pirates, unaccustomed to contend with disciplined men, attacked them at first as if they were equals, but the effect of steady fire, of guns and rockets, soon undeceived them, and they were defeated with considerable loss.

The expeditions of Captain Keppel had for many years a great effect: the pirate fleets disappeared from the coast, traders were unmolested, and fishermen could pursue their calling in security.

The chiefs came to Sarāwak and entered into engagements to give up piracy, and live at peace with the neighbouring countries. Sir James Brooke did his utmost at that time to impress upon them the misery they would bring on themselves and others by a continuation of their evil practices; and he appeared to have had fair success, as for some time one or other of the head men would warn him when any restless warriors planned a foray out at sea, and this timely notice would generally suffice to nip the affair in the bud.

The Government of Brunei considered this as a good opportunity to reduce the Seribas and Sakarangs once more to their allegiance, and to raise contributions from these districts. As this attempt might, however, have given these Dyaks and Malays an excuse to break off all relations, Sir James Brooke interfered, and persuaded the Borneon Government to let them alone, and permit them to continue to govern themselves.

Until March 1846 these people continued quiet, and many of them left those districts and removed to Sarāwak. One Sakarang chief even placed his two sons under Sir James's care, that they should be sent to Singapore to be educated; but the restless old pirate could not be quiet, and in spite of these hostages being in English hands, led out a fleet, which attacked Banting, the chief town of Lingga, and succeeded in gaining possession of a portion of the town. The defenders suffered considerable loss in killed, and many women and children were captured. Just as the pirate chief started on this expedition, one of his friends remonstrated with him, saying, "The Rajah will kill your children." He answered, "I know the white men better than that; they won't hurt my boys." And true enough, after a short time the lads were sent back to their father. The pirates next attacked Lintang, the chief town of Kaluka; and although they were beaten off with loss, they managed to destroy a portion of it.

In 1847 Sir James Brooke returned to England, and then the pirates threw off all restraint, and commenced their ravages on an extensive scale. In 1849 I made a list of the towns attacked, the villages destroyed, and the large trading vessels which were taken during the years 1847 and 1848, and the total of the destruction of life and property was appalling; but it would not serve any good purpose to endeavour to repeat it here.

In September 1848 we arrived in Sarāwak in the Meander, and then Sir James Brooke and Captain Keppel publicly pledged themselves to attack both the Seribas and the Sakarang as soon as the fine season of 1849 permitted boating operations.

As Mr Gladstone appeals to our treaty with Borneo, I add a letter from the Sultan of Brunei to show how his Highness interpreted its clauses:—

"From Sultan Omar Ali Safudin, the Ruler of Brunei and its dependencies, together with Pangeran Anak Mumein, and Pangeran Muda Mahomed; to Sir James Brooke, &c., &c.

"We have to inform our friend respecting the Dyaks of

Sakarang and the Dyaks of Seribas. Great is our distress to think of their doings, which are evil in the extreme—continually sweeping with destruction the coast of Borneo, and pirating on the sea, plundering property, and taking the heads of men. Exceedingly bad have been their doings hitherto, and in consequence of which, our subjects sailing on the high seas for the purposes of trade experience great difficulties. We sent Pangeran Surah to that coast, and he was attacked, likewise numerous *nakodahs*.[1] They attacked Nakodah Mahomed, and plundered his *prahu* of every article. On his way to Singapore, Mahomed Jaffir was attacked at Tanjong Serik; the pirates took the *prahu*, which the crew abandoned. Nakodah Matudin from Muka, bound to Singapore, was also attacked at Teluk Milanow.

"Such are the reasons for which we send this information to our friend, so that, if possible, he may check the doings of these Dyaks, and render it safe for our subjects seeking their livelihood at sea."

This was an official call on the part of the Brunei Government on the British to fulfil their part of the contract to put down piracy.

And it was time that something should be done, for during the first six months of 1849 these pirates attacked Sadong twice, as well as Sussang on the Kaluka, and Serikei, Palo, Mato, Bruit, and Igan. Almost all intercourse by sea ceased, as few who attempted to pass the mouths of the pirate rivers escaped unhurt. I calculated at the time that above 500 of the Sultan's subjects had been killed or taken captive between January and July 1849; and we knew that one large fleet had passed the mouths of the Sarāwak river to attack the subjects of the Sultan of Sambas.

It was during the height of this bloody work that Sir

[1] Merchants or owners of trading *prahus*.

James put to sea with the Sarāwak fleet, supported by four boats of the steamer Nemesis, to endeavour to check the pirates, while waiting the arrival of sufficient forces for what was hoped to prove the final attack.

Among other places which we visited was the river Kaluka, where I first saw what was the effect of continued piratical attacks on the prosperity of a country. Kaluka was famous in former times as a great commercial emporium, and as a most productive agricultural district.

This is what we saw.

After pulling a few miles up the river, we came to a spot where two branches meet, and here, stretched across the rounded point, were the ruins of the former town of Lintang. This had been a great trading place, and had lately held a population of above 6000, but having suffered severely from the attacks of the pirates, particularly that one in 1846 to which I have before referred, the inhabitants resolved to abandon their homes. About 1000, under their chief, Tuan Muda, came to Sarāwak, while the rest dispersed among the neighbouring districts. We then pulled up the left-hand branch to Sussang, the only town left in the district. It was stockaded at its lower end, and a few guns commanded the river. It was a wretched poverty-stricken place, the people being afraid to cultivate their fields, as they were constantly exposed to surprises. The houses, however, gave evidences of having formerly been of a superior construction, but were now rapidly falling to decay.

We then pulled for about 80 miles up the right-hand branch, and saw on the banks the ruins of one large town and several villages, but now the only inhabitants were those that lived in a strongly fortified hamlet. In no other part of Borneo had I seen more splendid open plains, or places better fitted for agricultural operations. They were now completely abandoned, though the continued groups of palm-trees showed how extensive the population had once been.

Never before had I been so struck with the irreparable mischief done by the piratical tribes, as when I saw this lovely country so completely deserted.

I have thus lightly sketched the proceedings of the pirates to show the mischief they did, and the extensive character of their operations. It is proved by Dutch and English authorities that the pirates of Seribas and Sakarang attacked the villages and the vessels of the subjects of the Sultans of Brunei, Sambas, and Pontianak, &c.; and Mr Boudriot, the Dutch official, told me that their archives were full of the records of the losses occasioned by these attacks; and I cannot but regret that Sir James Brooke did not follow up the idea thrown out by our unexpected supporter, and apply to the Government of the Netherlands for all information relating to the ravages of the Dyak pirates within the territories of Holland. He did not, as he was satisfied with the testimony contained in the 'Moniteur des Indes,' Timmink, and Earl.

That these pirates indiscriminately attacked the subjects of the Sultan of Borneo, not even respecting his envoy extraordinary, Pangeran Surah, is abundantly proved. Malays, Milanows, and Dyaks suffered to an almost equal extent; and the losses among the inhabitants of these lovely groups of islands lying between Borneo and Singapore were annually very great. They are subjects of the Rajah of Rhio, a Dutch-protected state.

I repeat that for the last thirty years it has been abundantly proved that the Dyak pirates of Seribas and Sakarang attacked the subjects of the Sultans of Borneo, Sambas, and Pontianak, as well as those of the Rajah of Rhio, besides every other person who came within their reach; and yet to this day we see Mr Gladstone repeating that this was intertribal war.

CHAPTER IX.

THE BATTLE OF BATANG MARAU.

1849.

I HAVE already described the rivers and districts inhabited by the Dyaks of Seribas and Sakarang. I need therefore only say that Sir James Brooke, knowing that the pirates living on the well-known rivers which bear their names had made every preparation to receive our expedition, determined to proceed 100 miles up the Rejang river, and fall upon them in the rear by the Kanāwit branch; but circumstances induced him to change his plan.

Our expedition was thus arranged: The Albatross was to stay in Sarāwak; the Royalist was to be anchored up the Batang Lupar, opposite the entrance of the Lingga river, to protect the women and children there, while the Balow warriors were away with us. The expedition was commanded by Commander Farquhar, a light-hearted, merry, earnest companion, with sound sense to guide him; Everest, eccentric, but exceedingly well read, particularly in poetry, and dearly loving an argument: as volunteers, Mr Urban Vigors, a very gentlemanlike, pleasant Irishman; and a Manchester solicitor, a droll fellow, with some good tastes.

Sir James Brooke, aided by his nephew Captain Brooke, by Grant, and by myself, as staff, and with his experienced

Sarāwak officers, commanded the native contingent, which was expected to exceed 100 war-boats and at least 3000 men,—the larger *prahus* carrying a crew of 70 men, while the smallest mustered 28.

On the 24th July 1849 the Nemesis started with the Royalist, Ranee, and seven English boats in tow, to rendezvous at the mouth of the river Morotabas, and then to take them to the Batang Lupar, to leave the Royalist at Lingga, and tow the boats to the entrance of the Kaluka, where they were to await the arrival of the war *prahus*.

We started in the evening, a slashing tide sweeping us along, so that in four hours we were at Morotabas,—the crews shouting and yelling, as *prahu* after *prahu* arrived, and let go their anchors, some running foul of each other, others carried towards the sea by the strength of the tide, and struggling to regain their places. At length the noise sank into a buzz, which gradually dying away, we set our watch and fell asleep. At 2 A.M. I was called to keep my watch, and looking around I was much struck with the curious appearance of the scene. The night was dark, and the scattered native boats had each a small fire, which threw a fitful light on the waters, like stars shining on the broad river. Now everything was still, with the exception of what perhaps rendered the silence more impressive—the distant but clear and distinct cry of the *jelātuk*, a bird of night. Then suddenly the strong stream would sweep a *prahu* from its anchorage, and loud cries of "Pull, pull!" would fill the air, and the rapid stroke of the oar show how difficult it was to stem the tide. Then again quietness, perhaps occasionally broken by the distant sound of a gong, or the low monotonous chant of the Koran, or the tinkling bell that marked the hours on board some of our boats. Such scenes were then both novel and interesting to me.

July 25, 1849, started for Sadong. The grey light gradually broke over the sky, and by degrees, few and far between, you would see a man rise, stretch, and look around him; then would come a little bustle of preparation, the anchor would be got in, and the men take their places at the paddles or oars, and gradually leave the river, and wend their way within twenty yards of the trees along this jungle-belted coast. At first there were few boats before us, but a dense mass of large and small Malay and Dyak *prahus* astern. Every now and then, however, a swift, long, and snake-like spy-boat would dart past, its light weight skimming over the waters under the impulse of the rapid movement of thirty paddles; then by comparison the heavy Malay war *prahu,* impelled by double banks of oars, would move by, leaving us with but few companions, as we were on the look-out for stragglers. So we continued, till on the 27th we rendezvoused off Kaluka, when it was settled that we should move back to the mouth of the Seribas, and cut firewood to enable the Nemesis to steam up the Rejang without danger of a short supply of fuel.

Whilst we were wooding, a message came from a respectable man living at Sussang, the capital of Kaluka, saying that the day before he was fishing on the sands with some followers, when they saw a fleet come out of the Seribas river. They instantly hid themselves, and soon afterwards counted ninety-four war *bangkongs* pulling northward towards the Rejang: a short time afterwards four more passed. Such apparently correct intelligence changed our plans, and it was immediately determined that we should endeavour to intercept the return of this fleet. The following arrangements were made: the Rajah and a large native squadron, supported by two boats commanded by Lieutenants Everest and Wilmshurst, were to anchor up the Kaluka, while the Nemesis, Ranee, five English boats, and a strong division of native *prahus*, were to anchor off the Seribas. It was expected that the pirates on their return, seeing the steamer, would dart for the Kaluka, and after proceeding up a few miles would be stopped by our division. The moment that Captain Farquhar perceived this movement, he was to follow with all his forces, and thus enclose the pirates between two fleets. Whilst we were anchored in these positions a message was received from Mr Crookshank, who had been driven by stress of weather into the Rejang, that a large fleet of the Seribas were off that river; that he had counted eighty *bangkongs*, and that a man at the mast-head had reckoned many more.

On the morning of the 31st, however, we fell down to the mouth of the river, preparatory to the steamer going next day to search for the fleet, as some appeared to fear that they were hid, or meditating an attack on Serikei, or would go home *viâ* the Rejang and Kanāwit; but the true reason was that we were but young warriors, and had not the patience to wait for the fruit to ripen. The Rajah in vain endeavoured to check this ignorant impatience, but at length gave way. As we left our anchorage, however, a second message was received that the Seribas fleet was near Serikei, and that an attack was intended on that town. To prevent this, it was decided that the Nemesis should start for that place the following morning.

That very evening, however, Captain Brooke and a party of us went to search for a wild pig or a deer along the sands, and we foolishly strayed for about three miles, when we turned. We had not retraced our steps above a mile when a spy-boat came pulling along the shore: suddenly all in her rose up, pointed towards the distant cape, then waved to us to return, shouting out, "The Dyaks are coming!" We scarcely credited the intelligence, but still walked rapidly back: the spy-boat arrived a little before us, and the news quickly spread. As we passed the Rajah's *prahu* we had to submit to a terrible scolding for having set so bad an example of discipline to the others,—a reproof which we fully deserved. We instantly got our *prahus* under way. Hurry, bustle, and confusion for some minutes reigned around, but silence having been restored, we moved down the river and formed a line across the entrance of the Kaluka. The Rajah fired a rocket to give the steamer notice, as the sun had now set. As no reply was given to this signal, another rocket was let off, and a faint shout was heard from the sea, which was answered by our whole fleet. A dead silence ensued for a minute, when from thousands of voices there burst a long, loud, deep yell of defiance, now sounding high, now low, as it was borne to us by the wind; a few guns were fired at us, and then the pirate fleet dashed away for the Seribas.

A rocket and a blue-light from the steamer told us that they were prepared, and then the deep booming of the guns announced the commencement of the action. Anxiously we waited that some of the fleet should come our way, when, after a short interval, we saw a dark line of war *prahus* approach our entrance. Everest and Wilmshurst, in their cutters, dashed forward to meet them, followed by a division of our light boats. A more picturesque scene could scarcely be imagined. The moon, dimmed by misty clouds, shed down a hazy light; the dark banks of the river, the breaking waves around the open sea before us, the distant booming of the heavy guns, the rockets curveting over the waters, the brilliant blue-lights of the steamer, the pursuing boats, the flashes of musketry, the yells and answering yells of enemies and friends,—forcibly struck our imaginations, as, rifle in hand, we waited the expected onset. For four hours the heavy firing told of a long, a prolonged fight. About midnight it almost ceased, and but an occasional report of a gun was heard.

We waited anxiously expecting intelligence till nearly 2 A.M., when the little Ranee came and brought us some news. It appears that when the pirates saw our fleet at the entrance of the Kaluka they thought that all was safe, as the steamer, being under the shore, was not at first perceived; but as they approached the mouth of the Seribas they were met by the Nemesis, whose heavy 32-pounders, loaded with round-shot, grape, and canister, scattered them in all directions. They tried the sea, but there the boats under Captain Farquhar drove them back; they tried the sandy point—there a large fleet of *prahus* poured in their fire upon them; they huddled in a confused crowd—all order was lost—a sort of *sauve qui peut* feeling took possession of their minds—and, to escape the fearful cannonade, they ran their war *prahus* on shore and escaped into the jungle, not before many of their boats had been taken or destroyed at sea.

One episode will show how daring some of these pirates were. A squadron of seventeen boats, commanded by an old chief named Lingir, saw the havoc that was done by the fire-ship, and determined to board her. The Nemesis was at that moment still, and Lingir and his gallant crews dashed at her; but when they were within fifty yards, they saw the monster begin to move, and rush full speed on them, and of the seventeen *prahus* only Lingir's one is known to have escaped. The brave pirates, however, whose *prahus* were sinking under them, jumped into the sea, swam towards the vessel with their swords in their mouths, and attempted to climb up her low sides; but they were beaten back into the water by the Sidi boys or blacks, who, seizing the billets of firewood on deck, hurled them at their fierce foes. I heard these anecdotes told by Captain Wallage of the Nemesis, who rightly fathomed their intentions—by the other officers on board—and also, some months afterwards, by Lingir himself.

August 1st.—A note from Farquhar to the Rajah to say that he had gone with the steamer up to the mouth of the Rembas to prevent any of the pirate fleet escaping, so in the morning we started with our division for the Seribas. As we passed along the shore, floating around us were very many evidences of last night's work—broken boards, half-sunk *prahus* baskets, mats, cloths, and every furniture of a boat; and when we neared the sandy point, we saw crowds of natives fishing for things in the water—and, as we could more clearly distinguish objects, a mass of war *prahus* left high and dry, the *debris* of the pirate fleet. When we landed, we found our men loading themselves with every description of spoil, the crews of our squadrons eagerly picking up what the others had thrown aside. Gangs of men were told off to destroy the large boats, and these were working with their axes to cut up the planks, preparatory to burning them. Some were launching and repairing the smaller and more serviceable boats, which were to be used as tenders to our heavy *prahus;* others collecting into groups, were telling of their exploits. Parties pushed into the jungle in search of enemies and plunder: now a man came dashing out, saying that he had seen Dyaks in one direction; then another, open-mouthed with news, told of a *rencontre* in which men on both sides had been killed—*rencontres* which only existed in the imagination of the relater, for by this time the enemy were miles away; of mangled bodies of women found headless, and hacked to pieces—the murdered, and, in their flight, the encumbering prisoners of the Seribas. I, a new-comer, could scarcely credit this story, and went with a party to verify the statement. We soon came to a spot where rough mats had been thrown over certain objects. One having been removed, we saw the dead body of evidently a young woman, naked, with head and breasts cut off, mangled—evidences, in fact, of an indecent and barbarous cruelty, too revolting to be described. Let us hope that these mutilations took place after death. We did not look further, the sight was too sickening; but we were told that four or five other bodies had been found of headless women and young girls. The mats that covered their remains we could see, but we did not approach them. These murderers were without pity.

It was indeed a scene of confusion, of hurry, of running to and fro, of proud exultation amongst the natives, for never previously had the pirates received such a blow. And justified were they in their opinion—a fearful retribution had fallen upon the marauders for all their massacring, their plunder, their cruelty, their bloodthirstiness. Scarce a man was present that had not lost through them a father, a brother, or a son, or who had not had some female relative borne away into captivity. The loss in war-boats was very great,—about 75 were lying on the sands. I measured one: it was over 80 feet in length, 9 in breadth, and some of its side-boards measured 60 feet. Its crew must have been over 70 men. In all, the pirates lost about 93 of their war *prahus*. We calculated at the time that the fleet consisted of 100 boats, with 4000 men as crews. We afterwards learned that there had been 105, and that 12 escaped up the river.

From a prisoner, and from the subsequent conversation of chiefs who had been present, we were enabled to trace the proceedings of this pirate fleet previously to the *rencontre*. It had left the Seribas the morning before we arrived there, and had continued its course to Palo, a small village to the north, famous for its salt manufacture. When they arrived opposite that place they were hailed, asked who they were, and what they wanted. The Seribas replied that they had not come to attack the place, and that they would do them no harm if no opposition were made. The terrified people said none should be offered. The pirates then went on shore, plundered them of all their stores of salt, took as much rice as they required, and, having selected two or three young women that pleased them, carried them off and proceeded to Mato, a town on one of the *embouchures* of the Rejang. They tried to surprise this place, but the inhabitants being better prepared than the people of Palo, repulsed their attack; but the Seribas managed to secure some heads, as they captured a detached house, and they seized likewise four young women. They then dropped down towards the mouth of the river to pick up fishermen and any trading vessels which might pass that way, and succeeded in securing a large one coming from Singapore laden with piece-goods and iron, and another going to Singapore with sago. The crew of one of these *prahus* got away to the shore, while that of the other was cut off before it could escape, and every one killed. We found in the pirate war-boats the cargoes of these two trading vessels, and our men divided a very handsome booty. They then started for Serikei, a large town up the Rejang, hoping with the aid of the tribes in the interior to take that place; but the Malays were well armed, and some feints were easily repulsed. Besides, the chief of this district, Sirib Musahore, carried on secretly an extensive trade with the pirates in the interior of the Rejang, and those Dyaks dared not offend him or their supplies of salt and iron would have been cut off. He was one of the intriguing, mischievous, half-bred Arabs, who did so much evil on the coast.

The Seribas and Sakarang Dyaks had long been expecting an attack, and had therefore confined their cruises to the neighbourhood of their own rivers; but this being the Ramadan, or fast month, they imagined that the Sarāwak Malays would not think of joining our expedition, and had therefore thought to strike a blow or two before our arrival.

This fleet was commanded by the brave old Pamancha, and almost every well-known Malay and Dyak leader had been present during the engagement, and they had induced many of the other Malays to join the expedition in order that they should manage their guns and supply them with musketry. Their surprise on seeing our fleet anchored at the mouth of the Kaluka was only exceeded by their terror on finding their way home blocked by the European force.

The dead bodies of the women we had seen were those of the captives taken at Palo and Mato. Whilst we were burning the *débris* of the pirate fleet, the Malay chiefs collected in Sir James's *prahu*, and proposed to him to occupy a neck of land between the Seribas and the Kaluka, and thus cut off the retreat of the mass of the pirates who were now in the jungle. But he said that perhaps they had been punished enough, an opinion in which many did not agree; but the Rajah was firm, and immediately decided to draw off the native boats. Had he followed out the advice given, few, if any, of the pirates would have escaped.

However, Sir James Brooke assured himself that the pirate *prahus* had been destroyed before he gave orders to all to push up the river and join the Nemesis off Rembas, and then continue our course to Paku higher up. It was arranged that the Nemesis should remain at this spot with all our heavy native war *prahus*, and that the rest of us should proceed in the light boats, and, accompanied by the European force, ascend the Paku branch and attack the chief town of that district. The river Seribas from its mouth to the Paku, a distance of about fifty miles, is broad and deep, with a tremendous tide, so that it took but a few hours to reach the steamer anchored there.

3*d.*—To-day we left our war *prahus* and got on board

some Dyak boats that we had selected from among the heap lying at the entrance of the river. The little steamer Ranee and the European boats led the way, and it was with the utmost difficulty that we could keep back the impetuous rush of native *prahus* that pushed ahead, the hindmost anxious to find themselves in the foremost ranks, and these again endeavouring to maintain their position — pushing, crushing, and running one into the other, the crews, some 3000 strong, all shouting at once, either "Back, back," or "On, on." A strong flood-tide swept up the river, and had not the steamer twice got aground we might have reached the town that night with great ease. No sooner was she off than away went her funnel. Her mishaps were without end, and caused much delay: it was a pity to have brought so unmanageable a boat up a winding stream. At length we arrived off the site of the old town, which had been destroyed by Keppel in 1843, and there the most unaccountable order was given to anchor, which lost us two hours of flood-tide, and caused us all our future troubles, and probably our loss of life.

The banks of the Paku are rather pretty, and afforded, as we advanced into the interior, some picturesque scenery, with lofty trees, high overhanging banks, occasional wooded hills; and the water around us was as clear as crystal. The fruit-trees along the banks are innumerable, and were loaded with produce, which unfortunately for us was not ripe. At the site of the old town were some remains of cultivation,—fruit-trees, and cocoa-nut and areca palms. We anchored here and passed the night, intending to start early in the morning. This gave a respite to the pirates, of which they were not slow to avail themselves. They cut down enormous trees and let them fall across the stream to obstruct or delay our advance, and give them time to carry away or conceal their valuables. While we were at dinner, one of our party incautiously sat on the gunwale of the boat, and leaning back overbalanced himself and fell into the river. We had a hearty laugh when he was pulled in, as, notwithstanding his unexpected bath, we found him holding on vigorously to his plate and knife in one hand and the grog-bottle in the other.

4th.—We started at daylight, and had not proceeded many miles when we found the felled trees, of which I have already spoken, stopping the way. Tedious and heavy work it was to cut through them: no sooner did we get past one than another presented itself. We continued this axing till eleven o'clock, now getting on a quarter of a mile, then stopped by a small obstruction,—advancing a few yards and a heavy tree to cut through; and these were placed in such positions that had the Seribas had the courage to collect with a dozen muskets, they could have inflicted on us very heavy loss. We at last came to a mighty tree that our axe-men could not get through, so returning a quarter of a mile to a cleared rising ground, we determined to march to the town; and all set to work to get their breakfasts.

This being over, and some delay occurring from the slow arrival of the native boats, which could only pass the obstructions one by one, time was given for the complete arrangement of the plan. The old Orang Kaya of Lundu, of whom I have often previously spoken, was with us, with his three sons, Kalong, Bunsi, and Tujang, and his brave followers. It was arranged that these should lead the way to remove obstructions and cut the path, while a strong party of Malays were to protect them. The English marines and blue-jackets were then to march in a body, and to be followed and flanked by the natives in a mass, while an adequate party remained to protect the boats.

On the left bank the forest came down to the water's edge, but on the right there was rather open ground, interspersed with tall trees and clumps of bamboo. Captain Farquhar landed his force and marched them to the top of the cleared hill, and there took up a strong position. Everything was in active preparation, some of us had returned to the boats to get completely ready for the march, when suddenly a distant yell was heard. A volley of musketry, shouts and cries, a bustle on the summit of the hill, and the English forming into line, told that something of importance had occurred. We rushed on shore. At first no explanation could be given of the alarm. Some cried "Kalong is dead," others Bunsi, others Tujang, some that all three were killed—when the return of a Dyak, bearing a wounded companion, gave the first intelligence. The wounded man said that they were ahead, when a party of the Seribas dashed from under cover of the bamboos, and killed a great many: he himself was severely wounded; he believed that all the young chiefs were dead, —he was sure that one was. The headless trunk of Bunsi, and the frightfully mangled body of Tujang, were now brought in; but Kalong, the eldest son, had escaped. Not knowing the strength of the enemy, and wishing to be prepared, the English were ordered to fall back a little and form a line along the summit of the hill.

This slightly retrograde movement began to produce a panic among the natives, who imagined that our men were retreating, and they commenced rushing to their boats. At first I thought that the enemy in overpowering force must be approaching; but the Rajah, speaking quietly to the men as they passed us, and laughing, observed, "Don't be afraid," and turning to us said, "Let us advance." We instantly pushed up the hill. This had an immediate effect,—the men turned and followed the Rajah in crowds.

Quiet being restored, we heard some account of the skirmish. The clearing party had advanced some hundred yards into the bamboo jungle, when Tujang, high-spirited but rash, pushed on ahead, and was followed by his elder brothers. Over-confident, they were almost unarmed, and without their fighting-jackets, and had advanced with a few men beyond immediate support. Tujang and Bunsi were stooping to pull out the bamboo spikes, when from behind a thick clump out dashed twenty of the enemy, and cut them down before they could draw their swords. Kalong, seeing his danger, sprang back and was saved; and the immediate advance of some Malays under brave Patah, who poured in a volley on the enemy, saved the wounded, and enabled them to recover the bodies of the slain.

It was a melancholy hour for the old Orang Kaya of Lundu. The father was but a little way behind when they fell. Proud of his sons, and especially fond of Tujang, he at first could only find vent for his grief in bitter reviling of those whom he accused of deserting his sons. He retired with his tribe to their boats, and sent Kalong to the Rajah to request permission to return to Lundu to bury his children. The surviving son came, and in a subdued voice said, "I have lost my two younger brothers." "Tell the Orang Kaya," replied the Rajah, "not to grieve; his sons died like brave men." A proud though faint smile of satisfaction was for a second visible, as praise from their great chief was indeed appreciated by them. Unwilling to allow his brothers' death to pass unrevenged, Kalong wished to remain with us; but the old Orang Kaya, bowed down by grief, begged him to return home.

One would have thought that the measure of his grief had been full, but another incident occurred which filled it to overflowing. The Ranee steamer had been left at last night's anchorage, and a cutter, bearing the dead body of a sailor who had accidentally shot himself, arrived at the moment the Lundu chief was passing. There was a little bustle on the steamer's deck; a rope caught the hammer of a musket left at full-cock by shameful carelessness; it went off; the ball passed between two of the officers, grazed a boy, struck the Orang Kaya's son-in-law, killing him on the spot, and finished by burying itself in the breast of a Malay. The old man, completely overcome, burst into tears, and holding up his fingers to the officers, could only say, "Three sons in one day," and continued his melancholy journey. From him the enemy could expect no mercy, and every Seribas that came in his way, during his passage down the river, was killed; many fell by his own hand.

The jungle being so thick, and it being necessary to allay the excitement among our native forces, it was determined to advance by the river next day to a point nearer the town. Some rockets were fired at the Seribas who appeared in the distance, and a few of the more adventurous who climbed into the trees were disturbed by a rifle-ball, but the rest of the day passed quietly. A picket was placed on shore, and a guard of natives. The enemy were constantly heard in the jungle, and volleys sometimes awoke us; but knowing that there could be no danger from which being awake would save us, we were but little disturbed by the firing, and slept well.

5th.—Starting at low water, we managed to pass under the enormous tree that had hitherto barred our passage, and soon reached the spot from whence we were to march on the town of Paku. When the whole force was collected, we found that Sir James had come to the very injudicious conclusion to keep all the English to protect the boats, and to allow the natives to march alone to attack the town. It was injudicious in many respects: it exposed our force to a check; it annoyed the English, who had borne the brunt of the difficulties; and it was unfair to the officers, who naturally sought opportunities of distinguishing themselves.

The native force, however, was well selected, and consisted of about 1500 men, the best armed of the fleet. They had been away nine hours when a dense smoke at the distance told of their success. They had arrived at the place without any opposition, as the pirates were cowed by their defeat at the mouth of the river; and in the skirmish of yesterday the latter had lost many men, including their Malay chief.

6th.—The whole of this day was spent by our native allies in organising expeditions against the villages around. The enemy made no effectual resistance: they even abandoned a strong stockade, threw their guns into the river, and fled at the approach of our men. The plunder obtained by our Malays and Dyaks was enormous, and in their eyes of great value; but to the Europeans it was useless, except a few things as curiosities. Some brass guns were secured, and some handsome rifles, that in their panic the Malays had left in their houses.

During these two days of enforced idleness we amused ourselves in cutting down the brushwood with our swords, bathing in the sparkling river, joining in the *jereed*-throwing of the natives, and trying our skill at Malay football.

During the great heat of the day Lieut. Everest and I passed our time in reading to each other Thomson's "Seasons" and "Paradise Lost." Sir James Brooke, wondering what could be interesting us so much, approached to join in what he called a "very pretty and appropriate amusement during a warlike expedition," until disturbed by the necessity of receiving reports from the different chiefs, as they came in from their forays.

7th.—Early this morning we made preparations for our return to the main river. Before leaving, however, Sir James Brooke wrote a long letter to the pirates, in which he pointed out to them the necessity of abandoning their marauding pursuits, as the English had determined to put down piracy. On our arrival at the Nemesis we found that one prisoner had been taken,—not, however, before he had wounded one of our men. They had to stun him with the blow of an oar before he could be secured. On board the steamer he was treated with great kindness, clothed, well fed, and, as usual, christened "Jack" by the sailors. He proved of the utmost service to the cause of peace by the accounts he gave of the kindness he had received. This was the only Dyak prisoner taken.

I need not enter into any particulars of the rest of the expedition. We went to the Rejang, ascended that river about 100 miles, attacked the pirates living on the banks of its Kanāwit branch, and then returned. Many of us thought that we should not have stopped until we reached the fort of Buah Ryah,[1] the great chief of the Kanāwit, and the most famous pirate of those regions; but Commander Farquhar considered that his men had been exposed enough, and wished to go back. Captain Brooke and some others of us had been ahead with the light division, and had come upon and skirmished with the enemy, and felt assured that on the morrow we should find the pirates in force. We returned late to headquarters, to find dinner over; so, removing to the other end of the village-house, we sat down to ours, and indulged ourselves in our discontent at the idea of returning. The one whom the Rajah subsequently called his "unruly child" raised his glass, and gave a toast aloud, "Oh for one hour of bonnie Keppel!" Directly Farquhar heard this, he jumped up

[1] The names adopted by some of the Dyak chiefs from the Malays are significant, and remind one of the North American Indian titles. For instance, "Buah Ryah" may be interpreted "the Chief Fruit" (or, "the Fruit of the Festival"); the name of another chief (now dead) was "Tongkat Langit," or "the Support of the Skies." Buah is the Malay for fruit; Ryah means festival, holiday, high; for instance, "Hari Ryah" (Hari is day) is the great day of festival at the end of the Ramadan; Tongkat means walking-stick, staff or supporting post: Langit, is the heavens, or rather the sky.

and came and tried to persuade us that he could not act otherwise. He was no doubt right, and he and his men had acted most gallantly and successfully under great difficulties.

Two or three incidents occurred during this expedition[1] which are perhaps worth relating.

One village was surprised by the Malays, and the inhabitants had only time to save themselves in the woods. A mother, being hard pressed, let go the hand of her son, a boy of ten, who was seized by our men. He was brought on board our *prahu*, and Sir James took him under his protection, had him clothed and well fed, so that before he had been on board many days he grew quite confidential. As we were coming down the river, near the site of his village, the little fellow asked, "Where are you going to take me?" "To Sarāwak." "I wish you would land me, and let me find my mother." Sir James hesitated, but being assured by the natives that a Dyak boy of ten could manage for himself, it was determined to land him where he had been picked up. He was loaded with presents, and with food for three or four days. To prevent his being annoyed, a Malay guard was left at the landing-place until the last Dyak ally had passed on. This little fellow remained three days alone in the jungle, but never wandered, and was found by his mother at the spot she had left him. This conduct, so different to what is customary in native warfare, had an excellent effect: as soon as possible after the expedition his mother sent him to

[1] About this time one of the Malay crew of the advance or spy boat got wounded, a barbed spear having been hurled at him from the river-bank as he lay in the boat. The barb entered his back, and was in such dangerous proximity to his spine that the naval surgeons were afraid to extract it, and so contented themselves with cutting off the shaft. On the return of the expedition to Sarāwak, Bishop McDougall undertook the operation, and to the astonishment of all (the patient included), succeeded in extracting the barbed spear-head, so saving the man's life.

Sarāwak, with presents to the white man who had been kind to her boy.

Confidence in the truthfulness of the white man is proverbial in Borneo. As we were breakfasting in a deserted village, voices hailed us in the distance to know who we were. They were told. They answered they wanted to have a talk: Sir James answered that he could not stop then, but if they would come down to the next night's resting-place they should be well received. Next morning they came; and for the first time in their lives they met white men and saw a steamer. These Dyaks remained ever after our firmest friends. This judicious mixture of severity and kindness always won the hearts of the Dyaks, and tended to increase their great respect for the English Rajah.

One of these chiefs said: "The inhabitants of our creek are determined to give up piracy, and will do so, but we cannot answer for the other tribes. If we give up cruising, shall we and our friends who think with us escape when you attack the refractory chiefs?" Sir James answered: "If you see a flock of sparrows devouring your rice, do you not try to kill them; and if by chance a harmless linnet should be among them, does he not run great risk from being found in such company? He may be killed; in a flock of birds it is difficult to distinguish between the mischievous and the harmless."

During our absence from Sarāwak the Meander had arrived, when, finding the work done, and his orders calling him elsewhere, Keppel sailed away, and we saw him no more in Borneo. Had his special boats been with us during this expedition, we might have pushed further up the Kanāwit, and by destroying Buah Ryah's fortified village, have given peace much earlier to the distracted districts of the Rejang.

CHAPTER X.

MR GLADSTONE AND THE RAJAH.

1849.

When I heard that Mr Gladstone intended to enter into a detailed and careful examination of the whole question of Sir James Brooke's dealings with the Borneo pirates, I wrote to him privately, offering to furnish him with every information in my power, as I did not wish to see my old chief's reputation exposed, to be tarnished by accusations founded on imperfect evidence, by a master in the art of making the worse appear the better cause. Subsequently I replied to the following queries posed by Mr Gladstone :[1]—

Query 1.—That the so-called pirates were not pirates in the proper sense of the term, as applicable to persons habitually infesting the seas and assailing European commerce.

"The Dyaks of Seribas and Sakarang were pirates in the proper sense of the term, as their fleets and squadrons frequented the coasts of Borneo, and the Natunas, Tambilan, and other groups of islands, during the whole of the south-west monsoon—that is, during the time that

[1] I may notice that I answered all these queries at a sitting, without having any papers or works to which I could refer. I was going out of town, and wished Mr Gladstone to have what I had written as soon as possible.

native war-vessels could keep the seas: that their cruising-grounds were from Tanjong Kidurong, north of the river Bintulu, to Pontianak, or about 400 miles in extent; whilst some of their vessels have pulled as far as Banjarmassin on the south coast.

"The Seribas and Sakarang pirates, though usually called Dyaks, were not entirely so, as in every large *prahu* was a contingent of Malays who worked the swivels and guns and used the muskets; and in many cases boats were almost entirely manned by Malays. In the latter expeditions the Dyaks had many guns and muskets of their own.

"The Seribas and Sakarangs were pirates in the proper sense of the term, as they attacked trading vessels under whatever flag they might be. They also ravaged the coasts of the Netherlands colonies, as was referred to by the Dutch officer, Monsieur Boudriot, who gave evidence before the Commission of Inquiry in 1854, and whose evidence was more clear and important than that published in the proceedings.

"In no text-book, and in no instructions issued by her Majesty's Government, are pirates defined as only those who attack European commerce. It is sufficient to prove that men attack peaceful commerce on the high seas, or on the coast, to constitute them pirates; and the pirates of Seribas and Sakarang attacked every vessel they met which they thought they had a chance of taking; so that the Seribas and Sakarangs were pirates in the proper sense of the term as applicable to persons habitually infesting the seas and assailing peaceful commerce.

"The attacks on the Seribas and Sakarang pirates were commenced by the present Admiral Keppel in the years 1843 and 1844. His proceedings were approved by the Admiral on the station (Sir William Parker), the Admiralty, and the Foreign Office. No instructions in a contrary sense were given during the intervening years to

1849, when the action in dispute[1] took place. On the contrary, the Meander, in 1848, was fitted out with special boats, suited to river service, to enable her to attack the Seribas and Sakarang pirates with greater certainty of success. The Government were therefore satisfied of their piratical character, or they would have issued instructions accordingly. As the Admiralty Courts had accepted the evidence adduced in Captain Keppell's case, it could not be supposed that it would be rejected in Commander Farquhar's. In fact, the Admiralty Court, presided over by that distinguished judge, Sir Christopher Rawlinson, decided that the Seribas were proved to be pirates, and admitted Commander Farquhar's claim for head-money."

I might have added that the Lanun and Balagñini, whom no one has ever doubted to be pirates, principally prey on native commerce.

Query 2.—That there was no broad, clear distinction in manners and morals between the allies whom Sir James Brooke assisted or led, and the enemies whom he attacked.[2]

[1] Mr Gladstone's queries relate in a principal measure to the battle of Batang Marau, which was described in the last chapter (chap. ix.)

[2] There is an ancient custom of the Seribas and Sakarang Dyaks which forbids the inhabitants of a village from going out of mourning * after there has been a death in it, until a head (that of an enemy if possible, but still a *head*) has been obtained, to be used as a propitiation to the spirits at the religious feast. But owing to the vigilance of the Sarāwak Government subsequently to the action of Batang Marau, the headmen of every long village-house from which there issued a head-hunting party was severely fined in gongs, brass guns, or sacred jars; and hence there gradually arose difficulty in carrying out their traditional custom. On a recent occasion, however, the present Rajah (Mr C. Brooke) assembled all the chiefs at Simangan Fort, on the Batang Lupar, and proposed to them to abandon this old law, or rather to substitute for "a head" "a good harvest." This suggestion was immediately accepted by the assembled chiefs. Can there be any stronger proof of the wisdom of Sir J. Brooke's

* It may puzzle Britons to know how nearly-naked savages go into mourning: it is by allowing their hair to go long, and to dispense with coloured waist-cloth or head-cloth or brass ornaments of any sort.

"The native contingent that acted under the orders of Sir James Brooke was composed of men as distinct in manners and morals from the pirates as it is possible for men to be. Although two branches of a Dyak tribe furnished a light squadron, the bulk of our native fleet consisted of war-boats manned and led by as respectable and as worthy traders and chiefs as any living in the East,—many of them as good men as ever existed, as the late Bandar of Sarāwak, and his brother the Datu Imaum, who served because it was necessary to clear the sea of marauders. It was the trading population who manned the fleet, as they are the good sailors of the coast. I knew them during fourteen years, and can declare that it was not possible to find a more marked distinction than that between our men and the pirates; and this testimony could be corroborated by probably every one who has resided on the coast, whether as her Majesty's officers, or as missionaries, merchants, or as Sarāwak officials."

Query 3.—That the action in and after which the slaughter took place was partly on and near the shore, partly in chase of several *prahus* which were in flight for the open sea—viz., seventeen *prahus*[1] which escaped Captain Farquhar, and when making for the Batang Lupar were raked abeam with grape and canister by the Nemesis, which drove many of them on shore badly crippled, when they fell an easy prey to the Dyak boats. Five of these, which were pursued by the Nemesis, were destroyed in detail out at sea, till there was not a living being on board.

"As this appears to be taken from an account given by the officers present, I have little doubt of its general cor-

policy, which has been so ably carried on by his lieutenants and successors, than this example of the moral power exercised by an Englishman in the midst of savages?

[1] *Vide* chap. ix. p. 179.

rectness. The Nemesis was quite justified in preventing any war-boats escaping to the Batang Lupar, as it would have exposed the peaceful inhabitants living on the lower reaches to death and pillage."

Query 4.—That there is no evidence that either the five or even the seventeen *prahus* fought in any way against the Nemesis or the Snake, which appear to have performed this operation; or especially as to the five, which were not doing anything except flying for their lives.

"It is a positive fact known to myself, to Bishop McDougall, and to every superior officer in Sarāwak, that Lingir, who commanded the squadron of seventeen *prahus*, made a determined rush at the steamer, and only failed in taking her by her suddenly going ahead full speed. Nothing would have been easier, had the pirates got alongside, than for them to have captured her. They were about 700 strong, and the very bravest and most daring of the fleet; while the Nemesis had not even her best Europeans on board, as they were away with the boats, leaving a few whites to point the guns. The rest of the crew were Indian Lascars and African Sidi boys. I heard Lingir himself some months afterwards tell his version, and I heard Commander Wallage, the day after the affair, explain how the steamer escaped capture. It must not be forgotten that these pirates have always shown themselves both daring and brave, but the inferiority of their arms renders an encounter with European regular forces quite hopeless. Lingir escaped home with his *prahu*, so that at least one of the seventeen got away.

"The Snake was a Malay boat manned by perhaps thirty men."

Query 5.—That there is no evidence that either the seventeen or the five *prahus* were warned or summoned to surrender.

"It appears to be forgotten that there were 105 pirate *prahus* in the action, manned by about 4000 men; that the affair took place between 8 P.M. and midnight; that the whole of the forces engaged were constantly shouting and yelling like demons; that until firing actually commenced the fleets were not within speaking distance; that these pirates never give quarter, and therefore never expect it — in fact, practically asking them to surrender was impossible, or it would have been done."

Query 6. — That the rate of pay per pirate slain or taken being £20 per head, the prize-money claimed was £3000, and the sum voted and drawn was £20,700. Thus it would appear that about 1000 persons must have been destroyed in the action by her Majesty's forces entitled to claim prize-money, not including the great slaughter by natives in the jungle, and principally in the operation just now referred to.

"The rate of pay was £20 per pirate killed, and £5 per pirate present. The naval forces were paid—

500 pirates killed, at £20, ...	£10,000
2140 „ present, at £5, ...	10,700
	£20,700

"Whilst I was H.M.'s Acting Commissioner from Feb. 1st 1851 to Aug. 11th 1855, I made the most minute inquiries on the subject, and, as I testified before the Commission, found that there were killed and present during the action, and that the payment therefore should have been for

300 killed, at £20, ...	£6,000
3700 present, at £5, ...	18,500
	£24,500

"The pirates landed on their own coast; and though many died of disease and fatigue, very few—perhaps not one, certainly not twenty—were killed by the natives on shore (I have been reminded that one was killed by the eldest son of the Orang Kaya of Lundu). They had a great dread of the prowess of the pirates, and did nothing but shout the watchword, for fear the Europeans should mistake and fire into them, the light squadron commanded by Mr Steel in the Snake excepted, as they cruised at the entrance of the river, to try and prevent the pirates getting up the Seribas.

"It must not be forgotten that subsequently we lived on terms of great familiarity with the chiefs of the pirate fleet. They used in after-years to come up to our houses and sit round our-dinner tables, and tell us stories of that eventful night, that destroyed for ever their love of piracy.

"I myself landed on the beach in the morning, and found that there had been no pursuit, except to pick up what had been thrown away by the pirates during their flight. One of these parties found the headless bodies of the young girl-captives whom the pirates had killed in their rage, and I went to verify the fact, and saw the sickening-spectacle.

"There were between seventy and eighty war *prahus* on or near the beach, some very fine, measuring 80 feet in length and 9 feet beam, carrying a crew of about seventy paddlers, besides the chiefs."

Query 7.—That we do not hear of any judicial proceedings for piracy against prisoners on this occasion, and are consequently altogether ignorant of the pleas they would have advanced in self-defence, and the colour which these pleas might have given to the case.

"There was only one prisoner taken, and he with great difficulty. He was of inferior rank, and was subsequently sent with a message to the pirate chiefs to the effect, that if they ceased pirating these attacks would cease. I never heard either a Seribas, Malay, or Dyak, or in fact any pirate, who denied his deeds: on the contrary, they gloried in their success, and the daring evinced in their distant cruises. In the Eastern Archipelago piracy was considered an honourable pursuit, worthy of a nation of warriors.

"When I say only one prisoner was taken, I mean in this operation. Subsequently a chief and several women and children fell into our hands."

Query 8.—That the Act 6 George IV., the treaty with Borneo, and the instructions given under the Act, appear—

a. To contemplate only the case of pirates in the sense of persons who assail British commerce.

b. To assume that pirates will be summoned and as far as possible taken alive, with a view to a regular trial, and to being judicially dealt with.

"I need not refer to the Act 6 George IV., but the treaty with Borneo being negotiated with the Sultan of Brunei and Sir James Brooke, referred to the pirates who infested his Highness's coasts, and destroyed the trade of his subjects, and was aimed as much at the numerous Seribas as at the better-armed Lanuns and Balagñini, and had no special reference to British commerce, but to commerce in general, and no instructions were given to Sir James Brooke to make such a distinction.

"When it is possible to summon pirates to surrender, it is done. In the action between the Nemesis and the Balagñini in 1847, a boat was sent off for that purpose, when the pirates immediately opened fire on her. On every occasion the pirates have commenced the action. In 1843 the Prime Minister of Brunei called upon Captain Keppel to aid in putting down the Seribas and Sakarang pirates, and himself sent the forces he could command.

"I might also have added that every attack made on the Seribas and Sakarangs by H.M.'s naval forces was at the direct written request of the Government of Brunei, and the one in 1849 under the provisions of our treaty."

Query 9.—That the proceedings of this expedition, as far as the seventeen *prahus* are concerned, appear to have been conducted in contravention of the spirit exhibited by public authority in the Act, treaty, and instructions.

"By no means: the seventeen *prahus* made a direct attack on the Nemesis, and were defeated. One at least of the *prahus* escaped, as Lingir, their chief, lived to give his own account of the affair; and most managed to reach the shore, where they abandoned their boats to Mr Steel's squadron."

Query 10.—That they seemed to have been founded on the principle that extermination was the proper method of dealing with these people, as with wild beasts or vermin, and that under the circumstances the saving of life was not desirable.

"On the contrary, had these pirates shown the slightest desire to surrender, every European officer present would only have been too glad to stay his hand. It was not, however, possible to take prisoners, as though vessels could be seen at night with the aid of blue-lights, it would not have been easy to distinguish a swimming man. Besides, the strong tides there would sweep out to sea or up the river—in either case, far away from the scene of contest—any one struggling in the water. Within that week, I do not remember to have seen more than three bodies floating in the river. Admiral Farquhar was essentially a humane man, and Sir James Brooke was as gentle-hearted as he was brave and good."

Query 11.—That it appears from the journals of Sir James Brooke that probably more than four-fifths of the crews of the *prahus* thus indiscriminately slaughtered were slaves who, whether there against their wills or not, were there without having any option, and were of the class of persons on whose behalf Sir James Brooke had at another time contended that some distinction ought to be drawn.

"There is some error here. Sir James Brooke may have made some such statement when referring to the Lanun, Balagñini, or Maludu pirates who came from the Sulu seas or Gilolo, but not when speaking of the Dyaks. The Dyaks have but a few female slaves—the young women who have been seized in their cruises. Occasionally a chief may have a few slave debtors, but their war *prahus* are manned by the fighting men of the tribe. Among the 4000 pirates who were present at the action on the 31st of July 1849, there may have been a few slaves attending on the Malay chiefs, but not to an appreciable extent. As a rule, they would be considered too valuable to be risked. The slaves in the Lanun *prahus,* on the contrary, are mostly captives taken during the cruise. The *prahus* start with the fighting crew and a few hundred slaves. As they advance, the captives take their places at the oars; and if their number become unwieldy, a chief will man a captured *prahu* from his own crew, and none of the chief's followers do more than steer and manage the sails, and pull in the swift boats to surprise unwary fishermen. These pirates rarely kill any but their European captives, as their object is to acquire slaves, taking them in one country to sell them in the next.

"In all this controversy mistakes have arisen from not paying sufficient regard to the person who gives testimony. As a rule, naval men know nothing of the countries in which they operate, and therefore make mistakes in writing their reports. Even in Sir James Brooke's journals it is necessary to remember that they were written from day to day, and that they are only first impressions—often wrong—and that they were published under the auspices of persons entirely ignorant of Borneo. In 1847 I corrected many sheets of Mundy's book, and one blunder of my own was the cause of a fierce controversy.

"I have not read the papers connected with this affair for years, as I did not intend to enter into the controversy in my life of Sir James Brooke; but in looking for Mr Boudriot's name, I noticed one of the errors to which I refer: Captain Wallage says that the pirates landed in a hostile country, whereas the action took place at the mouth of the river Seribas, and the pirates therefore landed in their own country. It was uninhabited, but quite familiar to them. It was almost an island, so that by pulling up a branch of the Seribas we could have held the neck of land between that river and the Kaluka, and thus cut off the retreat of the 3200 men who had landed (I calculate that 500 got up the river, being the crews of the *prahus* which escaped); but Sir James Brooke knew that the pirates would not surrender, and would thus either perish of starvation or be killed in detail. He therefore called on his forces to follow him to the attack of the pirate town of Paku, and thus enabled the mass of the pirates to get home. He said at the time to those who objected to his clemency that the blow struck would suffice to destroy the system,—and it did, as no pirate fleet has gone to sea during the last twenty-eight years, and the inhabitants of those rivers are now far more industrious than any others on the coast.

"When I arrived in Borneo in 1848, the commerce of the coast was dead, and the whole trade of Sarāwak in 1849 was, I think, about £10,000. No trading *prahus* could venture to sea except well armed and with numerous crews, forty to fifty men manning vessels which really required but eight to manage them — and they generally sailed together, to give each other support. There was insecurity everywhere; and during the first six months of 1849, about 500 people fell victims to the Seribas and

Sakarangs. No wonder that the traders and peaceful inhabitants turned out by thousands to punish these marauders as soon as the presence of the English forces gave unity and direction to the attack. The contrast is now most marked. Instead of pillage, massacre, and insecurity, we have perfect peace upon that coast. Instead of a population, estimated at 100,000, bestowing its best energies on an exciting but nefarious pursuit, we have them eagerly turning their attention to trade and agriculture. These old pirates are far more energetic than the others, and are eager for improvement, and they are now looked upon as the mainstay of the English; and even during the Chinese insurrection of 1857, when Sir James Brooke was a fugitive from his capital, they turned out almost to a man to defend his cause, and greatly aided in driving the insurgents over the frontiers.

"These are indeed results, and would atone for any severity which might have been exercised in their accomplishment. But I am certain that no undue severity was used: and after twenty-eight years' reflection on the subject, I can arrive at no other conclusion than that the punishment of the pirates was just and necessary; that no one was killed beyond what was requisite for the success of the operation, and that there was no pursuit of the flying foe after he had abandoned his war-vessels; and that Sir James Brooke and Commander Farquhar, to prevent the destruction of the defeated, called off the attention of the native forces as soon as possible after the action, and thus prevented any useless slaughter.

"I may conclude by saying that in 1849 Sir James Brooke, as Commissioner, laid before Lord Palmerston all the evidence on the subject; that after some months of consideration his lordship wrote to say that he approved of what had been done, and that should Sir James consider it necessary, he was to continue to act in the same manner.

It is in the hope that you may come to the same conclusion as Lord Palmerston did that I have written to you at this length."

I subsequently received from Mr Gladstone some further queries, which I think worth inserting, as they complete the whole view of the subject:—

Query 1.—In what sense and on what evidence are we to understand that the Seribas and Sakarang people were on this occasion engaged in piracy as distinguished from intertribal war?

"The English and native forces were assembled to attack the pirates of Seribas and Sakarang for piratical acts which extended over a series of years. It would have been immaterial whether they had committed acts of piracy or not on this occasion, but they did commit acts of piracy, for they captured a trading vessel coming from Singapore laden with piece-goods and iron, which goods were found in their *prahus,* and another vessel laden with sago just sailing for Singapore. The crew of one escaped on shore, whilst the others were killed. They also plundered the village of Palo, and tried to surprise the towns of Mato and Serikei. They took six young women at the village. Palo was almost the only salt-manufacturing village from which the pirates could get supplies, and therefore they did not destroy it."

Query 2.—It appears that they were not possessed of firearms. What weapons did they use?

"I am not aware on what authority this statement is made. I saw the flashes from their guns, and heard the report; and as most of the fighting Malays were in the fleet, and as all the Dyak chiefs have muskets, there must have been hundreds of firearms distributed in the 105 war *prahus* present. Any one who was present at the attack of Lang Fort, as I was, would have seen sixteen of our men knocked over by the first fire of the Dyaks, on which occasion they had no Malays with them. The arms of the Dyaks, as of our own men, were usually spears, javelins, and swords — and they are expert swordsmen. When fighting in the jungle the Dyaks seldom or never use firearms, as they are apt to get rusty and out of order; but in *prahus* they took all they had, and they have hundreds of brass guns, which are also used as currency. A man is said to owe another not so many dollars but so many *pikuls* (133¼ lb.) of brass guns. Some beautiful rifles were captured during this expedition."

Query 3.—Can any case be substantiated or alleged in which these tribes are known to have carried on sea-piracy, or assailed any British or European vessel? I find nothing on this head but vague allegations. Fifty-three merchants and inhabitants of Singapore, including six master-mariners, said there was no one among them who ever heard that any Dyak pirate, as distinguished from Malay, had been seen at sea?

"How could the Seribas and Sakarangs get to Pontianak and the other Dutch possessions except by sea? and the Dutch evidence is clear enough that they went and ravaged their colonies. Monsieur Boudriot's testimony alone proves that; and in the Dutch official reports on piracy, mention is made of the capture of one of their gunboats by Dyaks (under Rentab), in presence of her European consort, that vainly endeavoured to succour her. With the exception of Sir James Brooke's well-armed vessels, which went only to Sarāwak, no English or European merchantmen frequented this coast, as it was a favourite cruising-ground of the Lanun and Balagñini, as well as of the Seribas and Sakarang pirates. In 1854 we recaptured the guns of the Dutch vessel above referred to.

"If reference be made to the evidence given before the Commission in Singapore in 1854, it will be seen who these fifty-three individuals were, and why they signed this memorial. It would not matter if sixty instead of six master-mariners signed such a declaration, unless they knew something of what they were signing. Four-fifths of the European firms of merchants in Singapore signed in an opposite sense. The only pirates belonging to the coast of Borneo, south of the capital, were the Malays and Dyaks of Seribas and Sakarang. The master-mariners might give any name they pleased to them, but they were the only indigenous pirates. These Malays did not cruise alone: these pirates were not put down because they attacked British vessels, but because they destroyed commerce in general. I have answered these queries; but why they should be asked after the decision of the Royal Commission I do not quite understand. The Commission, after taking a Blue-book of evidence, decided they were pirates, which no one who understood the subject ever doubted."

Query 4.—Is it true that the *prahus* were propelled, not by sails nor oars, but by paddles?

"All these *prahus* can be propelled by the three means—all have masts and sails: they use oars when they are not on warlike expeditions, but paddles when required to move rapidly. On the night in question probably every mast and sail were taken in, and they trusted entirely to paddles, as then the men can sit closely together and urge the *prahus* on at double speed. All our own swift *prahus* were propelled by paddles, including Mr Steel's Snake. We never used oars, except when short of men, or in very heavy lumbering *prahus*. The men prefer the paddle to the oar." [1]

[1] After the extraordinary blunder into which Mr Gladstone was led by Mr Motley, he will be less inclined to trust to the evidence of master-mariners. The brig Amelia, of which Mr Motley was master, was, in 1852, capsized off the north-west coast of Borneo, when over sixty of the passengers and crew were drowned. The master and a few others, twelve in all, contrived to save themselves in the long-boat, and reached Bruit, one of the villages of the friendly and hospitable Milanows, and did *not*

Query 5.—Is there any indication whatever of any resistance by the five flying *prahus* which the Nemesis destroyed?

"I do not know; but I do know that they made no effort to surrender—no sign of wishing to give up; and it would have been a grievous error to have allowed these war-boats to escape into the Batang Lupar, whose peaceful inhabitants (on the lower reaches) would thus have been exposed to death and pillage. For the inhabitants of the coast, thinking themselves safe, as our fleet was out, went to sea on trading voyages; and some Dyak pirate *prahus*, returning from a foray in the Dutch settlements, caught a Sarāwak trading boat at sea, took her, and killed the whole crew. She was commanded and owned by Abang Hassan, nephew of the Datu Patinggi. To have allowed the five *prahus* to have escaped to the Batang Lupar would have been to sacrifice the innocent to save the guilty. I also understood that the five *prahus* were part of the squadron of seventeen which made the sudden and daring attack on the Nemesis."

Query 6.—Had these *prahus* no captives on board? If there were any in the five fugitive *prahus*, must they not have been destroyed with the rest of the people on board, all of whom perished?

"As far as my information goes, the whole fleet had no more than the six captives before referred to. I never heard of a besieging army stopping the bombardment of a town because there might be innocent women and children

arrive "*among these cruel pirates*," whose dwellings, instead of being on the coast, are *far* up the rivers Rejang, Seribas, and Sakarang. Mr Motley's printed account of the affair, in the Singapore newspapers in 1852, is strangely out of harmony with his gratuitous letter to Mr Gladstone in 1877. Mr Motley and his fellow-survivors were hospitably received and cared for by the officers administering the government of Sarāwak. Thus Mr Gladstone, to support his charges, has been led by an obscure writer, of whom he knows nothing, to print, in the columns of the 'Contemporary Review,' statements which are opposed to fact.

in it. It by no means follows, however, because Captain Wallage saw no living being on board of the five *prahus*, that the crews were all killed. Even when caught by a storm at sea, the Dyaks jump overboard and hold on to the sides, which lightens the vessel and gives her greater steadiness. It is so common an occurrence that the Dyaks are provided with bundles of a certain kind of bark, which, when wetted, gives out a liquid that poisons the water around, and prevents the sharks attacking them.[1] It was a cloudy night also, and objects were not distinctly seen: therefore the crews may have been in the water, holding on to the sides of the boats, which would explain the discrepancy in the calculation of the number killed—300 instead of 500."

Query 7.—In this night-operation and general firing, could it be known, with any certainty, whether the *prahus* had any captives on board or not?

"Not at the time; but we could be pretty certain that there were not many. The Seribas and Sakarang Dyaks never took men captives,[2] nor elderly women, nor any boy after ten or eleven years of age, nor any very small children. They said, even if they did not require heads, that men captives were dangerous, elderly women of no use, boys after ten or eleven would always be trying to escape, and that infants were troublesome. I have seen hundreds, I might say thousands, of the heads taken by the Seribas and Sakarangs—for instance, when we captured their villages on the Kanāwit—and among them were very many skulls, evidently of children. We had the curiosity to count the number of skulls in one village: it was a few over 300—more than 100 in one village-house! On that occasion we took eighteen villages."

[1] See my work, 'Life in the Forests of the Far East,' published in 1862.

[2] One male captive (a relative of his own) was once taken by Rentab, but he was reserved to undergo the most fearful tortures.

Query 8.—As regards the headless bodies of young girls, how is it known that they were killed by the pirates? If they had such persons in the *prahus*, was not this a strong argument against the slaughter wrought by the Nemesis?

"Even if there was no other evidence, it is not to be supposed that our men would kill their own friends and relations. Sarāwak had become as a city of refuge, and its population had increased from 1500 to 13,000 by refugees from the whole coast, who there sought shelter from the pirates. I saw only one body, and that was of a very young woman: her head, as well as her breasts, had been cut off, and a spear . . . Our men offered to show me the others who had been treated in the same manner; but the sight of one was enough. I, however, saw the mats which our people had drawn over the naked bodies; and there were five of these little heaps. I subsequently asked Lingir why they had killed these women. He said he was not there, which was true; but that their captors were afraid that they would escape, and that the men were so savage at their defeat that there was no restraining them. He allowed, however, that it was a wicked deed.

"There was no slaughter wrought by the Nemesis. Captain Wallage was ordered to do his best to destroy the war *prahus* of the pirates, and he did do his best, and only did his duty in doing so. That pirates were killed during the operation was the natural result, but he would have fired into no *prahu* that had shown any desire to surrender.

"The Seribas and Sakarangs perfectly understood the usage of the white flag, and when they desired to stop Captain Keppel's further attacks they hoisted it, a parley ensued, and the fighting ceased. Had they hoisted a white flag during this engagement, it would have been respected."

How little Mr Gladstone appreciated the information given will be understood by those who have read the article, "Piracy in Borneo, and the Operations of July 1849." He preferred the testimony of an ignorant skipper, or a casual visitor, to the evidence afforded by such men as Sir James Brooke, Captains Keppel, Mundy, Farquhar, and Mr Low, and all the other authorities in Borneo; and the result is, that he has not written a paragraph which does not contain an error.

Mr Gladstone has so imperfectly studied the question, as to confound in one confused mass the habits and customs of such people, for instance, as the Lanuns and Dyaks, as different in every way as Turks from Russians; but having a foregone conclusion to defend and to explain, he has in his article made mistakes which no one who had really studied the Further East could have fallen into.

Sir James Brooke acted on Mr Gladstone's words—"I have no right to sacrifice the interests of the sufferers to what I should call maudlin, artificial, and in the worst sense, sentimental pity." But Sir James's sympathies were with the victims, while Mr Gladstone's are for the pirates.

I had intended to have answered Mr Gladstone's article, paragraph by paragraph, and I believe would have shown that a little knowledge is a dangerous thing. But I have decided not to do so. I could not hope to convince Mr Gladstone, and the few who agree with him may rest in their unbelief. Those who cannot believe that a man may act from noble and generous motives, will admire Mr Gladstone's attempt to prove that both Sir James Brooke and the naval men engaged in these operations acted from sinister and unworthy motives, when they boldly and successfully cleared the seas from the most bloody pirates that ever infested them; while those who have followed Sir James's career with sympathetic appreciation will require no further evidence of the nobleness of his motives.

CHAPTER XI.

PEACE MEASURES — ATTACKS ON SIR JAMES BROOKE'S POLICY — MISSION TO SIAM.

1849-1851.

WE were all glad to find ourselves back in Kuching. Gradually our force dispersed. The natives went to their several homes; the Albatross, Royalist, and Nemesis to Singapore, carrying with them many of our party for a change of air and scene; and Grant and I alone remained with the Rajah. We had been above 160 days in boats and ships during the last nine months, and Sir James was completely exhausted. We were not, however, idle. By the arrival of the mails we found that a couple of clever newspapers in England were backing up Mr Hume in his attacks on our proceedings.

The European inhabitants of Sarāwak then met and resolved to present an address on the subject.

Sir James Brooke was of a very excitable and nervous temperament. The savage attacks to which he was subjected roused his anger, and did him permanent injury. He never was again that even-tempered gay companion of former days. He thought too much of these attacks, and longed to answer every petty insult and calumnious insinuation. The exposure to which he had been subjected during the last two years had told on his health, and fever and ague constantly prostrated him. He, however, occasionally diverted his mind by making excursions among the Dyak tribes of his own country, and the peace and contentment he ever found there had a soothing effect. Boxes of new books occasionally arrived, and brought us, among others, Macaulay's 'History of England.' This the Rajah began to read out to us; but it so interested him that he soon dropped that slow plan, and carried it off to his private apartments, and we saw it no more until he had finished the book. He was an excellent reader.

Ever and anon, however, the packet would come in; and, galled at length by some furious assault of Hume and Cobden, he one day sprang to his feet, and said, "I wish I had the two before me, sword in hand, on the sands of Santubong."

We passed all the autumn at Sarāwak, receiving deputations from the different pirate districts—all asking forgiveness, and promising to abandon piracy. The battle of Batang Marau, as the affair of July 31st was called, was the death-blow to piracy, as far as expeditions at sea were concerned; and although for many years the Dyaks continued restless, and attacked within the river-system various villages allied to Sarāwak, it was never again necessary to call upon the navy to punish them. It was indeed difficult to eradicate from their hearts that love of head-hunting on which their fame depended.

Many of these pirate chiefs had never before seen a European except in the undesirable position of an armed opponent, and their curiosity was insatiable. Sir James used to take them into his room, show them his swords, his uniform, every curiosity that could amuse them. He had a magic-lantern arranged for the evening, when the shouts of laughter would show that these wild children of the wood could understand a joke. The scene which amused them the most was a representation of a party of body-snatchers rushing from a churchyard pursued by skeletons who are pelting them with skulls. After seriously regarding this representation, a chief turned to Sir James and remarked, "Ah, I see that in former days you white men also took heads!"

The new and improved magic-lantern subsequently brought out was a *fiasco:* the natives naturally could make nothing of Scripture scenes or views of cities. The skull scene was wanting, and the rest of the fun was taken out of the show. Many a time, on returning at midnight after dining with a friend, I have found lights in the great hall, and Sir James sitting there with two or three natives, listening to their tales of distress, or hearing accounts of different tribes,—for it was at this hour that the truth would come out, and Sir James would obtain that intimate knowledge of what was passing in his country which often surprised those that lived with him.

When I appeared it was, "St John, come and have a glass of sherry and a talk," and there we would remain till two o'clock, discussing men and measures; for although at that time I was not looked upon with the same affection as some others, Sir James appeared to have an unlimited confidence in my discretion. It was at these times, alone together, that I endeavoured to calm his indignation, and lead him to despise attacks which were founded on ignorance, or suggested by the spite of such enemies as Mr Wise. I had but poor success, as by nature Sir James was impetuous. His feelings on the subject may be gathered from this sentence: "I am not the man to be bullied, or meekly to kiss the rod in such silly hands." In November he received several letters from Lord Grey—kind and considerate, as might be expected from so judicious a friend; and now the public and official declarations of Lord Palmerston might have satisfied him that he was warmly defended by the home Government. I believe he would have been satisfied, had not some zealous but injudicious friends at home kept up the excitement. I notice that the Rajah writes that a perusal of his journals would prove that he "had acted on large principles;" that "I have done justice and loved mercy, and that I have overcome difficulties without much heeding the personal consequences."

Among the measures which the Rajah at this time took to check the Sakarang Dyaks was to build a fort opposite the entrance of their branch of the Batang Lupar to stop the egress of any large fleet. We started from Sarāwak to aid the work, but were driven back by a heavy blow of wind which lasted several days; however, the people of the surrounding districts assembled, and in a short time built up a very formidable stockade, which was armed and prepared for a stout defence by Mr Crookshank.

Sir James Brooke's fame among the Dyaks was now at its greatest height, and parties walked from the far interior to see the noted chief. These visits were most pleasing to the Rajah, as they showed the effect of kindly and good government.

Fever and ague having, it would appear, obtained a firm hold both of Sir James and of his secretary Grant, the doctor strongly advised them to try a change of climate, and proceed either to the hills of Penang or to those of Ceylon; but before doing so, Sir James determined to revisit Labuan, where discord appeared to reign supreme. Sarāwak was at this time particularly unhealthy; it was not surprising that those exposed as we had been should suffer, but others as Mrs McDougall became so ill that her life was despaired of. The Rajah in one of his letters refers to it thus: "In this case the lady is so much loved and respected, so amiable and so clever, that we should indeed deplore her loss, and despair of readily making it good."

We left Kuching on the 11th of December, and reached Labuan on the 14th, to find everything in disorder. It is not my intention to enter minutely into the details of our stay: they would not be interesting to the public. Sir James Brooke on his arrival was astonished to find that the Lieut.-Governor was in open quarrel with all the principal officers, and that some charges had been made against him which affected his personal honour. The Governor began then to institute an inquiry; it lasted for a couple of months, and ended in the suspension from office of his *locum tenens*.

As I was actively employed during the whole time in preparing the papers for the Colonial Office, I had ample opportunities of studying the whole case, and I arrived at a different conclusion from my chief, but had not the influence to press my views successfully. I thought that the Lieut.-Governor had been both injudicious and intemperate, but that nothing was proved which could in any way affect his honour. Mr Napier was a superior man, of polished education and remarkable information, and the way that he turned the tables on a boastful doctor showed how accurate was his scholarship. But his temper was too irascible.

This separation of two old friends was very sad, and tended to embitter the discussions which hereafter arose on the subject of Borneo, and the part Mr Napier took in these was quite inexcusable.

Although the inquiry and subsequent proceedings lasted from the 14th December to 23d February, Sir James did not neglect other things. He visited Brunei, he did all he could to induce the Chinese and wealthy natives to come over to Labuan; but he failed, because they knew that, though he was Governor of the island, he seldom resided there. The influence of Europeans over natives is personal, and nothing but a prolonged residence of Sir James in the colony would have attracted the neighbouring population.

Glad enough were we to leave Labuan and get away from disagreeable discussions. Fever and ague having again attacked Sir James, he determined to proceed to Penang; so on the 23d February we left Labuan, called at Sarāwak, and reached Singapore on 3d March.

We found both friends and news: Mr and Mrs McDougall, grieving for the loss of their little boy; Captain Brooke recovered from his illness; news that the 'Times' and the 'Globe' were defending Sir James Brooke's policy, while the 'Daily News' and 'Spectator' attacked it. In Sir James Brooke's place, I should have been delighted to have found so much attention drawn to Borneon affairs.

We found also that Sir James Brooke had been directed to proceed to Siam and Cochin China to form treaties with these countries, if he thought it advisable; and he did me the honour to nominate me secretary to these missions. We found Singapore excited by the news of these diplomatic movements; and the merchants were delighted to know that there was a chance of the trade with these countries being soon placed on a proper footing.

After a fortnight spent in Singapore, where we were *fêted* by the Governor and the principal merchants, we left for Malacca and Penang. The Governor, Colonel Butterworth, was one of the kindest and most hospitable of men, and he placed his country-house at the Rajah's disposal.

The same day that we arrived at Penang we ascended the hills, till we reached the Governor's bungalow or country-house, situated 2200 feet above the level of the sea. It was a splendid situation for enabling us to recover our shattered health, as the whole party was ill. We were five,—the Rajah, Mr and Mrs McDougall, Grant, and myself; and when we began to gain strength, a pleasant time we had there. I think that most of us looked back on the six weeks we spent on the top of Penang Hill as one of the most agreeable periods of our lives. Sir James and a party rode out every day, while others walked along the well-laid-out paths that lead you to the neighbouring hills. We had a very pleasant society from the surrounding bungalows, the officers, both military and civil, visiting us; while two clever French priests delighted us with the intelligent and exact accounts which they gave us of both Siam and Cochin China—countries in which they had resided during many years.

Sir James enjoyed his stay in Penang. He used to sit for hours in a veranda from which he could gaze on one of the loveliest scenes in the world: the high mountains in the distance; Province Wellesley, with its smiling features and winding rivers; the placid sea between it and the cultivated plain of the island, lying distinct as a map 2000 feet below; the plain covered with formal plantations, whitewashed houses, and admirable roads networking the island. Then, when a squall came on, the clouds would come sweeping towards us; the distant prospect, from some atmospheric cause, would become for a moment doubly distinct, but soon obscured by a mass of vapour that enveloped us; then a rent in the clouds would show pretty valleys, where the sun for a moment shone, and the far-off southern plain, and the ocean surrounding all, gave us glimpse after glimpse of beautiful scenery. Sir James never tired of watching these varied prospects, and always spoke of the hill at Penang with enthusiasm, —for in those days we were all united.

Those uneventful six weeks soon glided away, and Sir James, having now nearly recovered from the fever, determined to return to Singapore and prepare for his mission to Siam. We left Penang on the 10th March, in a large

Spanish ship, the Magnolia. We had nothing but light breezes and calms, so that it took us seventeen days to reach Singapore, a distance of little over 400 miles.

Our voyage, however, was a pleasant one. We had lately received 'The Vestiges of the Natural History of Creation,' a book now forgotten, but then famous. How we discussed it! How interesting it was to listen to Sir James Brooke's disquisitions on the various disputed points! None of us were a match for him, either in brilliancy of illustration or in apparent logical deductions from the premises before him. I say apparent, because although very fond of argument, he had not a logical mind. At one time the Rajah reminded me of the students in 'Gil Blas,' who laid hold of the button of every passer-by in order to have an argument. He never tired, and what was more, he was never a bore, as he could amuse even when he could not convince. I always felt in the position of the man convinced against his will. I was silenced for the moment, but ready to return to the charge on the very next occasion.

We arrived in Singapore on the 27th of March, and immediately began to make preparations for our mission to Bangkok. Before we left for Penang, Sir James Brooke had sent the Nemesis to Siam, with a letter to the Minister for Foreign Affairs, to announce our coming; and soon afterwards the American envoy, Mr Ballestier, arrived there in the frigate Plymouth. He had utterly failed in his mission, having been received in Bangkok with scant courtesy: this was partly owing to the American commodore having declined to afford him any efficient aid,—another instance how sometimes private feeling mars the public service. Having succeeded in thwarting the Americans, the Siamese now prepared to thwart the British: nothing but the display of a powerful force could have induced them to submit.

By the mail which arrived on the 4th April we were cheered by the news of the debate in the House of Lords, and Sir James felt relieved by the friendly tone of Lord Grey's speech; and Henry Drummond, then M.P. for Surrey, wrote to a friend, "The attacks on Sir James Brooke are most unfair."

While we were waiting in Singapore, intelligence reached us from Sarāwak which was not without its importance. It appeared that the American envoy, Mr Ballestier, had visited that place, bearer of a letter from the President of the United States, addressed to Sir James Brooke as the Sovereign Prince of Sarāwak, and expressing a desire to enter into friendly relations; and Mr Ballestier informed Sir James by letter that he was intrusted with full powers, and ready on the part of the United States to sign a treaty with Sarāwak; and he added that he was instructed by the President of the United States to thank Sir James Brooke, "in the name of the American nation, for his exertions in the suppression of piracy," and to compliment him on his noble "and humane endeavours to bring his subjects and the neighbouring tribes of Malays into a condition of civilisation, which the President hopes sincerely will be successful in the end." It was very unfortunate that Sir James was absent from Sarāwak during this visit.

Sir James Brooke, being a British official as well as Rajah of Sarāwak, was placed in a peculiar position: he therefore sent the correspondence to Lord Palmerston, who replied that he saw no objection to his entering as Rajah of Sarāwak into diplomatic relations with the United States.

Never were the words more appropriate, "There is a tide in the affairs of men, which, taken at the flood, leads on to fortune." Had Sir James Brooke treated this letter from the President of the United States with the seriousness which its importance deserved, how many years would he have saved of heartburnings, of that hope deferred which maketh the heart sick! But he neglected it; he never answered the letter, and the recognition of the United States was lost.

We stayed above four months in Singapore, waiting for the naval force which should accompany us to Siam. We felt angry at the delay; but it was difficult to be angry with that pleasant old Admiral Austen, who, being the brother of the author of 'Pride and Prejudice,' won our hearts by such relationship. Sir James was a great admirer of Miss Austen's novels; he read them and re-read them. During quiet hours in Sarāwak he would read them aloud; and he filled me with the same admiration, which has continued to this day. During our stay in Singapore, Sir James received the most welcome despatches from home. "The Government has approved of all that I have done, and has directed me to follow out the policy. This is quite conclusive and satisfactory." Would that he had continued to think so!

We started for Siam the 3d of August 1850, on board H.M.S. Sphinx, Captain (now Admiral Sir Charles) Shadwell, and took the Nemesis in tow. No account of this mission has, I believe, been published; so that I may enter into a few particulars of it, without, I trust, fatiguing the attention of the reader.

Siam was, in 1850, governed by a king who had little claim to the crown, as he was an illegitimate brother of the true heir, but at the same time older, and perhaps more able. He had once been very fond of the English, but he had, during the last few years, become as inimical to foreigners as he had formerly been friendly. I believe this change arose chiefly from the ill-conduct of the principal English merchant in Bangkok, who excited the king's anger by a violent and unjustifiable attempt to force the

Siamese Government to pay an exorbitant price for a steamer. Since that quarrel the position of foreigners had become almost untenable. As we only heard the English account of the origin of these quarrels, I have no hesitation in saying how ill they behaved. Had we heard the king's account, we should have probably to speak even more strongly.

We arrived off the bar of the Menam river on the 9th August, and heard that the whole country was in a state of great alarm, as they feared that the obstinacy of the king would result in a war with the English.

The Sphinx remained outside the bar, and Sir James Brooke sent Captain Brooke and myself, in the Nemesis, to the village at the mouth of the river to announce our arrival and the object of our visit. We found the forts crowded with soldiers, and rather expected to be received with a volley; but prudence prevailed, and an officer came off. Having settled matters with him, we returned to the Sphinx, which next morning attempted to cross the bar, but unfortunately stuck on it. This, I believe, was the cause of our subsequent ill success. Had the large steamer been enabled to get in, the effect would have been complete; but her misfortune changed the mind of the king.

After many preliminary negotiations, it was settled that Sir James Brooke should land at Menam and meet the Prime Minister, who had just arrived from the capital. Two large boats came from the shore to fetch us. As Sir James stepped into one, the yards were manned; and as we pushed off, a salute of seventeen guns was fired from the ship, and admirably returned by a battery of field-pieces on shore.

At the landing-place Sir James was received by Siamese officers of rank, with a guard of honour, who presented arms with much smartness. Hundreds of others were squatted on the ground near the hall of reception—a new mat-building, built for the occasion. The Praklang, or Foreign Minister, was seated on a broad divan, partly reclining on numerous cushions, with his gold betel-boxes and spittoons, and his pipe and his tea-service arranged around him. He rose as we entered, and pointed to chairs, which were placed on the opposite side of the room, but far to his left—a position of marked inferiority. Sir James, however, disconcerted all these plans by advancing straight up to the divan and shaking hands with him; then taking a chair, he placed it right opposite to the Minister. We followed the example, shook hands, placed our chairs in a line with Sir James Brooke's,—and thus the first fight on etiquette was over, and no further attempt was made to lower the position of our envoy. The Minister was completely taken aback, but in a few minutes recovered his equanimity. He was an elderly man, dressed in great state, with a long robe and under-dress of rich gold brocade. His countenance was not unpleasing, but had sometimes an expression of harshness. He looked a Chinese ruler, lolling his huge body about, while his sons and his followers crouched around him; and when any one approached, it was by crawling on his elbows and knees—a humiliating fashion.

Lord Palmerston's despatch to the Praklang was borne in with all due solemnity, and read; then, after the usual complimentary questions and answers, we took our leave, having settled to proceed to Bangkok in state-barges, after some of us had been to the capital to inspect the house prepared for the Mission. We were greatly interested with our trip, particularly with the floating-houses in the capital. As the temporary building was not suitable, we accepted the offer of an English merchant, a Mr Brown, that Sir James should take his house, and leave the other to visitors and the escort of marine artillery which was to accompany us.

We stayed about six weeks in Bangkok, during which Sir James did all that was possible to negotiate a treaty; but the king and the Ministers were opposed to it, and our efforts were unavailing. But before we left, Sir James entered into a correspondence with the heir to the crown, who was an educated prince; and the Rajah felt that it would be more beneficial to our future interests to defer negotiations until his advent to the throne.

Bangkok is a remarkable city, but more remarkable for the number and size of its temples than for anything else. Few personal incidents occurred during our stay. Sir James suffered from the heat more than we did, as he almost entirely confined himself to the house, unwilling, amid a hostile population, to expose the Mission to an insult.

One day, however, we persuaded him to visit a temple near our house, when suddenly a drenching squall came on. We most of us ran to get out of the rain, while Sir James continued at his slow dignified pace towards the house, as if no pelting shower was coming down. He afterwards reproved us for having completely forgotten our dignity before a crowd of orientals. There never was such a country for mosquitoes as Siam. Our rooms were filled with them; and at night it was not possible to eat one's dinner without having fanners seated under the table to drive the swarms away. They bit through everything. One of our marines sought refuge in a huge water-jar, and was found with only his nose protruding; while another who sought to do the same thing in the river was drowned.

Sir James was one day seated in the veranda, when he drew my attention to a group of children swimming round a floating-house. Presently a little baby, who was at its mother's breast, turned, struggled from her arms, and threw itself into the river, where it swam like a fish. On being taken out, its mother presented it with a cigarette, which it quietly smoked, even before it could walk.

When we found that every proposition was refused, the Siamese Government declining to enter into a new treaty, we asked for boats and left Bangkok; and next day we found ourselves on board the Sphinx, and off to Singapore.

Sir James was delighted to hear, on his arrival, that some thousands of Chinese had passed over from Sambas into Sarāwak, and that at last the wished-for movement had commenced. As soon as we had got rid of the Siamese correspondence we started for Sarāwak in the Nemesis, and reached that place in October 1850, to find everything flourishing.

The arrival of this large party of Chinese was a very singular event. In the Dutch-protected state, Sambas, south-west of Sarāwak, there were two distinct parties of Chinese—those who sided with the gold-workers of Montrado, and those who, from their position, were under the influence of the Dutch. The latter were the agriculturists of Pamangkat, and some few small companies of gold-workers.

At the beginning of the little war in Sambas in 1850, the Chinese of Montrado obtained the upper hand, drove the few Dutch soldiers into the town of Sambas, and then set upon their allies. The Pamangkats fled for safety over the frontiers into Sarāwak, and there settled either as agriculturists or among the established gold-workers. The English officers did their utmost to aid them on their first coming, and above a thousand at one time were supported by the Government. They well repaid this attention, as in a few months they tripled the revenue.

We reached Sarāwak on the 24th, and were enabled to spend seven weeks there—not seven weeks of rest, as the Rajah had promised to visit some tribes of Dyaks, and examine the gold-workings of the Chinese in the interior.

Sir James had also a very momentous question to settle, which was the future government of the coast between Sarāwak and Brunei. The Sultan, the nominal ruler, possessed but little real authority, although his name was often used by plundering nobles to enable them to carry out their measures.

About a week after our return to Sarāwak, the Rajah discussed the matter with me. This was the state of affairs. The next district to Sarāwak was Samarahan, which in everything followed the lead of its neighbour; then Sadong, governed by an ill-conditioned Malay chief named Bandar Kasim, who was always in trouble; then the great river of Batang Lupar, with its several branches of Lingga, Undop, and Sakarang, entirely independent of the Sultan, as was the next great district, the Seribas. Kaluka kept up an occasional intercourse with the capital, but paid no revenue; and the majestic river of Rejang was peopled by tribes who owned no allegiance to any one.

Sir James had long thought that this state of things should be altered. He acknowledged that the lawful sovereign should receive some revenue; but how to obtain it was the difficulty. The Sultan had placed himself completely in Sir James Brooke's hands, and generosity as well as policy urged him to do what he could for him.

I believe that the suggestion came from me that the Rajah should propose to the Sultan to hand over to him the government of the six rivers, under the condition that the surplus revenue should be divided between the Governments of Sarāwak and Brunei. The idea was simple, and recommended itself; and in order to prepare the way for its execution, Sir James determined to establish better government in these districts. He removed Bandar Kasim, as he was proved to have made a foray into Dutch-protected territory, in which twenty-five men were killed.

He appointed good men to govern Lingga: he had already built a fort at Sakarang to check the pirates, and had established there his old Balidah acquaintance Sirib Moksain.

Seribas he was forced to leave alone for the present, as to establish a Government there would have been too expensive. He left the native rulers at Kaluka and the Rejang; but he determined to build a fort at the mouth of the Kanāwit to prevent the pirates getting down the main river to ravage the coast.

These different enterprises were likely to prove somewhat expensive, as the revenues to be received would for a long time be nominal.

We sat talking till late: the rest of the party had retired, and the native chiefs had long since glided from the room. During these discussions the Rajah was in his glory—he so fully understood his subject, was deeply interested in every phase of the question, would apparently never accept a suggestion, opposing it with all his energy, but pondering over it afterwards. Many a time has he said, some months subsequent to a hot discussion, "St John, you were right; I shall carry out your idea," when the idea itself had to be recalled to my mind.

At this time Sir James Brooke had around him in his own establishment a number of young officers, as Captain Brooke, Crookshank, Grant, Brereton, Lee, and myself, who were thoroughly devoted, and who looked up to him for counsel. But it is a singular fact, for which I have vainly endeavoured to find an explanation, that, except to one, he never gave a word of advice; never tried to direct our studies; never tried, except during our usual discussions, to lead us; never thought of our conduct.

Perhaps he trusted to his own example; for during the twenty years I knew him, his life was a pattern of goodness. His sentiments were always noble, and his conduct was that of a thorough gentleman. Much as we respected our chief personally, we did not accept his opinions as infallible: on the contrary, in politics, on religion, on literature, we had endless discussions, which, as the Rajah said, generally terminated in the battle of Waterloo.

Sir James Brooke discussed every subject in the same manner as he played the game of chess: he made a brilliant onslaught, which required the utmost wariness to oppose. With time, however, we learnt his method, and opposed coolness and caution to all his attacks, and our lively talks went deep into the night.

As I have before observed, complaints continually came in about the governor of Sakarang, and therefore the Rajah determined to send Mr Arthur Crookshank and Mr Grant to inquire into his conduct. On their return to Sarāwak, they were accompanied by a son of the famous Sirib Sahib, the great pirate-protector of former days, whose family was now anxious to settle under the shadow of the new fort. They also brought with them the clever, pleasant, but untrustworthy Sirib Moksain, whom they had found completely unsuited to the post in which the Rajah had placed him. It was then determined to invite Mr Brereton to leave the Labuan service and take charge of Sakarang as a sort of independent governor,—a most singular arrangement. Brereton, however, proved himself to be the right man in the right place; he had formerly been in the navy,—and he possessed the art of winning the confidence of the natives, and although too impulsive, was, on the whole, an excellent and genial governor.

As Sir James had been away about ten months, the Dyaks of various tribes begged him to visit them. But he was still weak from old attacks of ague, and therefore determined to confine his visit to some tribes that lived on the tributaries of the Morotabas river.

We first went to the Maradang Dyaks. As they were

in fact but a branch of the Sibuyows, I need not enter into particulars concerning them, but I wish to mention a curious conversation which took place. After dinner was over, we sat by the light of a few wood torches in the broad enclosed veranda of a village house, when one of the Dyaks present said that he should like the Rajah to give them some account of his visit to Siam. We were surprised to hear such a question from this almost naked savage, and we inquired what he knew on the subject. The chief of the tribe then said that many present were familiar with Siam. He for one had been sent by their Malay rulers to pull an oar in a Lanun pirate boat, and had often cruised in the bay of Siam, and showed by his remarks how well he had observed, and told us many a curious anecdote of their proceedings. No wonder the coast of Borneo was in those days considered unsafe when the regular rulers of the country thus encouraged the pirates.

In the morning we started on a visit to a tribe of land Dyaks, the Sentahs, who lived on a little hill on the Quop. At first the path was very muddy, being through old rice-fields, then over pretty hills, young jungle, and now and then across purling streams. A Dyak path, even after being improved, is difficult to follow, and this was a regular specimen. As I have said, it lay first through abandoned fields, which the late rains had turned into slush; then through low jungle,—old, slippery, rotten trees were lying across the path; then large clearings for planting dry rice, covered with felled trees and branches half burnt through, over and on which you are expected to walk; then steps down the sides of steep hills, either cut in the clay or made of a single tree with rough notches; then through deer-swamps, bridged over by small trees ranged in a line in the middle of the path,—most difficult walking—a slip on either side would plunge you into mire.

Every now and then we arrived at a deep ravine over which was thrown the stem of a young tree, slippery and often rotten—a queer specimen of a bridge. At last we reached the Quop hill, very steep and very tiring, as we climbed up a succession of steps cut in the slippery clay; but at length, where some clear water crossed the path, a halt was called and a bath determined on, after which we felt refreshed, though five hours in the broiling sun had fatigued many of our party.

The Rajah was received with an exuberance of joy by the people—for this was one of the tribes whose captured women he had forced Sirib Sahib to restore—and ceremonies of all kinds immediately commenced. The old women danced with a slow measured step, and passed their hands over our arms, and then rubbed their bodies, thinking they had drawn virtue out of us; they brought fowls to the Rajah, that he might wave them over the heads of the people and wish them all blessings,—cool weather for their crops, and fertility for their women; then cooked rice was brought, on which the Rajah was asked to spit, and then stuff it down the throats of these simple old creatures. At last the Rajah, overcome by his unusual fatigue, fell asleep, and I undertook to strike a freshly killed fowl against all the door-posts, and thus sprinkle a little blood there. These ceremonies, and the beating of gongs and loud singing, effectually drove drowsiness from my eyes, and I envied the Rajah his power of sleeping under almost any circumstances.

Glad were we all next morning to start early, though before doing so Sir James had to cut a ratan that let a flagstaff into a hole properly prepared, dug to receive it. The curious part of the ceremony was that, as the Rajah cut the ratan, a live chicken was thrown in and crushed by the falling staff, a sacrifice to their gods. The same ceremony once, it is said, took place in the Rejang district, when a young girl took the place of the chicken; but this probably is but a tradition. The chief of the Sentah tribe was a simple, kindly old man, and once came to the Rajah to tell him that a party of Malays had, during a visit to his village, discussed the subject of religion, and said that all those who were not Mohammedans would be hurled into hell and burnt. "What," said the old chief, "will become of the white men?" "They will be sent to hell too, and you Dyaks will be the firewood with which they will be burnt." "I thought," continued the old chief, "that this was too bad; that I who have always done my duty to my neighbours should receive such treatment in the other world. But, Rajah, is it true?" The Rajah told him that it was not true, and advised him to join the Christians; "and then," he said, "you will share our fate, which is not likely to be that stated by the Malays." The old man took the advice seriously, and shortly after he and many of his tribe were baptised. However, he ever after went by the name of "Old Fireworks."

Though still suffering, the Rajah determined to fulfil his engagements, and in a few days started to settle matters with the Chinese company established to work gold in the interior, and then laid down rules which, had they been strictly followed, would have prevented the Chinese insurrection that broke out many years later.

No sooner had he returned to Kuching than he was attacked by fever and ague, and he was brought so low that he listened to the advice of his medical friend, and determined to return to Europe on sick-leave. He could have no rest in Sarāwak, as no sooner had he settled the Chinese affair than he had to enter into an inquiry as to the truth of the accusations made against the Malay governors of Sadong and Sakarang; and finding the charges proved, he had, as I have said, to depose both of them, and establish fresh rulers.

Before returning to England, however, he determined to make a visit to Labuan, and have another look at the colony. We started in the Nemesis, and arrived there on the 19th December. The place was healthy enough, but there was no trade; the coal was left almost untouched, and there was little prospect of advancement. We were only three weeks in Labuan, and even during that time we paid two visits to the capital. Sir James suggested to the Brunei Government his plan for the management of the six districts near Sarāwak, and left them to discuss it among themselves.

Had Sir James Brooke been enabled to have stayed six months in Labuan, there would have been a start given to it which would probably have insured its success. One of the parishes or sections of Brunei, called the Burong Pingé, contains the real trading population of the capital. It is inhabited by rich men entirely addicted to commerce, who were at that time anxious to withdraw from the tyrannical government of the Sultan, Omar Ali. They had entered into negotiations with Sir James Brooke to leave the capital in a mass, and establish themselves in our colony; but, on account of the question of domestic slavery, they had decided to establish their new abode at the mouth of the Kalias river, which is opposite Labuan, and have their trading establishments under our flag. But when they heard that the Governor was about to start for Europe, they grew less warm in the project, and at last determined to give up the idea until Sir James's return to Borneo. We thus lost to the colony the active aid of 1500 of the best of the population of the capital—so true is it that in the East influence is personal.

Probably Lieut.-Governor Scott or Mr Low could have done everything they desired, but they were more used to the Rajah, and wanted to be under him.

Another attack of fever and ague warned Sir James no

longer to defer his departure, and on the 9th January we left Labuan, to reach Sarāwak on the 13th, and four days after the Rajah sailed for Europe. We accompanied him to the Santubong entrance, and in my journal I find the words, "Farewell, dear Rajah! may your visit to Europe prove as beneficial to you as we all hope." Mr Brereton, on being acquainted with the plans which had been formed for him, joyfully threw up the Labuan service and accompanied us to Sarāwak, preparatory to taking possession of his governorship of Sakarang. Every one regretted his departure from Labuan, as he had the art of winning affection.

CHAPTER XII.

SECOND VISIT TO ENGLAND.

1851-1853.

THAT visit to England from which we all expected so much proved an unfortunate one, as instead of seeking rest Sir James picked up the gauntlet thrown down by the Eastern Archipelago Company, and fought it out with them, to the loss of health, money, and time in my opinion, but it was a necessary fight in the opinion of Sir James and some of his friends.

The Eastern Archipelago Company had been formed to develop the resources of the Indian islands, and particularly to work the coal of Labuan and the antimony at Sarāwak. It had been pushed into existence by Mr Wise, Sir James Brooke's agent in England, nominally to aid in advancing the work of the Rajah, but in fact to supplant him. His secret project was unknown to the directors when the company was formed, as it was difficult to fathom Mr Wise's schemes.

But the fact was, that by inadvertence and unpardonable carelessness, some private letters written by Sir James from Sarāwak had been allowed to fall into the hands of Mr Wise, and in these he had noticed some energetic expressions about himself, when Sir James, irritated by what he considered dishonest attempts to impose on the public, declared that he would kick Mr Wise to Old Nick if he continued to mix his name up in such schemes, and expressed the opinion that "a friend was worth a dozen agents." Mr Wise, however, was cautious as to showing his discontent, and only whispered his insinuations to my father; but when Mr Brooke positively declined to have anything to do with his projected company, and refused to sacrifice Sarāwak to the other's greed for money, Mr Wise grew furious, and then it was that he burst out to my father in accusations against his employer.

Mr Wise was an able man, and as crafty as he was able. As a minute examination of his different projects would be of no interest to the public, it will be sufficient to say that he had for a long time attempted to launch in the market a gigantic scheme, and he took advantage of the excitement caused by the arrival of Sir James in England during 1847 to carry out his project. He thought himself secure of a lease of the Sarāwak antimony ore; but he had not yet obtained the grant of the right of working coal on the mainland of Borneo, as Sir James had thought it his duty, as a Queen's officer, to pass the concession he had obtained to the English Government.

Mr Wise looked upon this act as treason to himself, and was rendered furious on hearing that some genuine capitalists in the city were trying to obtain the concession. Upon this he thought of my father, and after a consultation he drew up a long memorial, which the former agreed to place personally in the hands of Lord Palmerston.

My father was an old acquaintance of Lord Palmerston, and I well remember the particulars of the interview. When he had explained to his lordship the object of his wishing to see him, he handed him the memorial to read. On seeing its length, Lord Palmerston started, and said, "St John, your friend is a d——d long-winded fellow!"

but with that admirable aptitude for work for which he was remarkable, his lordship read it through, and promised to do his best. What were the steps subsequently taken I do not know, but Mr Wise was given the concession, and immediately formed his company. Many of the directors were rich, but nearly all were inexperienced men, and Mr Wise was allowed to do as he pleased.

Some months later Sir James Brooke thought that he discovered errors in Mr Wise's accounts to a considerable amount: an explanation was demanded, but refused; and Mr Wise, finding that Sir James had for a long time been aware of his covert hostility, now threw off the mask, and attacked his old employer on every occasion.

Our proceedings against the pirates in 1849 furnished him with the necessary weapons. By garbled extracts, by untrue reports, by means which I know not, he managed to obtain the confidence of obstinate old Joseph Hume, who dearly loved a grievance, and attacks on Sir James were commenced both in Parliament and the press. To minds that were prepossessed it was of no use furnishing proofs of the character of the pirates, or to bring forward the judgment of the Admiralty Court. It was of no use for the House of Commons to approve Sir James Brooke's proceedings by increasing majorities; it was no use for Lords Palmerston, Grey, and Ellesmere to stand forward in his defence, nor for that hard hitter, Henry Drummond, to demolish Mr Hume's case in the House. Mr Wise and his faction were determined if possible to ruin Sir James Brooke. Stung by this injustice, the Rajah decided to carry the war into the enemy's camp, and attacked the Eastern Archipelago Company, and did not give up the contest until he had seen the seal of their charter torn off by the judgment of a high tribunal. But at what expense of time, money, temper, and health was this triumph obtained!

Sir James found, therefore, that whilst in England he had not only to fight the Eastern Archipelago Company, but that he had to answer the "persevering and malignant persecutions," as Lord Palmerston called them, of a small minority in Parliament. He therefore seriously thought of getting himself elected M.P., so as not only to be able to defend his own proceedings, but to develop the policy which he thought England should pursue in the Eastern Archipelago. It is a pity that he could not carry out his project, as his fiery eloquence, his honest purpose, his thorough mastery of his subject, would have won the House, ever ready to listen to a man who has really something to say, and says that well.

On July 10, 1851, Mr Hume brought forward his motion of inquiry into the conduct of Sir James Brooke; but as he really could bring forward no evidence whatever against him, his motion, after a triumphant reply by Lord Palmerston, was rejected by a large majority. Mr Gladstone spoke on this occasion, and bore his testimony to Sir James Brooke's noble character. It is a pity that he has lately allowed doubts to creep in as to the correctness of his former judgment.

News having reached England that the old king of Siam was dead, and that the friendly Chau fia Mungkut had ascended the throne, it was decided to send another Mission to Siam, and Sir James Brooke was directed to prepare for an immediate return to that country. His passage was taken in the October mail; but before he could start, news reached the Government through Col. Butterworth, Governor of the Straits Settlements, that the king of Siam would prefer the Mission being deferred until after the funeral rites of the late king were over. This was a real misfortune, as, had Sir James returned to the East, made an advantageous treaty with Siam, and perhaps opened Cochin China and Kambodia, his mind would have been diverted from the attacks on his Borneon policy, and his opponents would probably have considered it useless to persevere in their "malignant persecution."

But it was decreed otherwise, and instead of sailing for the East, Sir James went down to hunt with Harry Keppel.

In 1852 the Rajah had still his two questions on hand, —the lawsuit with the Eastern Archipelago Company, and to defend himself from the attacks on his Eastern policy. The friends and admirers of Sir James Brooke, to mark their opinion of his conduct, gave him a public dinner at the London Tavern on the 30th April, and on that occasion Sir James explained his policy in a lucid speech, which was much admired. "Do not disgrace your public servants by inquiries generated in the fogs of base suspicions: for remember a wrong done is like a wound received—the scar is ineffaceable. It may be covered by glittering decorations, but there it remains to the end." Prophetic words! the scar did remain to the end.

In the House, Mr Sidney Herbert stated that Sir James was engaged in mercantile speculations, and a correspondence ensued which was quite unworthy of Mr Sidney Herbert, who shuffled out of the question he had raised in a very mean way, as he had in reality spoken at the instigation of Mr Wise, and had no evidence whatever to support his disparaging assertions.

One satisfaction Sir James Brooke had, which was to see the seal of the charter of the Eastern Archipelago Company torn from that document, as it was found that the capital, which had been certified as subscribed, had not been subscribed. While the director, Mr Wise, was engaged in managing the "malignant persecution" in England, and pulling the wires by which Mr Hume, Mr Cobden, and their lesser allies were moved, their manager in Borneo was not idle. This was a Mr Motley, who was employed with a disreputable adventurer named Burns to endeavour to obtain a letter from the Sultan of Borneo, complaining of Sir James Brooke's conduct. But their schemes failed.

It is a curious circumstance that both these men, who were great authorities with Mr Hume in his endeavour to prove the gentler character of the natives of Borneo, died by their hands. Mr Burns was killed by pirates in Maludu Bay; and when we recovered the ship of which he had been supercargo, I discovered among his papers not only proofs of their having endeavoured to bribe the Sultan to complain of Sir James Brooke, but I found a very curious argument written out, whether it would not be justifiable on his (Mr Burns's) part to receive slaves in payment of goods. His conclusion was that he would be completely justified. And this is a specimen of the men who banded against Sir James Brooke. Mr Motley was massacred, with his whole family, by those mild inhabitants for whose good conduct Mr Gladstone is now ready to vouch. The third of this band of rogues, Mr Riley, the tavern-keeper, was drowned.

It will show to what height party spirit rose, when I say that Mr Motley accused the Rajah of having bribed the pirates to murder Mr Burns, Capt. Robertson, and others of the crew of the Dolphin, as in 1846 he had paid Mr Williamson's servant to drown his master. Can anything be more infamous than such accusations? The fact is, that H.M.'s officers in half-civilised countries are brought in contact with wandering ruffians whose only object is to make money,—honestly if they can, but at all events to make money; and when we endeavour to check their illegal acts, we are exposed to shameful abuse, which sometimes finds an echo at home.

Sir James Brooke had many interviews this year with Lord Malmesbury, Secretary of State for Foreign Affairs, and explained to him the position in which he found himself placed. He pointed out, what had long been apparent to us, that it was not for the good of the public service that he should hold the positions of Governor of Labuan, Commissioner to the independent princes, and Consul-General, whilst he continued Rajah of Sarāwak. Two of the three were compatible, but not all together. As Governor of Labuan, or as Rajah of Sarāwak, he could hold the position of Commissioner; but as he could not efficiently superintend two possessions distant from one another, he proposed, therefore, that he should give up the government of Labuan, and be appointed Minister, with two paid *attachés*. The governorship was given up, and everything was settled, even to his return to Siam to make a treaty with the new king, who appeared likely to prove very friendly to the English, when the Conservative party went out of office, and Lord Aberdeen came into power. Lord Stanley, however, then Under-Secretary, assured Sir James that Lord Malmesbury had so settled the affair, that his successor would not disturb the appointments.

I must go back a little to refer to another subject, to which I have already alluded. I have already mentioned that the American envoy had visited Sarāwak in 1850, with proposals to form a treaty. The Rajah's absence, and the necessity to submit this proposition to the British Government, as Sir James was an officer in their employ, prevented anything being done at the time; but in 1851 he received letters addressed by Mr Ballestier from Washington, informing him that the President of the United States would be happy to meet him if he would visit America. As Lord Palmerston had written to him to say that H.M.'s Government saw no objection to his negotiating a treaty with the United States, he had made up his mind to start when some private affair delayed his departure until too late. I have always regretted the frustration of this voyage, and fear that Sir James did not attach sufficient importance to it. It might have been of incalculable service, and have saved him from years of subsequent annoyance. In the spring of 1852, his private secretary, Mr Charles Grant, visited Washington, but as he had not been furnished with any powers, he could only act in his private capacity. Had Sir James been acknowledged an independent sovereign by the United States, with powers to sign a treaty of friendship and commerce, perhaps he would not have had to wait fifteen years for recognition by his own country, and Mr Cobden would have ceased to persecute a man whom the President had delighted to honour.

In the hope of silencing Mr Hume, or at least those of his friends who had minds open to conviction, Sir James, early in 1852, published a series of letters, addressed to Mr Henry Drummond, on the subjects in dispute, and so demolished Mr Hume in the opinion of many judicious persons, that he hoped that he should hear no more of the matter. In fact, many who had suffered by Mr Hume's malignant tongue were much pleased; and one day Lord Torrington, so unjustly treated by Mr Hume when Governor of Ceylon, meeting Sir James, said: "I am delighted with these letters; you have thoroughly avenged me."

This visit to England, which was to have been one of quiet in order to restore his health, resulted in much fatigue and anxiety. He was at work all the time. During this long stay, how vainly he sought repose! It was not then in his nature to indulge in it; he called for an easy-chair, but was never content to sit in it.

With 1853 there came into power the Aberdeen Ministry, a coalition of the Liberals and Peelites. It had the faults of all coalitions, it lived by compromise. Joseph Hume and his coadjutors thought that they had now their opportunity; and as a return for their support, they insisted that the inquiry that they had so long demanded should be granted.

It must not be forgotten that Sir James Brooke had himself proposed an inquiry, and in an interview with one of the Conservative Ministers it was fully discussed. Sir James refers to it in the following words: "At the close of our conversation, I asked him whether it would be better or not, for the sake of the public service, to have a Parliamentary Committee at once at the opening of the session. He said it was worth thinking about, and that he would consult Sir John Packington and Disraeli about it: that it would be the bold course; and a friendly and fair inquiry made at my request, would be very different from one granted to the hostile motion of Mr Hume."

The entry of the Coalition Ministry into power upset this plan, and Sir James might have expected mischief when he found that Lord John Russell refused to confirm Lord Malmesbury's appointments, but his confiding and simple nature did not. He, however, determined to resign his appointment as Commissioner and Consul-General, but waited until his arrival in Borneo to do so.

None of the friends of Sir James Brooke could object to an inquiry being instituted; but the concealment practised on this occasion was not justifiable. The secret history of the granting of the Commission, about which we shall hereafter hear much, may probably never be known; but Sir James Brooke was informed that Mr Hume got hold of Sidney Herbert, as the Minister who was evidently hostile to Sir James, and through him induced Lord John Russell to come into his views.

I am under great personal obligations to the late Earl Russell, but these cannot prevent my expressing the strongest sentiment of regret at the underhand way in which this affair was conducted. While Mr Hume was informed that the Commission had been granted, this information was carefully concealed from Sir James Brooke, and it was only by accident that he found it out. Instead of consulting with Sir James as to the manner and object of the inquiry, and enlisting his aid, Government, as I have said, kept the news concealed from him. He left England in April 1853, boiling with indignation, which was increased on his arrival in Singapore, when he found that Mr Hume had already written to the editor of a reckless local paper to announce the Royal Commission; so that Mr Hume had long been aware of what it was intended Sir James should not know before he left England. A most unfair and unjust proceeding!

No wonder that Sir James wrote,—"The Commission makes me so fiercely indignant, and I dislike it because it is derogatory to my position and humbling to my pride." And later on,—"I guard against the besetting sin of my temper by inheritance—the indelible impression of injury received, and the unforgiving spirit which such an impression produces." The sense of an injury received did indeed remain impressed on his spirit to his dying day.

I must here pause to notice an observation made by Mr Gladstone: "His" (Sir James's) "language respecting Mr Hume and Mr Cobden, two men of the very highest integrity, and by no means given to extremes as humanitarians, is for the most part quite unjustifiable." It makes one flush with indignation to read such a remark. Mr Hume, who had spent years in grossly calumniating and "malignantly persecuting" the Rajah, and who had not the manliness to acknowledge the falseness of his statements when proved to be wrong, is not worthy of respect. I have no respect for the man who, by false accusations, embittered the life of a noble officer, whose shoestrings he was unworthy to untie, and who wrecked a noble policy.

And Mr Cobden I judge from the following sentence, never retracted: "Sir James Brooke seized upon a territory as large as Yorkshire, *and then drove out the natives*, and subsequently sent for our fleet and men to massacre them." The insolence and ignorance here displayed are about equal, and yet Sir James is censured for resenting the accusation of having massacred the peaceful inhabitants of his own country. Verily, Mr Gladstone has two weights and two measures.

I may mention that in the autumn of 1852, whilst on a visit to Lord Ellesmere, the Rajah nearly closed his career in a very tragic manner. His cab-horse ran away, and at one moment there was great danger of the vehicle being tipped into a canal; but passing this danger the horse dashed on, and came into collision with a large stone gate-post, and the cab was shivered to pieces, while all the inmates escaped unhurt.

Before dwelling further on this subject of the Commission, I must briefly notice the events which had taken place in Borneo during the Rajah's prolonged absence. Captain Brooke had been left in charge of Sarāwak; Mr (now Sir John) Scott, of Labuan; and I, of the Commissioner and Consul-General's department. It is with Captain Brooke's proceedings, however, that we have principally to do. Captain Brooke was exceedingly well adapted to the trust confided in him: he was of a frank and noble nature, of just though perhaps not broad views, and of so sweet a temper that he gained the affection of those around him. He had no remarkable ability, and never pretended to have: but he managed the affairs of the country with great tact, improved its revenues, and kept everything quiet, so that during the two and a half years of his uncle's absence nothing occurred to disturb its tranquillity.

He visited every part of it, and made himself thoroughly acquainted with the natives and their wants. I accompanied him in most of these journeys. Sakarang was the only place that gave him any uneasiness. I have mentioned the establishment of a fort there under Sirib Moksain; of his having been withdrawn on account of the unpopularity of his administration; and of the substitution of Mr Brereton as Governor, but a Governor without pay. No better choice could have been made of a man likely to win over the wild warriors of Sakarang than Brereton. He was young, active, knew the language thoroughly, and was of pleasant and taking manners. His only support was his own talent, aided by twenty armed men from Sarāwak, and by a small Malay population which had gathered round the fort, under Abang Aing, one of the most faithful Malays we ever met. The position was very curious. The fort was surrounded by various tribes of Dyaks who were at war with each other—the Sakarangs, the Batang Lupars, and the Undops, generally in alliance against the Linggas, though not always. These lived on different branches of the same great river. This was intertribal war, not piracy. Although the fort had been erected principally with the object of preventing fleets passing down the river to the sea, Mr Brereton early saw that he must induce these tribes to make peace before any success could attend his government. He first won over Gasing, an influential Sakarang chief, already half a friend since an interview with the Rajah in Sarāwak; and with his assistance, and that of Abang Aing, he managed to bring about negotiations which appeared likely to terminate satisfactorily. It took him months of labour, but at length he was enabled to announce to the Sarāwak Government that he had succeeded, and invited Captain Brooke to come to Sakarang in order that the peace ceremonies should take place under his auspices.

Thus Mr Brereton restored peace and tranquillity to a river district inhabited, it is thought, by nearly 40,000 Dyaks, and from that time everything appeared to go on well, except in the far interior, where a famous pirate chief named Rentab refused to be bound by these engagements. Not to disturb the sequence of the story, I will defer referring to this until I have to describe the expedition undertaken by Sir James Brooke himself against this chief.

It was a great satisfaction for us to witness the success of this young officer, who laid the foundation of the permanent tranquillity and prosperity of this important province, so that those who succeeded him had for some time comparatively an easy task. They have had their troubles, but prosperity was too solidly based by Mr Brereton for things not to proceed well. He exercised a very great personal influence over these chiefs. In fact, whether it was natural or whether he had acquired it, his manner of treating them was more like his own great leader than that of any other officer in Sarāwak.

We had troubles also with the Chinese gold-workers, but Captain Brooke's vigorous policy soon brought them to order, and kept them quiet many years.

In the autumn of 1851 the English schooner Dolphin [1] was taken by pirates in Maludu, and after it was recovered an expedition was sent by Admiral Austen to punish the guilty. Captain Brooke and I accompanied it, but as the expedition remotely affects Sir James Brooke, I only mention the fact. I now return to the Rajah.

[1] It was in this ship that Mr Burns, of whom I have already spoken, lost his life at the hands of pirates.

CHAPTER XIII.

RETURN TO BORNEO.

1853–1854.

After a few days' stay in Singapore, Sir James started with his English servant and Grant for Sarāwak in a little sailing brig, the Weraff, and reached Kuching the first week in May. The whole population poured out to meet him—for however unjust others might be, in Sarāwak he was rightly judged, and this comforted him. When we met, we all noticed that the Rajah's face looked puffed, and that he was covered with what appeared to be harvest bumps; but there being no medical man amongst us, nothing was said, although the Rajah complained of a feeling of lassitude. In the veranda of his house a crowd met him, and I heard a Borneon noble say, "Why, the Rajah has the small-pox." I took no notice, as the word he used was one not common in Sarāwak; but on the morrow the Rajah felt weaker, and very soon kept his room and his bed. I sat at his bedside, holding his hand,—for it is a curious circumstance that when ill Sir James could obtain no sleep unless some one whom he liked held his hand. I have done so until all sensation had left my arm; did I move it, he awoke immediately.

It was not long before Sirib Moksain entered Sir James's bedroom and said, "Good God, the Rajah has the small-pox!" No sooner did the Rajah hear this than he sent for some experienced persons, and these confirming the fact, he instantly forbade those amongst us who had not had the disease from entering his room.

He had, however, two experienced and tender nurses, — Mr Crookshank, the police magistrate of whom I have often spoken, and Sirib Moksain, the dispossessed Governor of Sakarang, the half Arab who had announced to him what was his malady. Day by day he grew worse; his disorder, confluent small-pox of a virulent kind, broke out all over the body, and his attendants were in despair. We would creep into the room, expecting daily that it was the last look we were to have of our beloved friend, whose countenance now was not to be recognised, so swollen, so disfigured had it become.

The native population were deeply moved. Every day the chiefs would walk softly up to the house, and sit for hours in the outer rooms, to have the latest intelligence; the poorer classes, Malays and Dyaks, would crowd the verandas and wait patiently for news, until, in a lucid interval, the Rajah gave the order that the house should be placed in quarantine to prevent the infection from spreading. Nevertheless, every day the native ladies would send perfumed water to wash the invalid, while Sir James lay stretched on broad plantain-leaves, whose freshness cooled the fever that burned the skin.

In the mosques there were daily and nightly prayers for the sick chief—in fact, all the population suffered apparently as much as we did. The weather was fearfully hot—94° in the coolest rooms—and not a breath of air disturbed the drooping leaves: all this was against our patient.

At last there was a favourable turn, but the weakness continued for above three months. In August, however, he was enabled to embark again on board the Weraff, and proceed to Labuan and Brunei. In appearance he was so changed that, when he landed in our colony, his former subordinates were almost affected to tears at the sight of him, while he cheerfully consoled them, saying, "You will not like me the less for being a little uglier." In a letter, written in July 1853, to my father, I notice these words: "Sir James was very excited about the Commission; but since his severe illness he has been much more calm. He says, 'After such a chastening as God has given him, all that man can do appears nothing.' I have never before seen him so quiet and resigned, or so even in temper."

From Labuan we proceeded to Brunei to meet the Sultan. Changes had taken place in the capital. Omar Ali, the murderer of his relatives in 1846, was now dead, and his chief Minister, Pangeran Mumein, reigned in his stead. As, however, Mumein had no claim whatever to the throne, but was placed in that position as a makeshift, he felt uneasy at Sir James Brooke's coming—as, besides the consciousness of many intrigues, he feared that his powerful tributary would not acknowledge him.

The Weraff not being able to reach the city on account of the freshes, the Sultan sent down large boats to fetch us up to the palace, where he had provided a couple of rooms for us, behind the audience-hall. Although we came with a full retinue of servants, we were not allowed to provide our own table. Three or four times a-day the Sultan sent in dishes in abundance, prepared by the ladies of the harem: the curries, the stews, the rice-cakes, were really delicious; and Sir James, in his convalescence, ate of them with pleasure, particularly after the wretched dirty cooking of our little merchant-brig.

Our first interview with the Sultan was very interesting. Crowds of nobles were assembled, the two rival factions mustering strongly, and with them was the wily Makota.

Presently the Sultan came in, looking uneasy. We rose and advanced towards him. The Rajah, after the first salutations, turned round to the chiefs and complimented them upon the wisdom of their selection, and then addressing Mumein as Sultan, continued the conversation. The effect was immediate—the gloomy, uneasy look vanished, and all was cordiality.

The Eastern Archipelago Company, anxious to obtain some testimony against Sir James Brooke which might be used before the Commission, had employed their unscrupulous agent in Labuan to gather evidence. The ex-Lieutenant-Governor, smarting under the punishment he had received, had written to his former friends in Brunei to tell them all about the Commission. We were therefore deeply amused by the conduct of our old enemy Makota, who longed to refer to the pending inquiry, but did not dare.

Sir James soon relieved him by entering on the subject himself, and by explaining all. He committed, however, the natural error of confounding his feelings with facts, and describing the conduct of the English Government as more hostile than it really was. It was in vain for me to point out to the Rajah the impolicy of this: he was obstinate—and when obstinate, no one at the time could turn him from his view. This error, a few years later, had a most deplorable effect.

The real object of Sir James Brooke's visit was to carry on negotiations for the cession of the six districts in the neighbourhood of Sarāwak, which then yielded no revenue to any one. In fact, in their state of disorganisation, they did not pay their expenses. Sir James Brooke proposed to the Brunei Government a certain fixed sum, and half the surplus revenue. The fixed sum was small; but the revenues of Sarāwak at that time were very small also. Few difficulties were raised: the Sultan, happy in the assurance of Sir James Brooke's support, and desirous to get something where his predecessor had obtained nothing, agreed to cede the districts on the conditions proposed; and before we left, the necessary deeds were prepared and signed. A year's revenue was paid in advance, large presents were made to different nobles, and all ended happily. Even Sir James Brooke's most inveterate enemies could not say that this cession was obtained by force or intimidation. It was known that the British Government was considered hostile—that Sir James had now lost the support of the navy—that he was there in a small unarmed merchant-craft,—yet everything he desired was granted immediately.

When this serious business was over, Sir James turned his attention to the intrigues against himself personally, and thought it desirable to obtain from the Sultan and Makota the original letters written to them by the ex-Lieutenant-Governor of Labuan, as well as some other documents, in order to lay them before the Royal Commission. Sir James went about the business in a straightforward way, but made no progress. The letters were mislaid, perhaps lost—they could not be found—and day after day passed without any result. At last one of our party took the affair in hand, and with a better appreciation of Makota's character, called upon him, and plumply said, "I will give you so many hundred dollars if you produce the letters." He talked about his honour, his character being at stake; but he was assured that it was for his honour to expose an intrigue. Next day he slipped the letters into the hands of the negotiator; and the same evening, by his express desire, the bags of dollars were passed through a small window in our bath-room, which abutted on his house, and were received by his Excellency in person. Some years after, having to negotiate a convention with Brunei, and finding that, though settled, I could not get it signed, I sent to Makota, and told him it was useless to imagine that I was going to pass any bags through *my* bath-room window, as the English Government did not do business in that way; and it is creditable to his appreciative character that, on receiving the message, he laughed, and sent the convention, signed and sealed, next morning.

The Rajah enjoyed this stay in Brunei. He was tolerably quiet; he was regaining strength every day; he found that his influence had not declined,—on the contrary, he was looked upon as the final arbitrator in the disputes of the natives; he had obtained the documents requisite to enable him to introduce regular and legal government in the six districts, and he was satisfied.

I have often heard people talk of their influence among the people of Borneo. Many have acquired a certain amount, as Captain Brooke (the Rajah's heir), Mr Crookshank, Mr Brereton, Mr Charles Johnson,[1] and one or two other Sarāwak officers; but their influence was but reflected influence, acquired because they were relatives and officers of the old Rajah. Mr Low and I had some influence in Brunei, but no one more surely than Mr Low would confess that much of his influence was reflected; and I took care, wherever I went, to take the position once given me by Sir James, of his adopted son, and I found that, whether among Dyaks, Malays, or others, it was fully appreciated.

The loyalty shown by the Sultan and Rajahs of Brunei to their old friend and protector was really remarkable. Every temptation was laid in their way to induce them

[1] Lieutenant Charles Johnson, R.N., whose name often appears in my narrative, had at this time retired from the navy, and had joined his uncle, Sir James Brooke, in Sarāwak, and eventually became Governor of Sakarang and Seribas. He afterwards adopted the name of Brooke, as his elder brother, Captain Brooke (now deceased), had previously done; and he is now (1879) Rajah of Sarāwak.

to attack Sir James before the Commission—they were assured that rich Sarāwak would be restored to them; but they resented every offer. This conduct should not have been forgotten, but it was forgotten; and to the faithlessness Sarāwak has shown to Brunei I may refer hereafter. Everything being satisfactorily settled we left Brunei, and having called at Labuan sailed for Sarāwak, where we arrived in September after a favourable passage, to find plenty of correspondence, though nothing about the Commission.

The Malays were soon summoned to a meeting, at which Sir James gave a full account of his visit to the capital. The cession of the six rivers was received with a murmur of applause, and when the Rajah explained the purport of the Royal Commission, there were some genuine and hearty responses made by the chiefs. Very often they only follow the lead given, and utter what they think will please; but some of them now spoke out, and expressed their sentiments in no measured terms.

The announcement of the appointment of a Commission to inquire into the conduct of Sir James Brooke was, however, an unmixed evil. It disturbed the minds of the population, it raised ambitious aspirations in many a chief who had formerly been a zealous supporter, and laid the foundations of troubles which lasted many years.

We remained in Sarāwak several months quite undisturbed. To aid Sir James Brooke in his recovery, Captain Brooke had, during our absence in Brunei, built a pleasant little cottage, on a spur of the Serambo mountain. We had a lofty peak on our left, while on the other three sides the hill sloped steeply down 1200 feet to the plain and river below. In a ravine close by, rose a huge rock some 70 feet in length by 40 in breadth, somewhat in the shape of a mighty but very blunt wedge. The thicker end was buried in the ground; the centre, supported on either side by two rocks, left a cave beneath; while the thinner part, thrust up at an angle of 30°, overshadowed a natural basin, improved by art, in which the Rajah loved to bathe. A rill that glided from under the rock supplied us plentifully with cool, clear water. It was a beautiful spot, a charming natural grotto, in which we often passed the burning mid-day hours. As we sat there we could catch glimpses of distant mountains, of the plain below, and here and there a reach of the river, all these seen through noble trees, and brilliant vegetation of various kinds. What pleasant days we spent there! With our books, our writing, and our chessboard, we managed to let time slip imperceptibly away. We were surrounded by groves of fruit-trees, and at a little distance below us were three Dyak villages, to which we made constant visits. I do not think it necessary to enter into many descriptions of Dyak ways, or of the Rajah's visits to these tribes, as so much has already been published on the subject; but we saw many curious customs.

The Dyaks were naturally as much or more pleased than the Malays at the recovery of their Rajah; they could not believe but that if he died the old system of oppression and robbery would recommence; and they were right, as in 1853 the Sarāwak Government would not probably have continued to exist had its founder been removed, particularly under the depressing influence of the estrangement of the English Government from the infant settlement.

Deputations from all the tribes on the right-hand branch of the river came and pressed us to go among them, but the state of the Rajah's health would not permit it. Some chiefs, however, in our neighbourhood, as those of Bombok, Serambo, and Paninjow, would take no refusal; and as they were near, the Rajah used to make an effort to be present at their great festivals. I have entered so fully into all these ceremonials of the Dyaks, however, in my own work on Borneo,[1] that it would be but a repetition to dwell on them here.

In May 1854 news reached us that Charles Grant, resident of Lundu, was attacked with small-pox, and the Rajah immediately left Kuching in order to nurse him. Luckily it was but a mild attack, and he soon recovered. Whilst the Rajah was away, a most curious incident occurred. We were sitting at dinner, when the room was lit up as by a brilliant flash of lightning: a few seconds afterwards another bright flash and an awful explosion. The powder-magazine had been blown up. We ran down to the river-side, jumped into our boats, and were soon across, to find the stockade laid flat on the ground, and the sentinel lying near, dead but unwounded: the concussion of the air appeared to have crushed his life out. On inquiry, we found that a bright meteor had traversed the sky, and unluckily chosen the powder-magazine for its place of rest.

There was one subject that had for years occupied the Rajah's mind, and that was to bring the Dyaks of Sarāwak under the direct control of the Government. The plan which had been adopted in former years was a modification of the old system. The Rajah governed Sarāwak with the aid of three Malay chiefs, called Datus — the Patinggi, the Bandar, and the Tumanggong. Each family in every tribe of Dyaks was expected to pay about a bushel and a half of rice every year to the Government, half of which went to the Rajah, and half to the Malay chiefs. The Patinggi held in hand the tribes of the left-hand branch of the river, the Bandar those on the right, while the Tumanggong had sway over the tribes on the coast, and the fisheries. These three Malay chiefs, by

[1] Life in the Forests of the Far East.

custom also, had the right of the first trade with the Dyaks, which in former days had been forced upon the aborigines by every species of tyranny. When Sir James was appointed to the governorship, he endeavoured to restore the traditional customs, but he found that with all his watchfulness the Dyaks were exposed to much oppression. At length, when the revenues of Sarāwak permitted the payment, he decided on giving the different chiefs fixed sums per month, in lieu of all claims on the Dyaks.

The Patinggi had been the great offender in all this system of oppression, so that Sir James commenced with him, and commuted his revenue of rice into cash, giving him 50 per cent extra to cover the profits he had derived from trade. This was at first gratefully received, as he hoped privately to continue the forced trade; but when checked in this, he began to show signs of discontent.

I propose fully to explain what consequently occurred, as it is the keynote to troubles which lasted eight years, and which cost many valuable lives and much property.

I have before referred to our visits to the Patinggi's wives, when we saw his very fine-looking daughter, Fatima. This girl had her head turned by prosperity, and disdained to match herself with any Sarāwak gentleman. Unfortunately her father favoured her views, and fixed his attention on a certain Sirib Bujang, a harmless fellow enough, but brother to an intriguing chief, Sirib Musahor, of Arab descent. This man was the ruler of Serikei, and was suspected of being in league with the pirates of Seribas and Sakarang.

The Rajah strongly objected to this marriage, as he knew the danger of bringing the Arab influence into Sarāwak; but the chief pressed for his consent, and the ladies pretended to be in despair; so that, on the eve of his departure for England in 1851, his consent was given. In order that the marriage should be solemnised with great state, the Patinggi began making immense preparations, borrowed of his friends, and daily pestered Captain Brooke, then in charge of the Sarāwak Government, to afford him the means of meeting his expenses, as he said the honour of Sarāwak was at stake. Finding what he thus obtained to be inadequate, he began oppressing not only his own Dyaks, but those of Sadong, who had been temporarily confided to his care, and deep were the complaints which arose on all sides.

The marriage came off with great *éclat*. For days every house was covered with flags, and cannon fired from dawn to sunset; free tables for all friends were prepared, and the amount consumed per day was a never-ending subject of conversation. Sirib Musahor had accompanied his brother, and was open-handed in his liberality. We all liked him. He appeared to seek European society, and by the hour would stand listening to Mrs McDougall playing the piano; at other times he was a diligent attendant at our chess club, and many a tough game have we had together. His brother, the bridegroom, was very inferior to him in appearance, but was a better chess-player.

After the marriage festivities were over, we all noticed a change in the Patinggi's manner. He began to give himself great airs, talked more of his own importance to, and in, the Government, and continued his oppressive conduct towards the Dyaks. Captain Brooke visited most of the tribes to verify these complaints, and, finding them true, had often to fine some of the followers of the Patinggi. The chief, however, showed no signs of discontent, but became a greater beggar than ever for donations from the public treasury.

When, however, the Rajah returned to Sarāwak in 1853, and brought news of the Commission, it appeared from what was afterwards discovered that the Patinggi began to entertain ambitious views; and, in concert with Sirib Musahor, it was suspected, began to broach the idea of expelling the English from the coast and seizing the reins of power. On our return from Brunei in September 1853 the Rajah called the Patinggi before him, and pointed out to him how ill he had behaved during his absence in England towards the Dyaks committed to his care. The cases of oppression were indeed startling; but the Rajah was desirous of sparing an old servant who had been faithful in former days, and, as I have said, offered to take over his Dyaks and pay a certain sum as an equivalent for the revenue and forced trade, but leaving the open trade free to him and his family. The Patinggi appeared joyfully to acquiesce, but his evil advisers soon began to whisper to him that he had been degraded in the eyes of the people. From that time forward he began actively to conspire.

Early in 1854 the Rajah and Captain Brooke went with a squadron of war *prahus* to the Batang Lupar river to visit Charles Johnson at Lingga and Brereton at Sakarang. I stayed in Kuching. One day, whilst sitting alone in my little cottage, the eldest son of the Datu Tumanggong, Abang Patah, came in to have a talk. He was one of the best of the Malay chiefs—frank, loyal, honest, brave as a lion. He subsequently lost his life in gallantly defending the Rajah's Government.

I saw by his uneasy manner that he had something to communicate, so, after answering a few leading questions, he said, "It is no use beating about the bush; I must tell you what is going on." He then unfolded the particulars of a plot which the Patinggi had concocted to cut off the Europeans in Sarāwak. The Patinggi had confided his plans to the other chiefs, but they had almost unanimously refused to aid him, and had determined to keep a watch over his proceedings; but they had not the moral courage to denounce him to the Government. At length Patah said, " I have become alarmed. The Rajah and Captain Brooke are away together. The Patinggi is with them, with all his armed followers, and in an unsuspecting moment all the chief English officers might be cut off at a blow." I promised, as he desired, to keep his communication a secret from all but the Rajah, to whom I instantly wrote, giving not only Patah's story, but other indications which had come to my own knowledge. An express boat carried the letter to its destination. The Rajah read the letter, and without a word passed it to Captain Brooke. The latter, having also read it, said, " What do you think?" " It is all too true," answered the Rajah, to whom conviction came like an inspiration. They had noticed some very odd proceedings on the part of the Patinggi, but having no suspicions, had not been able to interpret some of his armed movements; but now it was clear that he was trying to get the Europeans together to strike one treacherous blow. Nothing, however, was said or done publicly. The faithful were warned to watch well, and a few judicious inquiries brought the whole story out.

On his return to Sarāwak the Rajah determined to bring the Patinggi to justice. On account of his relatives he spared his life, and he was permitted to depart on a pilgrimage to Mecca. The scene in the court-house where this occurred was very remarkable. It must always be remembered that, beyond half-a-dozen English officers, the Rajah had only his Malays at his back to aid him in deposing a native chief.

The following is the Rajah's brief account of what occurred on this occasion: " Yesterday was a great day in Sarāwak, as I publicly, before about five hundred men, accused the Patinggi of misgovernment and treason, and deposed him from his office. I demanded all his guns, *lelahs*, muskets, and powder, which were brought into the fort by the Bandar, the Tumanggong, and the Abang-Abang, whilst the deposed Patinggi was sitting in court. All passed off quite quietly, and the people were entirely with us, true indeed as steel, with the exception of course of a few, his immediate followers and dependants. It was a measure rendered imperative by the Patinggi's intrigues, and his secret hostility to the Government for having checked him in his course of misgovernment and malversation."

The next in rank, the Bandar, now succeeded to the chief influence among the Malays; and his brother, as Datu Imaum, was added to the list of trusty counsellors. These two brothers were remarkable men. The eldest was a mild, gentle Malay, as truthful and honest as any man I ever met, but to whom ill health rendered active exertion repugnant. His energetic brother, the Imaum, the head of the Mohammedan priesthood, has proved by his acts and conduct how wise was the Rajah's choice, for he is still the mainstay of English influence among the Malays.

CHAPTER XIV.

THE ROYAL COMMISSION—EXPEDITION AGAINST RENTAB.

1854–1856.

At this time the Rajah was visited by his old friend Mr Read of Singapore, who was accompanied by Colonel (now Lieutenant-General) Jacob. This visit greatly interested us all, as the keen-eyed Indian officer readily took in the salient points in dispute, and from a doubter, or rather a questioner, became one of Sir James's firmest friends.

Whenever we went to stay on the mountain of Paninjow we forgot all about the Commission, or referred to it only as a troublesome business that the British Government had brought upon themselves; but the even tenor of our way was at length disturbed by the arrival of the three volumes of the Rajah's letters, edited by Mr Templer. From this time forward chess and light reading were put aside, and all thought was devoted to the coming Commission. Manifold were the useless despatches written, —worse than useless, because to an indifferent person they looked like bad faith. But in a man like the Rajah there was no bad faith, but an earnest belief that he was furthering the cause of truth and justice in summoning as witnesses to appear before the Commission in Singapore such men as Lords Palmerston, Clarendon, and others. It may, however, have been done to show how incomplete would be an inquiry conducted at Singapore. Vainly I endeavoured to moderate this stream of letters, which must have wearied the Foreign Office, already sufficiently troubled by the communications of one of Sir James's indiscreet friends, who even intimated to Lord Clarendon that the life of the Rajah would be placed in jeopardy should the Commission hold its sittings in Singapore, as party feeling there ran so high. I read the copy of this despatch as sent to us; but I trust some judicious adviser erased that paragraph before it reached the Foreign Office. More than this, the English Government was assured that Sir James Brooke was a man so superior, that should England find itself in a moment of supreme danger, she must call upon him to return home and save her. Such trash as this could not but fail to weary those to whom it was addressed, particularly as it had nothing whatever to do with the case. At last the Government peremptorily declined to continue so useless a correspondence, which could only injure the Rajah's cause, as it was supposed to be instigated by him.

These letters from home exercised a very unfortunate effect on the tone of the Rajah's own communications. No man, however, was better aware than himself of his unfitness for European statesmanship. In Borneo he was unrivalled; in Europe he would have had many a rival, with more experience and better training.

Sir James had now passed nine months in quiet, looking after the internal welfare of Sarāwak, and had determined to take no active step before the meeting of the Commission; but he was forced at length to break his resolve, on account of the successive delays which arose in the definitive selection of the gentlemen who were to act as Commissioners.

The boldness shown by Rentab, the great pirate chief, at last determined the Rajah to organise an expedition against him.

Rentab was one of the most notorious and truculent of the Dyak chiefs. He had, as I have before stated, won fame during a cruise off the Dutch possessions in Sambas, where he had surprised some Chinese at work in their fields. Information of this act had been instantly conveyed to the Dutch officials, and they sent out a gunboat to intercept his return; and the Sultan of Sambas also equipped a war *prahu*, manned it with a select crew, in which many young nobles were included as volunteers. The two gunboats soon came in sight of the fast-pulling Dyak *bangkongs*, which, however, appeared in distress, as they seemed only to hold their own with their pursuers. The Malays, excited by the chase, gave way with a will, and soon left behind them their European ally. When Rentab saw that the Sambas boat was beyond all immediate assistance from the Dutch guns, he turned on his Malay pursuers, came down upon them at the full speed of his war-boats, and overwhelmed the Malays beneath a shower of spears. Not one escaped. Having secured their heads, their guns, and gold-handled *krises*, he abandoned his prize, and pulled off from the heavy Dutch gunboat, which was vainly endeavouring to come to the aid of its consort.

This act raised Rentab's name in the estimation of his people, and he spurned all proposals to give up piracy.

In addition to the fort at the entrance of the Sakarang river, which, as I have said, had been erected to prevent the egress of pirate fleets, the Rajah had, at a later period, established another lower down the river to protect the Lingga district from the attacks of the Seribas. The former fort was commanded by Mr Brereton, of whom I have already spoken; the latter by Mr Lee, a quiet, prudent, young officer,—the very aid one would wish to have in a moment of danger. Ever since the action at Batang Marau on the 31st July 1849, the pirate tribes had been divided into two parties—those who wished to give up piratical cruises, and those who desired to continue in their old habits. Of the latter party Rentab was a chief. During 1853 he had determined on a bold stroke, which was not only to pass Brereton's fort, but previously to attack those Sakarang Dyaks who sided with the English. Information having reached Mr Brereton of this intention, he proceeded with a small force up the river Sakarang, and established himself in one of the threatened village-houses. Guns were placed in position, and the place was fortified. Mr Lee came up from Lingga in his own boat, and strongly advised that they should only repel attacks until his allies the Lingga Dyaks arrived with several hundred men. This was agreed to. At that moment a large war-boat was seen coming down the river; a warning musket was fired, but the pirates continued to descend. Another *bangkong* followed; upon this a charge of grape was sent at them. Immediately there appeared to be great confusion in the foremost boat; the Malays shouted to follow and secure the prize, and in spite of Mr Lee's warnings, the Sakarang party under Mr Brereton dashed off in pursuit. Though convinced of the danger, Lee could not abandon his friend. He followed up in his boat. The enemy appeared scarcely able to get away, as the fresh was heavy and the pullers few; but no sooner had they drawn the Malays a couple of reaches from the fortified village, than they rose to their feet with a yell, and at the same time a dozen huge war-boats came sweeping down the river and were on the pursuers before they could fire their guns. A pell-mell ensued, boat against boat. Lee made a stubborn resistance: with his double-barrelled gun he kept the enemy at a respectful distance, until, wounded and surrounded, he attempted to board an enemy's *bangkong*, and was knocked into the stream, and as he rose his head was severed from his body. Mr Brereton was more fortunate, as he was dragged ashore by Abang Ain and some faithful Malays, who never deserted their English leaders. The force of the current had now driven the contending boats under the guns of the battery, and its defenders instantly opened fire, which caused considerable loss to the Dyaks, and forced them to retreat up the river.

We all regretted poor Lee—a gallant officer, who would in after-days have been of infinite use to the Sarāwak Government, as he was prudent in speech and cautious in action. Charles Johnson succeeded him at Lingga.

Rentab, though proud of his trophy, was forced to retreat, for, as I have said, his party had suffered heavily. This action roused both sides, and Gasing, the Sakarang chief and friend of the English, collected some thousands of men, marched into the enemy's country, and plundered and burnt twenty villages.

Rentab, however, was still, in 1854, looked upon as a leader of the pirate party, and was soon able to collect enough men around him to build and palisade a village on a hill near the river Lang (the Hawk), and establish a fortified position on the neighbouring mountain—Sadok. From these places he continually sent out expeditions, which harassed the friendly Dyaks, until at length they declared that the fort of Sungei Lang must be destroyed, or they would have to come to terms with the enemy.

Sir James Brooke put off as long as possible this expedition, as he did not like to act while the Commission was pending, fearing it might look like an open defiance of the English Government; but at last, roused by the cries of the people, he determined to undertake it. War-boats were prepared, contingents summoned from the neighbouring rivers, and at length in May we started. When we reached the Sakarang river, we found that we reckoned eight Englishmen and about 7000 Malays and Dyaks.

To prevent the Dyaks of Seribas collecting to the aid of Rentab, our old Datu Tumanggong was sent with a strong force to that river, while Mr Steel advanced up the Kanāwit from the fort at its mouth, to keep the thousands of warriors in that river from joining the enemy.

It was a formidable affair preparing this host for our long journey into the interior. We pushed up the Sakarang as far as our war-boats could reach; we then stopped, built a stockade, and ran up a house for Sir James Brooke, as his state of health would not permit him to march. He remained behind with a strong force to protect our *prahus* and provisions, and as a rallying-point in case of accident. Then we started, some to march, others to push ahead in our light tenders. We soon found that marching would neither suit the English nor the Malays, as the way was through tangled jungle and low abrupt hills, so that the parties would scatter, and be an easy prey to an enterprising enemy.

Most English leaders in Borneo that I have seen have a mania to show the Dyaks that they are as good woodsmen as themselves, and push on at a rapid pace, forgetting that all are not equally active, and that their followers are heavily laden. I have done as much jungle-walking as most men, but I always found it necessary to husband my own and my followers' strength for the first few days; then, being in condition, longer marches could be made, and there would be few or no stragglers. Our leaders on this occasion were not of that opinion, and the army broke into parties, so that at last we were forced to take to our light *prahus*, leaving the Dyaks to march, or to come on in such boats as they could seize on the banks of the river. They were soon all provided.

It was an exciting expedition. As we advanced into the interior, the woods were thronged with enemies, who at night would creep down and stab at the occupants of those boats who had been too lazy or too tired to erect fences on the banks for their protection. We and our immediate followers slept on shore in little huts, run up at the moment, as leaky as sieves, and the most uncomfortable places in the world. It was astonishing that amid so many experienced men nobody ever thought of a light waterproof tent,[1] such as I afterwards used, which cost ten shillings, and was made by one of my servants, and in which a dozen men have remained perfectly dry during the heaviest storms. To the want of such a tent I ascribe the death of one and the illness of nearly every member of our English party.

The night of our arrival at the landing-place from which we were to march on the village of Lang was an exciting one. Shouts and yells were heard in the woods around; spears were thrown, ghastly wounds had to be dressed, and much sleep was impossible. As each took his turn as sentry, we had leisure to meditate on our position.

The morning broke brightly after a heavy night's rain, and parties were told off to protect the boats. We advanced, but all was confusion. No one would lend a hand to haul up the gun until the English seized the ropes and began pulling away. The ground was slippery, and it was heavy work. At last we reached a spot where the village was visible. It looked formidable enough for our irregular forces to attack. On the ridge of a long hill a high stockade extended, enclosing on three sides half-a-dozen village-houses, which we could see were crowded with men. The opposite side to us was defended by a sort of precipice, up which it was possible to climb, but not in the face of an enemy. We now tried the range of the four-pounder howitzer, but from that distance its fire was evidently ineffective. As our wild allies had pushed ahead, we left the guns for the present and advanced to within a

[1] My tents were made of grey or white shirting, after the model described in Galton's 'Art of Travel;' they then received three coatings of linseed, dried on separately.

hundred yards of the stockade, meeting, as we ascended, men carrying the dead and wounded who had fallen during an effort to surprise the village. This had a damping effect on many of our followers.

While we stood consulting as to the best method of attack, the enemy aimed a gun at us, and the ball rushing over our heads, caused a most undignified duck, which, however, only occurred once. The place was attacked with rifles, gun, and rockets, but at six o'clock it was still untaken, and then we observed banking upon the horizon the heaviest of black clouds, which foretold a fearful storm. We all preferred the risk of an assault to passing out a night in such weather, and Captain Brooke determined to attack. Whilst he was gathering the party together, three men, old Panglima Usman and two fortmen, walked up to the place, and climbing over the stockade, pushed open a narrow door, upon which we all rushed forward, and in a moment had possession of the fort. A few shots were fired at us, and the thing was over. We could not well understand how it was that the Panglima had been able to do this with impunity, but he knew the men with whom he had to deal. The Dyaks have a horror of rockets, and each time they saw Mr Crookshank prepare one, they threw themselves on the ground, and remained there till the dreaded engine of death had passed over. The Panglima took advantage of this, and was not seen till, crowning the stockade, he fired his carbine at them, and the pirates, thinking that this was the head of the attacking column, fled, and when we hustled through the opening, we could only get a few shots at the enemy, as they rolled down among the bushes on the side of the precipice.

As the day closed we had observed hundreds of men collecting on each cleared hill behind us, and lining as well the edge of the jungle: as these Dyaks were all dressed in red cloth jackets the effect was singular, as they looked at the distance like English soldiers. There were thousands in all who were waiting to see the result of the attack: if it succeeded they would retire; if not, they would have fallen on us during the night.

At seven o'clock the rain descended in torrents, and right glad were we to find ourselves housed, though the buildings rocked under the weight of the thousands who crowded into them.

We spent a few days there punishing the followers and abettors of Rentab, who himself had retired to his other fort in the mountains, and then we returned to our boats. We ought to have attacked Rentab in Sadok, his stronghold, but natives do not like to undertake two attacks during the same expedition. The loss of Lang did not prevent Rentab holding firm in his eagle's nest at Sadok: and it took constant expeditions during eight years before it was ultimately captured. The descent of the river was rapid; the heavy rains which had deluged the country every night had raised the level of the river several feet, and as we dashed along at the rate of at least fifteen knots an hour, we felt that one touch on a hidden snag would probably send us all to eternity. But there was no help for us: on we went — there was no stopping, as 200 boats of every size were behind us, and our men pulled with a will to get clear of the rest.

We were glad to meet our chief again, who had been tortured by conflicting rumours; now it was news of our defeat before Lang, then of a fresh that had carried away all our boats—but as no *débris* had passed before the stockade, he comforted himself.

When we reached Sarāwak we found news; the Commission had arrived at Singapore, and a vessel of war had come to offer a passage to Sir James and those whom he wished to take with him. He had to go alone, as both Grant and myself were down with fever. Brereton was ill with dysentery,[1] and Captain Brooke and his brother were both suffering.

How impatient we were to get well! At length I was sufficiently recovered to bear the fatigue, and was carried down to a boat, and taken to the mouth of the river, where Grant and I embarked on board a merchant-ship, and on our arrival at Singapore we were both sufficiently recovered to land without aid.

The Rajah was indeed glad to see us, as he was almost alone, and wanted our assistance. How derogatory to his position and humbling to his pride did he feel this Commission to be!

What shall I say of the Commission? The two gentlemen sent down to conduct it were very different. Mr Prinsep, the chief, was incapable—the mental malady to which he soon after succumbed showed itself too often; and the Hon. Mr Devereux could alone do anything, and endeavour to control his colleague. He was an able, sarcastic man, well fitted for the work. But the results were most unsatisfactory. The Eastern Archipelago Company had nothing to say to which the Commissioners could listen; Mr Wood, the editor of the 'Straits Times,' was astonished to find himself called upon to prove the case for the enemies of the Rajah—and as the silly man knew nothing, he could only involve himself in a cloud of absurdities. The Lieutenant-Governor of Labuan tried to bring on his case, but that was beyond the scope of the inquiry. The only curious incident which occurred was the stepping forth of a Dutch civil officer, Monsieur Boudriot, of whom mention has already been made, and who said that, being on his way home from Java on sick-leave, he had accidentally attended the Commission, and

[1] Mr Brereton died shortly afterwards, and Mr Johnson was appointed ruler of Sakarang and Seribas, in addition to Lingga.

he begged to offer himself as a witness. This gentleman's evidence of itself would carry conviction to impartial minds, for he had held high positions on the coast of Borneo, and knew the Sakarang and Seribas Dyaks to be savage, inhuman wretches, and undoubted pirates.

Sir James had early retired from the Commission, as Mr Prinsep had permitted Mr Wood to take almost an official position during the inquiry.

The result was what might have been expected. As no specific accusations were brought against the Rajah, no specific answers to them could be prepared. Mr Prinsep reported favourably as far as the inquiry went, unfavourably on a subject on which he made no inquiry,—a very hazy result. Mr Devereux did his best to render his report satisfactory, and this was the only one which was seriously considered by Government.

Sir James Brooke himself did not manage his part well. He was anxious to prove the complicity of the Eastern Archipelago Company in all the intrigues which had brought about the Commission, and he wearied the Commissioners and the public with a useless and tiresome examination of Mr Motley, the agent of the Eastern Archipelago Company in Labuan. Nothing could be got out of the man, as he knew little, and had only been playing the part of the frog in the fable. We tried ourselves to induce the Rajah to confine himself to two issues, which were really important: first, whether the Seribas and Sakarang were really pirates; and if they were, had undue severity been exercised in suppressing them. To Mr Grant and myself these were the only important subjects; and we endeavoured to impress our chief with our ideas, but with only imperfect success.

No man perhaps felt the absurdity of the whole inquiry more than Mr Devereux. He vainly inquired, What are the charges, who are the accusers? and probably had he not been hampered with an impossible colleague, he would have closed the Commission at once, saying, "There was nothing to inquire about." As, however, the Commission was there, we exerted ourselves to send in witnesses; and even without going to Borneo, we found among casual traders and residents witnesses enough.

In speaking of Mr Devereux's probable opinions, they are but impressions made on my mind, as I should remark that out of the Commission I never once spoke to him. Sir James Brooke, from mistaken delicacy, begged us not to accept invitations, but to keep ourselves retired until the inquiry was over, so that we never met the Commissioners except in the court-house. I have ever regretted our following this advice, as, had we been enabled to converse frankly and unofficially with the Commissioners, we might have greatly influenced the result; for we knew more of the question than perhaps all the others put together, and so sensible a man as Mr Devereux would not have been unduly influenced by our suggestions.

In speaking of Sir James Brooke, I wish to present him exactly as he appeared to myself, and neither to conceal nor palliate his errors and faults. I watched him closely during the course of this Commission, and I thought that I detected in him the same impatience of opposition which I have ever observed in those who have lived much alone, or in the society of inferiors, whether of rank or intellect. Sir James had lived much alone, or with those to whom his word was law, so that he had had rarely the advantage of rubbing his ideas against those of his equals, and therefore treated as important subjects matters which, to others less interested, were but trivial. I dwell on this, as it appeared to me to be the cause of the Commissioners not having been able to report in a manner which would have fully shown to the world the importance of the Rajah's work in Borneo. Sir James Brooke did not direct the inquiry to the real issues, therefore it failed. The Commission was at last closed, having been conducted, as the Rajah truly says, without dignity and without propriety.

It was a pity that the Commission did not visit Borneo, as an inquiry conducted entirely in Singapore could not be satisfactory. Sir James Brooke did not consider it his business to send witnesses to Singapore; but had the Commission come to Sarāwak, he would have given them all the aid in his power. As it was, the evidence adduced satisfied all those who were not moved by a carping spirit. It is obvious, however, that an inquiry conducted 400 miles from the spot where the events occurred, and where the best evidence was to be obtained, could not be complete.

The Commissioners left, and we looked out for a way to get back to Borneo. We all embarked in the little schooner that had brought us from Sarāwak, but we met the strength of the N.E. monsoon, and were driven back to Singapore. Sir James Brooke landed, while we continued on board the Maria, and at length reached Sarāwak, where the Rajah soon after joined us, having been offered a passage in H.M.S. Rapid, Captain Blane.

Now commenced a really quiet life. Sir James was free from the anxieties caused by the coming of the Commission, and devoted himself to the happiness of the people. He no longer held a public appointment, as he had, *de facto*, resigned his position as Commissioner, and I was acting in his stead. This was perhaps the happiest time he ever spent. The country was progressing; no signs of discontent had yet shown themselves (except the affair of the Patinggi); he could live in the capital or in his country cottage as he felt inclined, and he returned to a course of chess and pleasant reading. We worked together at Staunton's Problems, and he fancied that he discovered an error in one of the most famous: we both studied the matter earnestly, and though I tried hard, I was fully convinced that an error had been discovered. We ought to have sent it to the 'Chess Chronicle.'

After our 11 o'clock breakfast Sir James attended court, and decided the pending cases, and then returned home at two. I never failed to know of his arrival by hearing a cheerful voice calling to me, "Now, St John, lunch and chess; we must work out that problem;" or he would quietly carry off my Staunton, and study in his room the Scotch or some other opening, and then, proposing a game, win half-a-dozen running, laughing heartily all the time at my surprise, and perhaps vexation, at being beaten so easily, for I was generally the better player, and could only be regularly defeated by surprise. It was at this time that he wrote, "I am chess mad."

We had at this time in Sarāwak the famous naturalist, traveller, and philosopher, Mr Alfred Wallace, who was then elaborating in his mind the theory which was simultaneously worked out by Darwin—the theory of the origin of species; and if he could not convince us that our ugly neighbours, the orang-outangs, were our ancestors, he pleased, delighted, and instructed us by his clever and inexhaustible flow of talk—really good talk. The Rajah was pleased to have so clever a man with him, as it excited his mind, and brought out his brilliant ideas. No man could judge the Rajah by seeing him in society. It was necessary to get him at his cottage at Paninjow, with his clever visitor Wallace, or with his nephew Charles, the present Rajah, who was full of the crudest notions, the result of much undigested reading, but who could defend his wild thoughts cleverly, pleasantly, and gaily; or with Mr Chambers, the present Bishop, who was too firmly convinced of the truths he preached to be offended by finding others differing from him, and too earnest not to feel assured that in the end he would convince his opponents. With these, either at our house in Kuching or in our mountain cottage, we were ever in discussion, and our discussions were always either philosophical or religious. Fast and furious would flow the argument, till somebody would observe, "It is just eleven," when the Rajah would instantly arise, wondering that the interest of the discussion had made him pass by his hour of rest, which was now usually ten o'clock, as fever and ague and small-pox had weakened him too much to permit him to pass the night in talk.

In the morning we would catch each other looking in the library for the authorities, and perhaps the arguments, with which to support another discussion in the evening.

I am afraid we thoroughly wearied many of our party. Captain Brooke thought that such discussions led to no result. Grant was very determined in his opinions, but did not care for the subjects discussed; while Fox, a later volunteer in the Rajah's service, though ever an inquirer, was trammelled by his education, and would at first touch with diffidence subjects which the more trained mind of Mr Chambers feared not to handle. Bishop McDougall (of whom I have already spoken), when he joined in these discussions, did not, by his arguments or tone, give encouragement to the inquirers. There was, however, no irreverence in our treatment of sacred subjects, so that no one else was offended. Not that we were *always* discussing, for Sir James Brooke was one of the few men I have known who could enjoy the luxury of silence.

In one of the Rajah's letters I find him laughingly referring to these discussions: "There was a tremendous argumentation on the being and attributes of the Deity. Charlie [Johnson] was there in high feather, the room resounded with their voices, and the energetic bangs on the table made the glasses ring. Everybody was an atheist and pantheist by turns. Charlie and St John collared Chambers with hard names, and then everybody sat upon poor Charlie, who said that God was everywhere and nowhere at the same time, or words to that effect. Then the company roared at St John for his heterodoxy, fiercely contested my definition; but at last it was discovered that everybody meant the same thing, that everybody said it in a different way, and half-a-dozen times over, and that we were proper and very orthodox all the time."

I have said how devoted the Rajah was to his duties in court. Some cases came before him in the spring of 1855, which made me seriously doubt his efficiency as a judge. The fact was, that we thought him much too tender-hearted to be intrusted with duties which required some sternness, as he had no jury to aid him in difficult cases. I am writing of what we thought at the moment. Two Senah Dyaks had killed two others under circumstances which we considered showed both treachery and premeditation, and many of us insisted that they deserved death. The Rajah, however, refused to condemn them, alleging their youth, the possibility of provocation, and that an example was not necessary; he therefore sentenced them to the highest native punishment short of death—the infliction of a heavy fine. Some of us were very angry, and in the evening we had a fine discussion, which terminated by the Rajah's getting excited, and saying, "It is my will that it should be so." I forgot myself so far as to answer, "That's the tyrant's argument." Captain Brooke became quite alarmed, and said to me next morning, "You will ruin yourself with the Rajah if you argue in that fashion." However, it was not so. I would have expressed my regret, but the Rajah would not let me—"No, St John; we were both in the wrong." Referring later on to some of our discussions, he wrote: "You are *an arrogant fellow by nature*, but you have been less so as Consul-General than as Private Secretary. I have always looked upon you as a child of my own—a saucy and turbulent one, but not the less dear or valued on that account."

With regard to the sentence, we all disagreed with it, thinking that it would lead to a revival of the old practice of head-taking: but the Rajah was right; the punishment he inflicted had a deterrent effect, and I never heard of a similar crime being committed in Sarāwak during the rest of my stay in Borneo.

In October 1855 two of the principal members of our party left us for England—Captain Brooke and Charles Grant. It was a great loss. To me the departure of Brooke was especially so, as we had been inseparable companions for seven years, and he was one of the most likeable of men. During that long period of daily intercourse I had ample opportunities of judging him, and thought him quite worthy to be the Rajah's successor; for though he had not his talents, he had found out the way to win the hearts of the Malays—a talent which his brother Charles did not possess in the same degree. The latter, however, could manage the sea Dyaks well, and he lived so much among them that he actually appeared to acquire their manner and expression, which, in effect, showed an indifference to what other Europeans liked; but underneath there was plenty of gaiety and much fun, and not without a certain amount of undeveloped talent. At that time, however, we scarcely did justice to Charles Brooke's capacity, as he had had few opportunities of showing what was in him. In 1859, however, he came to the front, as I shall have occasion to show.

Our two friends had scarcely been gone a week when Sir James Brooke received a despatch from Lord Clarendon announcing that his resignation of the posts he held under Government was accepted, and adding a cold approval. At the same time the Rajah was informed that I was appointed Consul-General in Borneo. The Rajah was deeply disappointed. After all the promises that had been made by Lord Clarendon as to the intention of the Government to make amends to him should the Commission report favourably—to receive a calm, almost a contemptuous, approval of his conduct, was too bad. What was the value of such approval unless it was intended to afford him some support, after having lowered his position both with native and European? The Rajah was indignant, and in his indignation wanted me to refuse the appointment, and take service under him. Lord Clarendon, it is said, named me Consul-General as a mark of friendship towards the Rajah; but he would not look on it in that light. And yet, had the Government been really hostile, and appointed such a man as the Hon. Mr Edwards, the Governor of Labuan, to be Consul-General, what trouble he could have caused, as we shall see he did afterwards whilst temporarily holding that post.

I could not well refuse the appointment offered me by Lord Clarendon; but I said that if the Government wished me to do anything hostile to Sarāwak, I would immediately resign. As this did not satisfy the Rajah, and as he insisted that a British Consul-General could not live in Sarāwak until the English Government recognised it, I reluctantly agreed, and wrote to Lord Clarendon to propose that I should take up my residence in Brunei.

There can be no doubt, however, that our Government behaved in a most shabby manner to the Rajah. After nearly ruining his reputation and prestige by sending out a Commission to try him, as the natives believed, they did nothing for him,—they simply dropped him; and I heard it afterwards alleged by a gentleman now high in office, that it was on account of the violence of the Rajah's language. Some allowance surely should have been made for the natural irritation arising from the injustice committed. He suspected the Peelites of being his enemies, and, if we may judge from Mr Gladstone's late writings, not without reason.

Our local politics were now disturbed by the news which came of the fall of Sebastopol. "Whilst victory was trembling in the balance," the Rajah wrote, "our hearts here palpitated as to the capture of Sebastopol;" and later on, "My heart beats high at the glorious news of the fall of Sebastopol. How deeply is our love of country entwined about our heart-strings! It astonishes me." Our Malays were as excited as we were, and crowds came when the mail arrived to hear the last intelligence from Roum. Those who saw them could not doubt as to the interest taken by Eastern Mohammedans in the Sultan of Turkey. The Rajah sent the chiefs a fat bullock to aid in the feast which was given in honour of the victory of the allied armies.

A new question now arose—that of judicature. When the Commission was first appointed in 1853, Sir James Brooke would have resigned his post as Commissioner, but the Government requested him not to do so until the report of the Commission was known. When, however, he read the very unfair instructions issued by Lord Clarendon, but which had the appearance of having been drawn up by Mr Hume, he instantly sent in his resignation. Lord Clarendon, however, would not accept it, and I continued to act for him as I had done from the 1st February 1851. Suddenly there arrived an Order in Council settling the question of jurisdiction by ordering all cases in which British subjects were concerned to be sent for trial to the nearest British colony. As this was not only contrary to our treaty with Brunei, but was a direct attack on the sovereign rights of native states, Sir James Brooke refused to allow it to be put in practice in Sarāwak, and said that he would only submit to force.

I had a meeting with the native members of the Sarāwak Council, and the good sense of the meeting was summed up in a few words by the Datu Bandar. "The Sarāwak courts," he said, "have tried cases for the last fourteen years in which both natives and Europeans have been interested. Has any one complained of these decisions?—if not, why should there be any change?"—or, "Cannot we leave this question alone?" the famous saying of Lord Melbourne.

I drew up a scheme which I thought would work well: that the courts should continue to try all cases; that the laws under which British subjects were tried should be recorded; that no punishment more severe than that allowed by English law should be inflicted; and that in serious cases the judge should be assisted by a jury. I added, by desire of the Rajah and for my own special benefit, that the Sarāwak Government would allow the Consul-General to sit as one of the judges in every case in which British interests were concerned.

By the treaty with Brunei that had been negotiated by Sir James Brooke himself in 1847, the English representative was intrusted with great powers. All disputes, whether between British subjects or between them and other foreigners and natives, were to be decided by him, and all crimes were also to be tried by him, without permission to the local authorities to interfere in any way. I saw that this would not work even in the capital; so then I proposed a mixed court, which was ultimately accepted both by the English and Brunei Governments, and it answered very well.

As soon as it was possible to receive an answer, I heard from the Foreign Office that they approved of my plan concerning Sarāwak.[1] It was, however, afterwards modi-

[1] "FOREIGN OFFICE, *February* 9, 1856.

"SIR,—Lord Clarendon desires me to write you a line to say that he has not been able to come to a decision upon the various questions of jurisdic-

fied, at the suggestion, I believe, of Earl Grey, who saw a chance of future difficulties if the Consul-General were to be one of the judges, and this part of the scheme was very judiciously omitted.

It was a pleasant duty for me to have to address a despatch to the Sarāwak Council, informing them that her Majesty's Government had no desire whatever to interfere with them, or to prevent their choosing what government they pleased; and then I added that the Government accepted the proposed plan for settling the jurisdiction question. I was not, however, allowed to ask for an *exequatur*.

I never saw the Rajah more pleased than when this question was settled. He appeared thoroughly happy, and for a time I thought that he had forgiven the British Government.

But I am anticipating. Before this was arranged, two other questions came up for settlement—the Bishopric and the Borneo Company.

I have scarcely mentioned the mission, as it really had no perceptible influence on the history of Sarāwak. Internal discussions had early broken out, and the house divided against itself could do no good. However, in 1855 Mr McDougall was named Bishop of Labuan, but the Rajah declined to allow him to act in his dominions unless he received letters-patent from the Sarāwak Government constituting him Bishop of Sarāwak. All objections were soon removed, and in the autumn of 1855 Mr McDougall paid a short visit to Calcutta, where he was consecrated, and was thenceforward Bishop of Sarāwak as well as of Labuan. He returned in time to join our

tion in Borneo, &c. These questions have been under consideration, and Lord Clarendon hopes the question of jurisdiction may be settled in accordance with your suggestions, which seem to offer a practical mode of solving the difficulty. (Sd.) WODEHOUSE."

Christmas dinner. I shall have something to say about the mission further on. Attached to the mission was a school, which greatly interested the Rajah, and really it was the most useful thing about it. The children all appeared to have an affection for the Rajah, as children always have for those who are themselves loving in disposition.

The Borneo Company, in 1855, began to be talked of. Coal had been found in several places in Sarāwak, and this, with the antimony, sago, and gutta-percha, tempted some capitalists to propose forming a company to develop the resources of the Rajah's territories. At the head of the scheme was a gentleman "much reputed in the city," who talked a great deal of philanthropy; but nearly all connected with the scheme were sound men of business. The Rajah saw in this scheme a means of detaining me in Sarāwak, and was most anxious to induce the directors to name me their managing director in Borneo. As long as I was with the Rajah, and surrounded by friends, I listened to this; and, to increase the inducement, the Rajah offered to add to the managing directorship the government of Sadong, a district in which the company were to work coal. Luckily I did not get the offer of the appointment, or I might have been weak enough to have accepted it, and have had to go through the same ordeal as their last managing director, and appeal to the English courts to have my rights respected; but the Rajah took exception to the appointment made by the directors, and expressed his views in very forcible language. He had so high an opinion of my talents, that I verily believe he thought that, like Lord John Russell, I might command the Channel Fleet or be Archbishop of Canterbury, much more manage a mining and trading company. There was, however, a thought latent in all this that had not come much forward. The Rajah was beginning to think that the Borneo Company, judiciously managed, might ultimately develop into another East India Company, and absorb not only Sarāwak but the north of Borneo, and he thought that my being director might facilitate this; but the appointment made frustrated this plan.

In 1855 began troubles in foreign politics which continued for six years to agitate the country. I must briefly explain these, as much resulted from them. In the north of the Rejang (the boundary of Sarāwak) were several rivers in the territory of the Sultan of Brunei which were famous for their sago, and the most important trade of Sarāwak was with these districts. There lived in the principal of these places (Muka) two chiefs, named Ursat and Matusin, who were rivals. Ursat governed in the name of the Sultan, while apparently Matusin was employed to watch him. Matusin was popular, being out of power, while the exactions of Ursat were on the usual scale of Malay rulers. One day Ursat, sitting in his veranda surrounded by his people, jeered at Matusin as he was passing down the river. This roused all the Malay pride in the breast of the latter, and he determined to attack his adversary. On returning home he passed Ursat's house, and saw him still there, but almost alone; he sprang ashore with his followers, dashed into the house, cut down Ursat and those near him, burst into the women's apartments, and killed every woman and child he met. He, in fact, "ran amuck."[1]

This act roused all Ursat's friends, and Matusin was quickly expelled, and his wife and children fell victims to the fury of his enemies. Among those who aided in the latter operation was Sirib Musahor of Serikei, a district of the Rejang, in Sarāwak territory. For this interference he was heavily fined by the Sarāwak Government, and, in disgust, retired with all his followers into the Sultan's territory. I went in H.M.S. Grecian, Captain Keane, to

[1] In Malay, *amok*.

inquire into this affair, as I wished to take the latest news to the capital, for which place I was bound; but meeting some people off the bar of the river Muka, I heard that the whole place was deserted, and so continued my voyage. Whilst we were in Labuan, the Rajah came up in the Jolly Bachelor (sailing gunboat), and, leaving the Grecian, I joined him, and returned to Brunei. I may notice that at this time Dr Treacher was administering the government of Labuan, and he determined to show that all Government officers were not hostile to his former chief, so, on his landing in our colony, the Rajah of Sarāwak was received with a royal salute of twenty-one guns. As the British Government were at that time considered hostile, it required some courage thus to distinguish Sir James.

I never spent a more curious month than this. The Sultan had run up a very neat mat-house alongside the palace in which we took up our residence. We were three—the Rajah, Mr Low, and myself. The Rajah had come for a week's visit, but he soon found that this would not be sufficient. The whole capital was in intense excitement: we found one party arming against the other party—men standing with lighted matches by their guns ready to fire into a hostile neighbour's house; everybody was preparing for a fight. Immediately after our arrival the Rajah had an audience with the Sultan, who threw himself on his generosity not to desert him in this crisis; and Sir James replied that if they wished he would remain at the capital until everything was satisfactorily settled.

I never had a chance of witnessing a more startling proof of the personal influence that the Rajah possessed than what followed. Here was he in a 30-ton boat, with a crew of a dozen natives, the despised of his own Government, but the absolute dictator in Brunei. At his suggestion a general unloading of guns took place, the armed retainers of the Pangerans returned to their homes, and quiet was established. The different nobles explained to him their grievances, their wants, and wishes, and he set to work to put some life into the place. In this he was well seconded by Mr Low, who is one of the ablest rulers of Malays that I have ever met; and in sending whom to Perak, the Government have sent the right man.

The Rajah soon found that the nobles were anxious that the ancient titles should be restored, and that the Sultan should, previous to his own coronation, name the four great *wuzirs*, or ministers; and this the Rajah managed to arrange, and to settle to every one's satisfaction who should fill the offices. He did not expect much good from his work, but it has stood the test of time; and now, twenty-four years after, the same forms exist.

Makota, fortunately, was absent, so that he could not mar the negotiations. He had gone down ostensibly to settle affairs in Muka, but, in fact, to plunder for himself and the Sultan.

It was really interesting to watch the Rajah during this month,—how he received deputations from the Sultan, from the nobles; how the trading class would come when the latter were absent; how the poor and the distressed flocked to him for relief, and how many he aided. There he was in his glory, He worked to introduce vitality into this wretched Court, and how pleased these nobles looked when he had induced the Sultan to name the four great *wuzirs* that recalled the memory of their ancient splendour. It was a continual levee—from morning to night our house was crowded. I must not, however, dwell too much on Brunei politics, as they could not be made interesting; but the Rajah succeeded in reconciling the hostile factions which had remained separated since the murders of Muda Hassim and Bedrudin in 1846.

We returned to Sarāwak in the Jolly Bachelor,—a long passage, beating against the S.W. monsoon four days, trying to double Sirik point; but at length we arrived in Kuching, and the Rajah sent his nephew Charles to Muka to try and settle matters, but nothing permanent was effected, so that troubles in all the Milanow countries ensued, and the Sarāwak Government, from that time forward, deemed itself authorised to interfere in their internal affairs. As these countries were in the Sultan's territories the policy was doubtful.

In August 1856 I was to leave Sarāwak to take up my permanent residence in Brunei; but before doing so, I determined to have another good look at the interior of Sarāwak. I started with my friend, Fox, to go as far as the frontiers of Sambas, visiting, by the way, the Chinese gold-workings. During this trip two things struck me forcibly—the number of the new arrivals of Celestials from Sambas, which the gold-working Company had constantly denied, but whom we saw by hundreds crowding the sheds and houses of the Chinese town at Bau; and the second was a conversation I had with one of the principal officers of the Company. He asked me what had been the result of the Commission,—whether it was true that the Rajah was no longer friendly with the British Government; whether the Queen had ordered the navy no longer to protect Sarāwak; and finally, whether I, the Consul-General, was about to leave the country for good? I soon found that these subjects had been greatly canvassed among them, and they acknowledged that their friends in Singapore had written to them about these affairs. I tried to remove their unfavourable impressions, but my efforts were useless. The Rajah had announced that the British Government was no longer friendly to him, and the fact remained that I was about to quit the country. The evil results of the Commission met us at every step.

On our return to Kuching we endeavoured to press on the Rajah the importance of weakening the power of the gold Company, which, with its offshoots, evidently consisted of several thousand men, blindly subservient to their chiefs. The Rajah saw the danger, but decided to put off active measures until the return of his nephew, Captain Brooke, then in England, who was expected in a few months.

Although the Rajah was very satisfied with the favourable settlement of the jurisdiction question, which was more than he expected, still he thought that Lord Palmerston should do something to restore his prestige—"personally, I have received no amends." This cold treatment produced at length much irritation against the British Government, and it showed itself in every way. I do not pretend to say that Sir James Brooke had not every right to be angry with the treatment he had received; but he showed it in a most impolitic manner. The abandonment of the interests of Sarāwak was the theme of every conversation, whether with foreigners, Chinese, Malays, or Dyaks, and the result was most disastrous. Every turbulent chief, every one who had lost by the introduction of good government, began to think that the Rajah was no longer invincible, and that perhaps it was possible to expel the English from the country, and establish the old order of things.

One of the reasons why Sir James Brooke used this language must be explained. In 1855 and 1856 the Dutch were at war with the Malay chiefs of the interior, and the latter were continually sending to the Rajah for succour. His answer was invariable: The Dutch Government is a powerful one—in the end it will prevail; so the best thing you can do is to get, as soon as possible, the most favourable terms. One of these chiefs who came to see us took the advice, went to Pontianak, told the Resident he came by the Rajah's wishes, and got excellent terms from the Dutch authorities. However, during these disturbances there was much trouble on the frontiers, and the Chinese, Malays, and Dyaks often crossed the border to revenge the injuries they suffered from tribes under Netherland rule. The Rajah endeavoured to prevent this, and used to say, "As I am abandoned by the British Government, the Dutch may take this opportunity of attacking us if you do these things." His words, however, had a different effect from what he intended.[1]

At last August arrived, when I had to leave Sarāwak, and Captain Drought, in the Auckland, came to fetch me away. It was a sad parting for both of us; but we thought that arrangements would be made which would bring about my return. In a few weeks I heard—"I often feel the loss of your society; the free communion of so many years is not readily supplied; but I do not consider it as lost, for we may look forward to your return to Sarāwak." He complained that he could never get an argument; it was all "yes" and "no."[2]

So weary did he, in fact, feel, that he determined to take a change to Singapore. Mr Crookshank had just returned from his leave, with his bride, who, the Rajah said, "shed a brightness on the old house;" and therefore, leaving the government in his hands, he started for Singapore in the Spartan, Captain Sir William Hoste.

From Singapore I soon received volumes of correspondence, which all, relating to local politics, would not be interesting to publish; but he could not see me alone in the wicked old capital without forcing his affectionate advice into my ear.

Whilst in England, his nephew, Captain Brooke, and

[1] I notice in the midst of all his cares at this time, his sending £25 to John Briggs of the Yew Tree, Reigate, in remembrance of some kindness received in childhood.

[2] Most of his other old friends and followers were at this time in England, or at out-stations.

his friend and follower, Charles Grant, were both married—the former to a sister of the latter; so the Rajah occupied himself in choosing furniture for the newly-married people. He still wished for my presence in Sarāwak; but as that could not be, he insisted that one of my family should have the first vacancy in the Government service in his territory. "I want ——, who has claims upon us through his brother, who is a firm friend, and was a firm friend in the hour of adversity." I insert these lines, as it shows how the Rajah never forgot those who were true to him.

In Singapore the Rajah lived a very quiet life. He declined all parties, and visited only among intimate friends. He was detained longer than he expected; but at length the vessel he was waiting for—the Sir James Brooke, the first steamer sent from England by the Borneo Company—came in to Singapore, and soon after he started in her for Sarāwak.

All was not quiet here: the Kungsi, or Chinese Gold Company, had been behaving with so much violence, that Mr Crookshank, who was in charge of the Government, thought it advisable to man the stockades, and to send for Mr Charles Johnson to bring up a force from Sakarang. This awed the Chinese, and quiet was restored; but I notice, in one of Mr Johnson's letters to me, he says, "I really hope that the Rajah will pull down that high tone of theirs, or they will do us a mischief one of these days." This was also the opinion of Mr Crookshank and many others; so that, to guard against surprise, fifty men were directed to sleep in the forts.

When, however, the Rajah arrived, he summoned the chiefs of the Kungsi, fined and punished them for their illegal acts, and seeing them submissive, he dismissed the guard, believing it impossible that so wild a scheme as attacking him could ever enter into the Kungsi's plans.

The Rajah's last letter to me before the Chinese insurrection is dated February 14 (four days before the attack), and consists of twenty-eight closely-written pages—a sort of epitome of coast politics; but it winds up with these words: "Congratulate me on being free from all my troubles. The Chancery suit has been concluded by a decree of the Court on Wise's annulling the paper I gave him."

The tone of his correspondence at this time was jubilant. Everything was going on well—the calm before the storm.

CHAPTER XV.

THE CHINESE INSURRECTION.

1857.

CHINESE colonists are the mainstay of every country in the further East; but they carry with them an institution which may have its value in ill-governed countries, but which in our colonies is an unmitigated evil—I mean their secret societies. A secret society is generally instituted under the form of a benevolent association, but secretly the members are banded together to obey no laws but their own, to carry out the behests of their leaders without question, and to afford protection to each other under every circumstance. If a member commit a crime he is protected or hidden away; if he be taken by the police, the society secures him the ablest legal assistance, furnishes as many false witnesses as may be required, and, if he be convicted, pays his fine, or does all in its power to alleviate the discomforts of a prison. Should the society suspect any member of revealing its secrets, or from any cause desire to be rid of an obnoxious person, it condemns the individual to death, and its sentence is carried out by its members, who must obey their oath. On these occasions the mark of the society is put on the victim to show who has done the deed. In our colonies we have not been able to put them down.

For many years the Chinese attempted to form secret societies in Sarāwak, but the Rajah's vigorous hand had crushed every attempt, and it appeared as if success had attended this policy,—and so it had, so far as the Chinese of the capital were concerned; but in the interior, among the gold-workers, the Kungsi stood in the place of a secret society, and its chiefs carried on an exclusive intercourse with their fellow-countrymen in Sambas and Pontianak, the neighbouring Dutch possessions, and with the Tien-Ti-Hué (Heaven and Earth Secret Society) in Singapore.

When Mr Fox and I made our long tour in the interior among the Chinese settlements, we became convinced that smuggling was being carried on to a great extent, for, however numerous might be the new-comers, the opium revenue had a tendency to decrease.

At last it was discovered that opium was sent from Singapore to the Natuna Islands, and from thence it was smuggled into Sarāwak and the Dutch territories. It was proved that the Kungsi had been engaged in this contraband trade, and it was fined £150—a very trifling amount, considering the thousands they had gained by defrauding the revenue—and measures were immediately taken to suppress the traffic,—which, together with the punishment of three of its members for a gross assault on another Chinaman, were the only grounds of complaint that they had against the Sarāwak Government.

But these trivial cases were not the cause of the Chinese insurrection in Sarāwak, as before that date all the celestials in the East had been greatly excited by the news that the English had retired from before Canton, and that the Viceroy had offered £25 a-head for every Englishman slain. The news was greatly exaggerated. It was said that we had been utterly defeated by the Chinese forces, and now was the time, they thought, when they could expel the English from Sarāwak, and assume the government themselves. The secret societies were everywhere in a state of great excitement, and the Tien-Ti sent an emissary over from Malacca and Singapore to incite the gold-workers to rebellion, and used the subtle but false argument that not only were the English crushed at Canton, but that the British Government was so discontented with Sir James Brooke that they would not interfere if the Kungsi only destroyed him and his officers, and did not meddle with the other Europeans or obstruct trade.

It was also currently reported that the Sambas Sultan and his nobles offered every encouragement to the undertaking; and the Chinese listened much to their advice, as these nobles can speak to them in their own language, and are greatly imbued with Chinese ideas. To explain this curious state of things, I may mention that the young nobles are always nursed by girls chosen from among the healthiest of the daughters of the gold-workers. And I may add that about that time there was a very active intercourse carried on between the Malay nobles of Sambas and Makota, the Rajah's old enemy and the Sultan of Brunei's favourite minister, and that the latter was constantly closeted with an emissary of the Tien-Ti-Hué of Singapore, to which I am about to refer.

To show that it was not mere conjecture as to the Tien-Ti sending emissaries abroad at that time, I may state that on the 14th February, four days before the insurrection in Sarāwak, a Chinese named Achang, who had arrived at Brunei from Singapore a few days previously, and had a year before been expelled from Sarāwak for joining that secret society, came to my house to try and induce my four Chinese servants to enter the Hué, and added as a sufficient reason that the Gold Company of Sarāwak would by that time have killed all the white men in the country.

At Bau, the chief town of the Chinese in the interior, the secretary to the Kungsi showed a letter from the Straits' branch of the Tien-Ti-Hué to a Malay trader named Jeludin, urging them to work against the foreigner. I mention these facts to show the ramifications of these secret societies, which in every country where they exist are the source of endless trouble and disorder.

During the month of November 1856, rumours were abroad that the Chinese Gold Company intended to surprise the small stockades, which constituted the only defences of the town of Kuching, and which, as no enemy was suspected to exist in the country, were seldom guarded by more than four men each; but, as I have before mentioned, Mr Arthur Crookshank, who was then administering the government, took the precaution to man them with a sufficient garrison, as it was said that during one of their periodical religious feasts several hundred men were to collect quietly and make a rush for the defences which contained the arsenal. On Sir James Brooke's return from Singapore he instituted some inquiries into the affair, but could obtain no further information than such as vague rumour gave; but still, such experienced officers as Mr Arthur Crookshank, and the chief constable Mr Middleton, were not satisfied, as they felt that there was mischief in the air.

I was sitting reading one day in my veranda in the consulate at Brunei, when a Malay hastily entered and said: "I have just arrived from Singapore. Whilst detained by very light winds we approached a schooner coming from Sarāwak, and one of the crew called out to me, 'The Chinese have risen against the Rajah, and killed all the white men.'" He knew no more. This, coupled with what I had previously heard, made me very uncomfortable. In a few days a short letter from a friend told me part of the catastrophe, but it was not for two months that I had the full particulars in a letter from the Rajah himself.

It appears that when the Kungsi saw their professions of loyalty were believed, they began to prepare for hostile operations; and on the morning of the 18th February 1857 the chiefs assembled about 600 of their workmen at Bau, and, placing all the available weapons in their hands, marched the force down to their chief landing-place at Tundong, where a squadron of their large cargo-boats was assembled. It is known now that until they actually began to descend the river none but the heads of the movement knew its true object, so well had the secret been kept. To account for their preparations, it was given out that an attack was meditated on a Dyak village in Sambas, whose warriors had in reality lately murdered some Chinese.

During their slow passage down the river, a Malay who was accustomed to trade with the Chinese overtook them in a canoe, and actually induced them to permit him to pass, under the plea that his wife and children lived in a place called Batu Kawa, eight miles above the town, and would be frightened if they heard so many men passing and he not there to reassure them. Instead of going home he pulled down as fast as he could till he reached the town of Kuching, and, going straight to his relative, a Malay trader of the name of Gapur, who was a trustworthy and brave man, told him what he had seen; but Gapur said, "Don't go and tell the chiefs or the Rajah such a tissue of absurdities," yet he went himself over to the Bandar and informed him; but the Datu's answer was, "The Rajah is unwell; we have heard similar reports for the last twenty years,—don't go and bother him about it. I will tell him what your relative says in the morning." This great security was caused by the universal belief that the Chinese could not commit so great a folly as to attempt to seize the government of the country, considering that they did not number above 4000, while at that time the Malays and Dyaks within the Sarāwak territories amounted to 200,000 at least. It is strange, however, and was an unpardonable neglect of the Bandar not to have sent a fast boat up the river to ascertain what was really going on. Had he done so, the town and numerous lives would have been saved, and punishment only fallen on the guilty.

But shortly after midnight the squadron of Chinese barges pulled silently through the town, and, dividing into two bodies, the smaller one entered a creek above the Rajah's house called Sungei Bedil, while the larger continued its course to the landing-place of the fort, and sent out strong parties to endeavour to surprise the houses of Mr Crookshank the police magistrate, and Mr Middleton the chief constable, whilst a large party was told off to attack the stockades. Strange as it may appear, none of these bodies were noticed, so profound was the security felt, and every one slept.

The Government-house was situated on a little grassy hill, surrounded by small but pretty cottages, in which visitors from the out-stations were lodged. The Chinese, landing on the banks of the stream just above a house in which I used to reside, marched to the attack in a body of about a hundred, and, passing by an upper cottage, made an assault on the front and back of the long Government-house, the sole inhabitants of which were the Rajah and a European servant. They did not surround the house, as their trembling hearts made them fear to separate into small bodies, because the opinion was rife among them that the Rajah was a man brave, active, skilled in the use of weapons, and not to be overcome except by means of numbers.

Roused from his slumbers by the unusual sounds of shouts and yells at midnight, the Rajah looked out of the

Venetian windows, and immediately conjectured what had occurred. Several times he raised his revolver to fire in among them; but, convinced that alone he could not defend the house, he determined to effect his escape. He supposed that men engaged in so desperate an affair would naturally take every precaution to insure its success, and concluded that bodies of the insurgents were silently watching the ends of the house; so summoning his servant he led the way down to a bath-room, which communicated with the lawn, and telling him to open the door quickly and follow close, the Rajah sprang forward with sword drawn and revolver cocked, but found the coast clear. Had there been twenty Chinese there he would have passed through them, as his quickness and practical skill in the use of weapons were unsurpassed. Reaching the banks of the stream above his house, he paused as he found it full of Chinese boats; but presently hearing his alarmed servant, who had lost him in the darkness, calling to him, he knew that the attention of the Chinese would be attracted, so diving under the bows of one of the barges he swam to the opposite shore unperceived, and, as he was then suffering from an attack of fever and ague, fell utterly exhausted, and lay for some time on the muddy bank, till, slightly recovering, he was enabled to reach the Government writer's house.

An amiable and promising young officer, Mr Nicholets, who had but just arrived from an out-station on a visit, and lodged in a cottage near, was startled by the sound of the attack, and rushing forth to reach his chief's house, was killed by the Chinese, his head severed from his body, and borne on a pike in triumph as that of the Rajah; while Mr Steel, the governor of Kanāwit on the Rejang, and an experienced officer, quietly looked through the Venetians, and seeing what was passing slipped out of the house, and soon found himself sheltered by the jungle; and the

Rajah's servant, whose shouts had in reality drawn the Chinese in his direction, had to display very unwonted activity before he could reach the protecting wood and join Mr Steel.

The other attacks took place simultaneously. Mr and Mrs Crookshank, rushing forth on hearing this midnight alarm, were cut down—the latter left for dead, the former seriously wounded. The constable's house was attacked, but he and his wife escaped, while their two children and an English lodger were killed by the insurgents.

Here occurred a scene which shows how barbarous were these Chinese. When the rebels burst into Mr Middleton's house he fled, and his wife following found herself in the bath-room, and by the shouts was soon convinced that her retreat was cut off. In the meantime the Chinese had seized her two children, and brought the eldest down into the bath-room to show the way his father had escaped. Mrs Middleton's only refuge was in a large water-jar; there she heard the poor little boy questioned, pleading for his life, and heard his shriek when the fatal sword was raised which severed his head from his body. The fiends kicked the little head with loud laughter from one to another. They then set fire to the house, and she distinctly heard her second child shrieking as they tossed him into the flames. Mrs Middleton remained in the jar till the falling embers forced her to leave. She then got into a neighbouring pond and thus escaped the eyes of the Chinese, who were frantically rushing about the burning house. Her escape was most extraordinary.

The stockades, however, were not surprised. The Chinese, waiting for the signal of attack on the houses, were at length perceived by the sentinel, and he immediately roused the treasurer, Mr Crymble, who resided in the stockade, which contained the arsenal and the prison. He endeavoured to make some preparation for defence, although he had but four Malays with him. He had scarcely time, however, to load a six-pounder field-piece, and get his own rifle ready, before the Chinese with loud shouts rushed to the assault. They were led by a man bearing in either hand a flaming torch. Mr Crymble waited until they were within forty yards; he then fired and killed the man who, by the lights he bore, made himself conspicuous, and before the crowd recovered from the confusion in which they were thrown by the fall of their leader, discharged among them the six-pounder loaded with grape, which made the assailants retire behind the neighbouring houses or hide in the outer ditches. But with four men little could be done; and some of the rebels having quietly crossed the inner ditch, commenced removing the planks which constituted the only defence. To add to the difficulty, they threw over into the inner court little iron tripods, with flaming torches attached, which rendered it as light as day, while they remained shrouded in darkness.

To increase the number of defenders, Mr Crymble released two Malay prisoners, one a madman who had killed his wife, the other a debtor. The latter quickly disappeared, while the former, regardless of the shot flying around, stood to the post assigned him, opposite a plank which the Chinese were trying to remove. He had orders to fire his carbine at the first person who appeared; and when, the plank giving way, a man attempted to force his body through, he pulled the trigger without lowering the muzzle of his carbine, and sent the ball through his own brains. Mr Crymble now found it useless to prolong the struggle, as one of his four men was killed, and another, a brave Malay corporal, was shot down at his side. The wounded man begged Mr Crymble to fly and leave him there, but asked him to shake hands with him first, and tell him whether he had not done his duty. The brave Irishman seized him by the arm and attempted to drag him up the stairs leading to the dwelling over the gate; but the Chinese had already gained the courtyard, and, pursuing them, drove their spears through the wounded man, and Mr Crymble was forced to let go his hold, and with a brave follower, Daud, swung himself down into the ditch below. Some of the rebels seeing their attempted escape, tried to stop Mr Crymble, and a man stabbed at him, but only glanced his thick frieze-coat, and received in return a cut across the face from the Irishman's cutlass, which was a remembrance to carry to the grave.

The other stockade, though it had but a corporal's watch of three Malays, did not surrender, but finding that every other place was in the hands of the Chinese, the brave defenders opened the gates, and, charging the crowd of rebels sword in hand, made their escape, though they were all severely wounded in the attempt.

The confusion which reigned throughout the rest of the town may be imagined, as, startled by the shouts and yells of the Chinese, the inhabitants rushed to the doors and windows, and beheld night turned into day by the bright flames which rose in three directions, where the Rajah's, Mr Crookshank's, and Mr Middleton's houses were all burning at the same time.

It was at first very naturally thought that the Chinese contemplated a general massacre of the Europeans, but messengers were soon despatched to them by the Kungsi to say that nothing was further from their intention than to interfere with those who were unconnected with the Government; which refinement of policy shows that the plot had been concocted by more subtle heads than those possessed by the gold-workers of Bau.

The Rajah had as soon as possible proceeded to the Datu Bandar's house, and being quickly joined by his English officers, endeavoured to organise a force to surprise the victorious Chinese—but it was impossible. No sooner did he collect a few men than their wives and children surrounded them, and refused to be left,—and being without proper arms or ammunition, it was but a panic-stricken mob; so he instantly took his determination with that decision which had been the foundation of his success, and giving up the idea of an immediate attack, advised the removal of the women and children to the left-hand bank of the river, where they would be safe from a land attack of the Chinese, who could make their way along the right-hand bank by a road at the back of the town.

This removal was accomplished by the morning, when the small party of English under tho Rajah walked over to the little river of Siol, which falls into the Santubong branch of the Sarāwak river. At the mouth of the Siol the Rajah found the war-boat of Abang Buyang with sixty men waiting for him, which was soon joined by six other smaller ones and some canoes· for no sooner did the Malays of the neighbouring villages hear where the Rajah was, than they began flocking to him. He now started for the Samarahan, intending to proceed to the Batang Lupar to organise an expedition from the well-supplied forts there. On their way they rested at the little village of Sabang, and to the honour of the Malay character I must add that during the height of his power and prosperity, never did he receive so much sympathy, tender attention, and delicate generosity as now when a defeated fugitive. They vied with each other as to who should supply him and his party with clothes and food, since they had lost all; and if to know that he was enshrined in the hearts of the people was any consolation to him in his misfortunes, he had ample proofs of it then. No wonder that, in reading these accounts, the 'Daily News' should say, "We have sincere pleasure in proclaiming our unreserved admiration of the manner in which he must have exercised his power to have produced such fruits."

When morning broke in Kuching there was a scene of the wildest confusion. The 600 rebels, joined by the vagabonds of the town, half stupefied with opium, were wandering about discharging their muskets in every direction loaded with ball-cartridge; but at eight o'clock the chiefs of the Gold Company sent a message to the Bishop of Labuan requesting him to come down and attend the wounded. He did so, and found thirty-two stretched out, the principal being from gunshot-wounds; but among them he noticed one with a gash across his face from the last blow Mr Crymble had struck at the rebels; and before his arrival they had buried five of their companions.

Poor Mrs Crookshank had lain on the ground all night desperately wounded, and with extraordinary coolness and courage had shammed death, while the rebels tore her rings from her fingers or cut at her head with their swords: then her life was saved by her mass of braided hair. Early in the morning a servant found her still living, and went and informed the Bishop of Labuan, who with great difficulty persuaded the Kungsi to allow him to send for her. She arrived at the mission-house in a dreadful state.

It was evident that in the intoxication of victory the Chinese aimed now, if not before, at the complete domination of the country, and summoned the Bishop of Labuan, Mr Helms, agent for the Borneo Company, and Mr Ruppell, an English merchant, and the Datu Bandar, to appear at the Court-house. The Europeans were obliged to attend the summons. The Malay chief came, but with great reluctance, and contrary to the advice of his more energetic brother; but he thought it expedient to gain time.

The Chinese chiefs, even in their most extravagant moments of exultation, were in great fear that on their return up the river the Malays might attack their crowded boats and destroy them, as on the water they felt their inferiority to their maritime enemies.

It must have been an offensive sight to the Europeans and the Malays to witness the arrangements of the Court-house on that day of disaster. In the Rajah's chair sat the chief of the Gold Company, supported on either side by the writers or secretaries, while the representatives of the now apparently subdued sections took their places on the side benches. The Chinese chief then issued his orders, which were that Mr Helms and Mr Ruppell should undertake to rule the foreign portion of the town, and that the Datu Bandar should manage the Malays, while the Gold Company, as supreme rulers, should superintend the whole and govern actively the up-country. During this time the Europeans could see the head of Mr Nicholets carried about on a pole to reassure the Chinese that the dreaded Rajah had really been killed. The Chinese chiefs knew better, but they thought to impose upon their ignorant followers.

Everything now appeared to be arranged, when the Bishop of Labuan suggested that perhaps Mr Johnson might not quite approve of the conduct of the Chinese in killing his uncle and his friends,—for most of them supposed the Rajah dead, and the head of Mr Nicholets was there as proof. At the mention of Mr Johnson's name there was a pause, a blankness came over their faces, and they looked at each other as they now remembered apparently for the first time that he, the Rajah's nephew, was the resolute and popular ruler of the Sakarangs, and could let loose at least 10,000 wild warriors upon them. At last it was suggested, after an animated discussion, that a letter should be sent to him requesting him to confine himself to his own government, and then they would not attempt to interfere with him.

They appeared also to have forgotten that there was Sadong under Mr Fox and Rejang under Mr Steel, who could between them bring thousands into line, and that Seribas also was panting for a field of exertion. All this appeared never to have occurred to them before undertaking their insensate expedition.

The Chinese were very anxious to have matters settled at Kuching, as with all their boasts they did not appear quite comfortable. They were not only anxious to secure the plunder they had obtained, but the leaders knew that the Rajah was not killed, and what he might be preparing was uncertain. They therefore called upon the European gentlemen and the Malay chiefs present to swear fidelity to the Gold Company; and under the fear of instant death they were obliged to go through the Chinese formula of taking oaths by killing fowls.

Next day the rebels retired up-country unmolested by the Malays, and a meeting was at once held at the Datu Bandar's house to discuss future proceedings. At first no one spoke; there was a gloom over the assembly, as the mass of the population was deserting the town, carrying off their women and children to the neighbouring district of Samarahan as a place of safety, when Abang Patah, son to the Datu Tumanggong, addressed the assembly. He was a sturdy man, with a pleasant cheerful countenance, and a warm friend to English rule, and his first words were: "Are we going to submit to be governed by Chinese chiefs, or are we to remain faithful to our Rajah? I am a man of few words, and I say I will never be governed by any but him, and to-night I commence war to the knife against his enemies."[1]

[1] Shortly after Sir James Brooke became ruler of Sarāwak, a case came before him in which Patah was concerned. He was then a very young man, and had contracted a gambling debt of £3; but by adding interest, his creditors quickly made it amount to £30. He could not pay this,

The unanimous determination of the assembly was to remain faithful to the Rajah, but they were divided as to the course to be pursued. Patah, however, cut the knot of the difficulty by manning a light canoe with a dozen Malays, and proceeding at once up the river, where he attacked and captured a Chinese boat, killing five of its defenders. In the meantime the women and children were all removed from the town, and some vessels were armed and manned, but imperfectly, as the Chinese had taken away the contents of the arsenal, and the principal portion of the crews of the war-boats were engaged in conveying the fugitives to Samarahan.

Patah's bold act was well-meaning, but decidedly premature, as the Malays, being scattered, could not organise a resistance, and urgent entreaties were made to the Rajah by injudicious people to return and head this movement. He complied, as he could not even appear to abandon those who were fighting so bravely for him; but he knew that it was useless; and arrived at Kuching to find the rest of the English flying, the town in the hands of the Chinese, and smoke rising in every direction from the burning Malay houses.

It appears that when the news reached the Chinese that the Malays were preparing to resist their rule, they determined to return immediately and attack them before their preparations could be completed. They divided their forces into two parties, as they were now recruited by several hundreds from the other gold-workings, and had forced all the agriculturists at Sungei Tingah to join them —in fact their great cargo-boats could not hold half their

and was in despair, as a noble to whom the debt had been transferred threatened to make his father pay it. Sir James having settled the case for him, won the young man's heart, who ever remained one of his most trusty and trusted followers. It was this Patah who gave the warning of the Patinggi's plot, *vide* p. 258.

numbers, so that one body marched by a new road which had been opened to the town, while the other came down by the river.

As soon as the Malays saw the Chinese boats rounding the point, they boldly dashed at them, forced them to the river-banks, drove out the crews, and triumphantly captured ten of the largest barges. The Chinese, better armed, kept up a hot fire from the rising ground, and killed several of the best Malays,—among others Abang Gapur, whose disbelief in his kinsman's story enabled the rebels to surprise the town, and who to his last breath bewailed his fatal mistake; and one who was equally to be regretted, our faithful old follower Kasim. The latter lingered long enough to see the Rajah again triumphant, and said he died happy in knowing it. Notwithstanding their losses, the Malays towed away the boats, fortunately laden with some of the most valuable booty, and secured them to a large trading *prahu* anchored in the centre of the river. Having thus captured some superior arms and ammunition, they could better reply to the fire of their enemies who lined the banks.

In the meantime the Rajah arrived opposite the Chinese quarter, and found a complete panic prevailing, and all those who had preceded him flying in every direction. Having vainly endeavoured to restore order, he drew up his boat on the opposite bank to cover the retreat, and after a sharp exchange of musketry-fire he returned to Samarahan to carry out his original intention.

He joined the fugitives, and his first care was to see to the safety of the ladies and non-combatants and wounded, and to send them off under the care of the Bishop and others to the secure and well-armed fort of Lingga. He now felt somewhat relieved, as he knew that there the fugitives would be in perfect safety, as they were surrounded by faithful and brave men who could have defended the fort against any enemies. There were no enemies in Lingga, except such as existed in the imaginations of the terror-stricken fugitives from Sarāwak, who had not yet recovered from their panic.

The Rajah prepared on the following day to take the same route, in order to obtain a base of operations and a secure spot where he could rally the people, and await a fresh supply of arms. It was sad, however, to think of the mischief which might happen during this period of enforced inaction, particularly as the Datu Bandar and a chosen band were still in Kuching on board a large trading *prahu,* which was surrounded by lighter war-boats. Here was our gentle Bandar, a man whom no one suspected of such energy, showing the courage of his father; Patinggi Ali, who was killed during Keppel's expeditions, and directing constant attacks on the Chinese whenever an opportunity offered. Thus harassed, the rebels were dragging up heavy guns, and it was evident that the Malays could not hold out for many days, particularly as there was now nothing to defend; the flames which reddened the horizon, and the increasing volumes of smoke, told the tale too well that the town was being completely destroyed.

With feelings of the most acute distress, the Rajah gave the order for departure, and the small flotilla fell down the river Samarahan, and arriving at its mouth, put out to sea eastward; when a cry arose among the men, "Smoke! smoke!—it is a steamer!" and sure enough there was a dark column rising in the air from a three-masted vessel. For a moment it was uncertain which course she was steering, but presently they distinguished her flag: it was the Sir James Brooke, the Borneo Company's steamer, standing right in for the Morotabas entrance of the Sarāwak river. The crew of the Rajah's *prahu* with shouts gave way, and the boat was urged along with all the power of their oars, to find the vessel anchored just within the mouth.

"The great God be praised," as the Rajah said. Here indeed was a base of operations. The native *prahus* were taken in tow, and the reinforcements of Dyaks, who were already arriving, followed up with eager speed. What were the feelings of the Chinese when they first saw the smoke, then the steamer, it is not necessary to conjecture. They fired one wild volley from every available gun and musket, but the balls fell harmlessly; and when the English guns opened on them, they fled panic-stricken, pursued by the rejoicing Malays and Dyaks.

Early that morning a large party of Chinese had crossed from the right to the left bank to burn the half of the Malay town which had previously escaped; but though they succeeded in destroying the greater portion, they signed their own death-warrant, as the Malays, now resuming the offensive, seized the boats in which they had crossed the river, and the Dyaks followed them up in the forest. Not one of that party could have escaped. Some wandered long in the jungle, and died of starvation; others were found hanging to the boughs of trees, preferring suicide to the lingering torments of hunger. All these bodies were afterwards discovered, as they were eagerly sought for. The natives said that on every one of them were found from five to twenty pounds sterling in cash, besides silver spoons or forks or other valuables, the plunder of the English houses.

The main body of the Chinese on the right bank retired in some order by the jungle road, and reached a detachment of their boats which they had left at its terminus, and from thence retired to Balidah, opposite Siniāwan, the fort famous in Sarāwak history, which the Rajah had besieged on his first arrival, and which after the insurrection became the headquarters of Mr Grant, a resident of Upper Sarāwak.

Thus was the capital recovered. The Rajah established

his headquarters on board the Sir James Brooke, and the Government soon began to work again. The land Dyaks, who had been faithful to a man, sent and requested permission to attack the enemy. This being accorded, the chiefs led their assembled tribes, and rushed in every direction on the Chinese, driving them from their villages, and compelling them to assemble and defend two spots only, Siniāwan and Bau, with Tundong, the landing-place of the latter town. The smoke rising in every direction showed them that the loss they had inflicted on others was now retaliated on them. The Gold Company, in their blind confidence, had made no preparations for an evil day, and it was well known that their stock of food was small, as everything had been destroyed except their own stores at the above-named places, and these were required to supply all those whom they had forced to join them from the town, and their whole agricultural population.

The harassing life they led must soon have worn them out without any attacks, as they could no longer pursue their ordinary occupations, or even fetch firewood or water, without a strong armed party, and as the Dyaks hung about their houses, and infested every spot. It soon became a question of food, and they found that they must either obtain it or retire across the frontier into Sambas. They therefore collected all their boats, and made a foray eight miles down the river to Ledah Tanah, and there threw up a stockade, in which they placed a garrison of 250 of their picked men, under two of their most trusted leaders. They put four guns into position to sweep the river, and, armed with the best of the Government muskets and rifles, they not only commanded the right and the left hand branches, but felt secure from a direct attack from the main river. Parties were sent out to plunder the Dyak farmhouses, and one bolder than the rest attempted to scale the mountain of Serambo to destroy the Rajah's country-house there; but the Dyaks barred the passage with stockades, and by rolling down rocks on the advancing party, effectually defended their hill. These Chinese were very different from those we see in our British settlements: many of them were half-breeds, having Dyak mothers, and were as active in the jungle as the Dyaks themselves.

To check the Chinese and afford assistance to the land Dyaks, the Rajah sent up the Datu Bandar with a small but select force to await his arrival below the Chinese stockade; but the gallant Bandar, on being joined by the Datu Tumanggong and Abang Buyong, and a few Sakarang Dyaks, dashed at the fort, surprised the garrison at dinner, and carried it without the loss of a man. The Chinese threw away their arms and fled into the jungle, to be pursued and slain by the Sakarang Dyaks. Stockade, guns, stores, and boats—all were captured; and, what was of equal importance, the principal instigators of the rebellion were killed.

As soon as the few that escaped reached Siniāwan, a panic seized the Chinese, and they fled to Bau, where they began hastily to make preparations to retire over the border. The Rajah, who was hurrying up to the support of the Bandar, hearing of his success, despatched Mr Johnson with his Dyaks to harass the enemy: these, together with the Sarāwak Malays, to whom most of the honour is due, pressed on the discomfited Chinese, who, fearing to have their retreat cut off, started for Sambas. They were attacked at every step, but being supplied with the best arms of the Government, they were enabled to beat off the advance-parties of their assailants, and retire in fair order along the good road that leads to Gumbang on the Sambas frontier. Still this road is very narrow, and every now and then the active Dyaks made a rush from the brushwood that borders the path, and spread confusion and dismay; but the Chinese had every motive to act a manly part, as they had to defend above a thousand of their women and children, who encumbered their disastrous flight.

At the foot of the steep hill of Gumbang they made a halt, for the usual path was found to be well stockaded, and a resolute body of Malays and Dyaks were there to dispute the way. It was a fearful position: behind them the pursuers were gathering in increasing strength, and unless they forced this passage within an hour, they must all die or surrender.

At last some one, it is said a Sambas Malay, suggested that there was another path further up the hill, which, though very steep, was practicable: this was undefended, and the fugitives made for it.

The Sarawak Malays and Dyaks, seeing too late their error in neglecting to fortify this path also, rushed along the brow of the hill and drove back the foremost Chinese. Their danger was extreme; but at that moment, as if by inspiration, all the young Chinese girls rushed to the front and encouraged the men to advance, which they again did; and cheered by the voices of those brave girls, who followed close, clapping their hands, and calling them by name to fight bravely, they won the brow of the hill, and cleared the path of their less numerous foes. Whilst this was going on, another column of Chinese surprised the village of Gumbang, burnt it to the ground, and then crossed the frontier. They were but just in time, as the pursuers were pressing hotly on the rear-guard, and the occasional volleys of musketry told them that the well-armed Malays were upon them; but they were now comparatively safe, as they all soon cleared the Sarāwak frontier, and although a few pursued them, the main body of the Malays and Dyaks would not enter Dutch territory, and halted on the summit of the Gumbang range.

The miserable fugitives, reduced to 2000, of whom above a half were women and children, sat down among the houses of the village of Sidin, and many of them, it is said, wept not only for the loss of friends and goods they had suffered from the insensate ambition of the Gold Company, but because they must give up all hope of ever returning to their old peaceful homes.

That Company, which on the night of the surprise had numbered 600 men, were now reduced to a band of about 100, but these kept well together, and being better armed than the others, formed the principal guard of the Tai-pe-Kong, a sacred stone which they had through all their disasters preserved inviolate.

Several times the assailants, who mistook it for the gold-chest, were on the point of capturing it, but on the cry being raised that the Tai-pe-Kong was in peril, the men gathered round and carried it securely through all danger. But here at Sidin, all immediate apprehension being over, the discontent of those who had been forced to join the rebels burst forth without control, so that from words they soon came to blows, and the small band of the Company's men was again reduced by thirty or forty by the anger of their countrymen.

Continuing their disorderly retreat, they were met by the officers of the Dutch Government, who very properly took from them all their plunder and arms, and being uncertain which was their own property, erred on the safe side by stripping them of everything.[1] Thus terminated the most absurd and causeless rebellion that ever occurred, which, during its continuance, displayed every phase of Chinese character,—arrogance, secrecy, combination, an utter incapability of looking to the consequences of events or actions, and a belief in their own power and

[1] The Dutch officers sent back to Sarāwak everything which they considered was public or private property.

courage which every event belied. The Chinese never have fought even decently, and yet, till the very moment of trial, they act as if they were invincible.

I think that this insurrection showed that though the Chinese always require watching, they are not in any way formidable as an enemy; and it also proved how firmly the Sarāwak Government was rooted in the hearts of the people, since in the darkest hour there was no whisper of wavering. Had the Chinese been five times as numerous, there were forces in the background which would have destroyed them all. Before the Chinese had fled across the border, thousands of Seribas and Sakarang Dyaks under Mr Johnson had arrived, and the people of Sadong were marching overland to attack them in rear, while the distant out-stations were mustering strong forces which arrived only to find all danger past.

I almost believe that it was worth the disaster to show how uniform justice and generous consideration are appreciated by the Malays and Dyaks, and how firmly they may become attached to a Government which, besides having their true interests at heart, encourages and requires all its officers to treat them as equals. The conduct of the Malay fortmen, of Kasim and Gapur, the generous enthusiasm of Abang Patah, and the gallant rush at the Ledah Tanah stockade by the Bandar and his forces, show what the Rajah had effected during his tenure of power. He had raised the character of the Malay, and turned a race notorious for its lawlessness into some of the best-conducted people in the world.

I may add that the results of the Chinese insurrection were very curious in a financial point of view. Though above 3500 men were killed or driven from the country, yet the revenue from the Chinese soon rose, instead of falling, which proves what an extensive system of smuggling had been carried on. The breaking up of the Gold

Company was felt by all the natives as a great relief. It is worthy of remark that while the Chinese were still unsubdued in the interior, boats full of their armed countrymen arrived from Sambas to inquire if Sarāwak was not now in the hands of the Kungsi, and were proceeding up the river to join them, when they were met by the Malays, driven back, and utterly defeated.

The Dutch authorities behaved with thorough neighbourly kindness on this occasion; for as soon as they heard of the rebellion of the Chinese, they sent round a steamer and a detachment of soldiers to the assistance of the authorities. Fortunately by that time all danger was past, but the kindness of the action was not the less appreciated. H.M.S. Spartan also, Captain Sir William Hoste, came over to Sarāwak, but I fear that his instructions were less generous: he could aid in protecting British interests, but not the Sarāwak Government.

Whilst struggling with all these difficulties, the Sir James Brooke, which had been sent to Singapore for supplies, now returned, bringing a large party to join the Rajah, —his nephew Captain Brooke and his wife, Mr Grant, Mr Hay, a new recruit of whom the Rajah says, "A gentlemanly man, young, of good family and the right stamp." I wish the Rajah had generally sought these qualifications in his recruits. There came also a lot of people connected with the Borneo Company, including Mr Harvey the managing director, Mr Duguid the manager in Sarāwak, and others. In giving me an account of the arrivals, the Rajah says,—"Our domestic intelligence is of the best and pleasantest. Brooke's wife is a sweet, sensible, but playful creature, . . . charming in manners;" and who that knew her would not re-echo these words?

When the news of the Chinese insurrection reached Seribas, all the chiefs were anxious to go to the succour of the Government; and while many were thus away in Sarāwak, our old adversary, Rentab of Lang Fort reputation, attacked the villages of our friends. The Rajah therefore determined to punish him, and started for Seribas himself to support the well-intentioned; and Captain Brooke visited the Rejang, while Mr Charles Johnson was ordered to attack Sadok, the chief's new stronghold, with his Malays and Dyaks; but as usual when he had but wild warriors to rely on, he failed, as they were not steady enough when serious fighting might be expected.

I went down to Sarāwak by the first opportunity, and reached it in July, to find everything proceeding apparently as if no insurrection had occurred. Though the Malay town had been burnt down, yet the inhabitants had soon recovered their energy, and had built their houses again, which, though not so substantial as the former ones, still looked very neat. Some things were missed in the landscape: the handsome Government-house with its magnificent library had disappeared; Mr Crookshank's and Mr Middleton's houses were gone, and with the exception of the Rajah, they were the principal sufferers, as the Chinese had had no time to destroy either the church or the mission-house, or the Borneo Company's premises; and although they all suffered losses from pilferers, yet they were trivial when placed in comparison with that noble library which was once the pride of Sarāwak. I may notice that some friends and others sent out a large number of books to replace it; but it appeared to me that every one had sent out a lot of unbound books he did not want, and in comparison to our pride they looked like a collection of rubbish.[1]

I have never seen a more perfect library than that of

[1] A "grace" was passed by the Senate of the University of Cambridge for sending copies of works printed by them, "as a testimonial of sympathy for your recent loss, and of admiration of your character."

Sir James Brooke—perfect in everything except the classics in the original, which would have been useless, as few of us had kept up our knowledge of them. Otherwise it was admirable: the best historians and essayists, all the poets, voyages and travels, books of reference, and a whole library of theology—books on every side of the question; and I well remember a sneaking parson from Singapore, who came on a visit, examining the library, and when he found the works of Priestley and Channing alongside of those of Horsley and Pye Smith, going away and privately denouncing "the Rajah as an infidel and an atheist, or, worse still, a Unitarian." Besides the loss of the library, the Rajah was at the same time deprived of all the records of his previous life,—for he had collected his journals and papers, and these shared the fate of his books.

I found, as I had expected, that the loss of worldly goods had had little effect on my old chief, who was as cheerful and contented in his little comfortless cottage as he had ever been in the Government-house. His health, which before was not strong, had been wonderfully improved by his great exertions to endeavour to restore the country to its former state, and I never saw him apparently more full of bodily energy and mental vigour than during the two months I spent in Sarāwak in 1857. Everybody took his tone from his leader. There were no useless regrets over losses, and it was amusing to hear the congratulations of the Malay chiefs: "Ah, Mr St John, you were born under a fortunate star to leave Sarāwak just before the evil days came upon us." Then they would laughingly recount the personal incidents which had occurred to themselves, and tell with great amusement the shifts to which they had been put for want of every household necessary. There was a cheerfulness and a hope in the future which promised well for the country.

I found that the deserted gardens around the town had

been in part reoccupied, for already there were Chinese here. In order not to interrupt the narrative, I have not before noticed that during the height of the insurrection, when the rebels had only been driven from the town a few days, news came that several hundred Chinese fugitives from the Dutch territories had crossed the borders towards the sources of the left-hand branch of the Sarāwak, and were seeking the protection of the Sarāwak Government. Though harassed by incessant work, the Rajah did not neglect their appeal, but immediately despatched trustworthy men, who safely piloted them through the excited Dyaks, who thought that every man who "wore a tail" should now be put to death. No incident could better illustrate the great influence possessed by the Rajah over the Dyaks and Malays, or his thoughtful care of the true interests of the country during even the most trying circumstances.

When the insurrection was completely over, the Rajah sent Sirib Moksain to Sambas with letters for the Dutch authorities. As the Sirib had been at one time in charge of the Chinese in the interior, he knew them well, and he said it was distressing to see the unfortunate agriculturists, who had by force been made to join the rebels, lamenting their expulsion from the country. They begged for permission to return; and subsequently many did, and established themselves in their old quarters.

I could not bring myself to visit the interior, and witness the ruin of so much prosperity.

CHAPTER XVI.

THIRD VISIT TO ENGLAND.

1857–1860.

A LITTLE later on the Rajah said, in a letter to his nephew, Captain Brooke—"I have sometimes thought that since the earlier days the bonds of sympathy between the native and European have been slacker." I quote these words, as they were but the reflex of my own thoughts during my visit in 1857. I did not as yet see any sign of want of sympathy, but there was want of intercourse. In the earlier days, every evening after dinner, the chiefs would assemble in the great hall, sit amongst us, and conversations were freely carried on as between equals. But when the ladies arrived that was all changed; after dinner the ladies retired into the drawing-room, where the gentlemen soon followed, or remained impatiently waiting for the natives to go. This they soon observed, and gradually they left off coming. No wonder the bonds of sympathy between the native and European became slacker. However, this is more for the future.

The Rajah, in the joy of his heart at the restoration of peace, determined to pardon two political offenders—Sirib Musahor, who had retired from Serikei, and the Datu Haji (the old Patinggi), who had been sent away to Mecca, but was then a pensioner in Malacca. Sirib Musahor had at first been injudiciously treated, but as he showed some desire to return to Serikei, he was permitted; and the Datu Haji was recalled apparently by the wish of the people, but I believe strongly against their real judgment, for they knew him to be a dangerous man, capable of revenge.

During my visit in 1857 I particularly noticed one thing in the Rajah, that though when in society full of mental vigour, yet when alone he showed a loss of buoyancy, a tone of melancholy in public matters, as if all ambition was dead within him. "I weary of business," he said to me. Just before I reached Sarāwak the news had arrived of the Indian Mutiny, and he was full of it. He turned "clammy with agitation when he first heard of it." How true is the ring of the following words! "I felt then, annoyed and disgraced though I have been, that I was an Englishman, and the ties and feelings which men have wantonly outraged are planted too deep to be torn up."

Mr Duguid, manager of the Borneo Company, having volunteered to give me a passage to Brunei in the Sir James Brooke, I was well pleased to return thus to my post, and the Rajah decided to accompany me. Before we started, however, we were invited to be present at the wedding of Mr Chambers, then missionary at Lingga, and now Bishop of Labuan.

On our way to Brunei we called in at Sadong to view the Borneo Company's coal-mines. The Rajah had a good look at them, but came back probably as wise as he went. This attempt to open a coal-mine in Borneo was a costly failure—rumour said £20,000—and all because, to save a few hundreds, the opinion had been taken of a practical miner who could have developed a real seam, instead of obtaining a report from a scientific engineer, who, by boring or other means, would have discovered if a workable seam existed. That is the way companies are too often managed.

We arrived in Brunei, and I lodged my guests in my mat-walled consulate, and the Rajah opened negotiations about the sago rivers.[1] Very little appears to have been settled, except that the Rajah was requested to see that right was done, whatever that might mean. He went to Muka, and thought that he had made an impression on the rival factions; but there was a blood-feud—the son wishing to slay the murderer of his father—and no hard or soft words could have any effect as long as Sarāwak supported that man of violence, Matusin.

The Rajah now decided to return home to England. He had many reasons publicly. He said he wanted to discuss matters with the Borneo Company; but in his heart he wished to enter into negotiations with the British Government, and see what they would do for Sarāwak. He wrote, when nearing Egypt, "I experience no pleasure when approaching England, nor pain either."

What an important year 1858 might have been for Sarāwak had the Rajah known how to secure his advance step by step! but he wanted to clear all obstacles at a bound, and failed.

In a letter which I wrote to my father, March 24, 1858, I find a sketch of the Rajah's character which is not without interest. "You say neither his friends nor his enemies have judged him right: perhaps so,—he has many injudicious friends. Take him for all in all, however, he is the man of the fewest faults that I have ever met with. Hasty judgment sometimes, and often hasty speech, are two faults which perhaps produce his third—that is, great impatience. In manner he is often absent and careless,

[1] In Sarāwak the term *river* is used instead of *district*, for the principal town of each district is generally on the banks of its main river, and it is by water and not by land that it is approached. The "sago rivers" comprise those of Oya, Mulla, Mato, Bruit, &c., and on their banks the sago-palm flourishes.

which, not being understood by strangers, offends them; but I will add what he is—he is a man of noble thoughts and noble actions, generous, generally most considerate; affectionate, and therefore beloved by all who are intimately acquainted with him. In conversation and argument brilliant when in happy spirits, playful when playfulness is required, earnest and sincere on all great subjects. I am not drawing his portrait, I am but touching on a few traits."

As soon as he reached England, the Rajah felt that the tone towards him was changed for the better. The Ministers received him cordially, and Lord Clarendon, thanks to Lord Grey's intervention, showed himself earnest in his desire to do something for Sarāwak. Both he and Lord Palmerston offered a protectorate: had the Rajah accepted at once, the affair might have been so advanced that no change of Ministry would have unsettled it. Unfortunately, however, the Rajah's views were now altered, and he began to wish to recover the private fortune he had expended in Sarāwak,—for the fact was, he was in reality a pauper: he had but his pension of £70 a-year. With very good reasons he urged that if England had a monetary interest in Sarāwak, it would be inclined to look better after it; but it is not the less to be regretted that he did not jump at the proposition of a protectorate, which would have so changed the status of Sarāwak as to have insured its prosperity.

Every kind of interest was shown. Mr Labouchere asked if Sarāwak would not take over the Indian mutineers; to which the Rajah replied,—"Hindoos? yes—but not Mohammedans;" and a small party was sent, who, I may say, behaved exceeding well. Later on came the question of a naval station in one of the ports of Sarāwak; but to this the Rajah would not listen—why, I do not understand, as it would not only have insured efficient protection, but would have been the first step towards that gradual absorption of the country which was so desirable.

On February 21st the Rajah went to a levee, and her Majesty spoke to him most graciously, asking after his health; and the Prince Consort shook him cordially by the hand. Indeed, the Royal family ever showed great interest in his career. The Ministers, he said, were remarkably friendly in their manner and tone. He was referring to Lords Palmerston and Clarendon, and he was particularly pleased with his reception at one of the evenings at the Prime Minister's.

Now came a change of Ministry, for Palmerston, being defeated on the Conspiracy Bill, retired to give place to Lord Derby; and the Rajah had to begin his work over again. How was it possible, however, for him to succeed in any negotiations with the British Ministers, when he could write of them in a few weeks as "base, truckling, and suspicious"? Evidently the Rajah was not meant for the work of European diplomacy.

On hearing of these things, I wrote to Captain Brooke: "The Rajah's life appears destined to be one eternal warfare; a Ministry turn and listen to him, and promise much, and would have performed sufficient, and they are ousted. Then came a cool Ministry, to whom Borneo is a bore; and the Rajah, irritated, writes to them in a very improper strain, 'holding them responsible for the lives and properties of British subjects in Sarāwak.' They shrug their shoulders and laugh, for they know they are not responsible, and would not care if they were."

Then came a tone of utter despondency, although, at the same time, he was vigorously backing his friends who were trying to agitate the country to support the Rajah, and thus influence the Ministry. Meetings, dinners, and speeches, however, did no good, as Lord Derby decided to have nothing to do with Sarāwak; but the Rajah notices,

"Lord Stanley has been more friendly than any other Minister." On a temperament so nervous as the Rajah's, this excitement was doing irreparable mischief. To add to his troubles, his nephews in Borneo were writing him letters on family and other matters which made him feel "dreadfully hurt and humiliated," and the cry from his heart could be understood—"Oh, I am weary, weary of heart! without faith, without hope in man's honesty;" and while working heart and soul for what he conceived to be the good of Sarāwak, he had to hear that his nephews were "horror-stricken at the idea of Sarāwak being sold into bondage." Thus were his nerves kept at the highest pitch of tension.

"On the 21st of October, after making a brief speech in the Free-Trade Hall at Manchester, I felt a creeping movement come over me. I soon knew what it was, and walked with Fairbairn to the doctor's. Life, I thought, was gone, and I rejoiced in the hope that my death would do for Sarāwak what my life had not been able to effect." Thus, sixteen days after the event, he described his first attack of paralysis, which effectually stopped his campaign. Then came the great deputation to Lord Derby, which, though powerfully backed, was a signal failure.

One of the wisest letters ever written to him, the Rajah received about this time from Mr Grant of Kilgraston, Captain Brooke's father-in-law, urging him to accept the protection offered by the British Government without haggling about terms.

I must now briefly glance at the events which passed in Borneo in 1858. In Sarāwak itself things were quiet enough; but in the sago rivers of the Sultan's dominions there were continual disturbances brought about by the desire of the Sarāwak Government to back up Pangeran Matusin, and by the constant interference of the Sarāwak *nakodahs* or traders in the internal affairs of those countries. As a rule, the Sarāwak *nakodahs* were very good men, but they were proud of their country, and wished to introduce the same system in these districts as that which prevailed at Sarāwak. So far, there was little to be said; but, like all merchants, they wanted a monopoly.

Captain Brooke, who was administering the government of Sarāwak, heard much of the evil doings in the sago rivers, and believing all the stories told by the *nakodahs*, proposed to the Rajah to seize these districts, and declare them independent of Brunei. Without reflecting, he fined people in the Sultan's dominions, and kept the fines. This high-handed policy greatly enraged the Sultan, and he begged me to remonstrate with the Rajah and Captain Brooke against this interference in his territory. This I did privately, and it is rather amusing to notice how these remonstrances were taken. The Rajah writes:—

"*Sept.* 7.—" St John seems to think that you are inclined to carry matters with too high a hand."

"*Nov.* 6.—I received a letter from St John, of which I did not quite like the tone and spirit; but I know he is a good fellow, and at heart our friend. I must add, that St John thinks that I, as well as yourself, carry matters with too high a hand towards Brunei. Do not notice it, because it is done in all kindness."

Before, however, Captain Brooke received this, he, being frankness itself, wrote to me to acknowledge that his policy towards Brunei had been too high-handed; but explained that he had been deceived by the highly-coloured reports of his officers in the Rejang. I have noticed these details, as they were the forerunners of troubles which brought about a great crisis in the history of Sarāwak.

In November 1858 there occurred an event which produced a complete change in the policy of the Court of Brunei. Makota, the Serpent, the Sultan's favourite minister, the Rajah's first enemy in Sarāwak and in his heart ever Sir James's enemy, was killed by the inhabitants of the interior whilst engaged in seizing a lot of young girls for his harem. The uproar that ensued is indescribable.

Makota, as I have often remarked, was one of the worst and most oppressive of the Borneon chiefs. Early in the month of November 1858 he started for a Bisayan village called Awang, a day's pull from the capital, and immediately commenced his exactions. He fined each of the chiefs for some imaginary breach of etiquette; and as the fines came in slowly, set his followers to beat the Bisayas, and went so far as to apply torture to several. Even this did not rouse the people. At length, as if he were courting his fate, he assembled the heads of the villages, and told them that they must each furnish him with one of their daughters for his harem, and at the same time directed his men to seize the girls. Still no resistance was made, and the young things were dragged off to his *prahu*.

When night came on, the chiefs met, and it was agreed to surprise the noble, and rescue their daughters. At midnight they came down quietly to the bank of the river, saw a light burning in the boat, and immediately attacked the sleeping crew. The surprise was so great that the men jumped into the water without any resistance, and Makota sought to escape with two followers in a small canoe; but one of the girls, seizing the light, held it aloft, and shouting to those on shore, drew attention to the flying noble, who was immediately knocked out of his boat by a well-directed stone, and he was drowned, being the only Malay I ever knew who could not swim.

On first receiving the news of the death of his favourite minister, the Sultan was furious, and swore he would exterminate the whole race of the Bisayas; but we used our influence to calm this anger, and but comparatively light punishment followed.

The year 1858 was marked by a great revival of Lanun and Balagñini piracy. Among others, a Spanish vessel was taken in the Sulu seas by Panglima Taupan of Tawi-Tawi: a young girl, the daughter of a Spanish merchant, was the only one on board not massacred. Taupan took her for a wife; and, as I wrote at the time—"Alas for the chivalry of the British navy! Sir ——, who was present when this information was given, said it was a Spanish affair, not ours." Another fruit of the Commission,—officers dared not act.

In December Captain Brooke lost his wife, and in her he lost his mainstay, for her calm judgment would have been of infinite service to him in the trials which were to come. Hearing at the same time of his uncle's attack of paralysis, he decided to return home,—a most ill-advised step.

The year 1859 was a most unfortunate one for everything connected with Sarāwak. When the Rajah lay on his bed of sickness, his friend, the present Sir Thomas Fairbairn, knowing the wretched state of the Rajah's finances, proposed to a few intimates that something should be done to relieve him from this state of distress: he thought that a Memorial Fund might be raised, and a sufficient sum collected to place the Rajah in a position independent of Sarāwak. Whether it would have succeeded or not had the project been carried out with equal zeal by all those who pretended to feel an interest in it, it is not possible to say, but it did not succeed, only about £9000 having been raised.

Rightly or wrongly, the Rajah always laid the blame of this failure on the shoulders of some members of the Borneo Company, whose interest he thought it was to keep him dependent on them. I notice how gradually he became more and more convinced of this, as I read his correspondence during the year 1859. It is not worth while, however, to quote these letters, as the whole thing has passed away, and the principal actors in this affair are now quiet in their graves. But it is necessary to notice it, as it partly accounts for the feeling of suspicion with which this Company was afterwards viewed, and which was heightened by the following circumstance.

The Chinese insurrection having completely disordered the Sarāwak finances, the Rajah's Government was compelled to borrow £5000 of the Borneo Company, and in the height of his distress they pressed him for repayment in a manner which made him write, "Their conduct is discourteous and avaricious."

Thus pressed by the Borneo Company to pay a debt due by the Sarāwak Government—harassed by the doubts arising as to the loyalty of some of his friends—with discouraging news from Sarāwak, and prostrate with sickness—the Rajah found himself in a most painful and distressing position. "A friend in need is a friend indeed," and at that moment one appeared who generously advanced the money owing to the Company, and thus enabled him to clear off the debt.

It was a mistake on the part of the Borneo Company thus to estrange the Rajah by trying to snatch the fruit before it was ripe. Had they shown that they were capable of governing the country, it is highly probable that, England failing him, he would have been glad to associate them with him in the government; but without the aid of the existing rulers of the country, I can imagine the failure that would have resulted.

The Rajah's correspondence during this year with her Majesty's Government was not pleasant, and ended apparently in complete estrangement. Fortunately public officers are not over-sensitive, and the unpleasantness was afterwards forgotten.

In 1859 the affairs of Muka were again brought prominently forward, as civil war had been renewed between Pangeran Matusin, the murderer of Ursat and his family, and Pangeran Dipa, the son of the victim. Mr Charles Johnson, then in charge of the government of Sarāwak, thinking that he must also pursue a high-handed policy, went there, and fined the Sultan's envoy, but had to carry off his *protégé* Matusin, whom the people would not have. Some glimmer of the truth as to the bad conduct of this Pangeran appears to have dawned on the mind of Mr Johnson, but he was so full of meaningless alarms about the intrigues of Borneon Rajahs, that his judgment was warped. He, too, kept the fine that was raised in the Sultan's territory, and refused to give it up, and went so far as also to propose to his uncle to seize the Sultan's territory, and appropriate it to Sarāwak.

When the Sultan heard of this insult of fining his envoy, he was very provoked, and sent for me to remonstrate on the subject, reminding me that the previous year the Rajah had written, "If we have wronged them [the Rajahs], or encroached on their territory, we are ready to make you, in your character as Consul-General and representative of England, the mediator between us."

Acting on this, I did remonstrate in a private letter, but to little effect, as the following extracts will show: "*March* 30.—St John's letter vain as usual." "*Aug.* 14.—St John's interference is unwarrantable, the result of vanity, the tool of Makota" (forgetting that Makota had been killed the previous year). But mixed with these I find: "St John's view is sensible;" "St John is a valued old friend and a true one." The letter which the Rajah calls "vain as usual," contains the following passage, to which he refers: it was so true that I cannot avoid inserting it. The Rajah had accused the Borneon nobles of being "weak, perfidious, and oppressive." I remarked in reply, "I will not enter again on the relations between Brunei and Sarāwak, but I must say, however 'weak, perfidious, and oppressive' the nobles may be towards their own subjects, they have been true to you at a time when their friendship was of some value. I have not forgotten, nor should you forget, their conduct at the time of the Commission, when there was every temptation to turn against you. I utterly deny that they have ever 'delayed, shuffled, or intrigued' in any matter in which you have been concerned; and all the accusations against them are founded on the *putana* (false reports) of the Sarāwak *nakodahs*, who delight as much in making mischief as the Pablats do here at the present moment. I do not think that you can bring a charge against them which I could not easily refute. Remember I am speaking of their conduct towards yourself and Sarāwak, not towards their own subjects." No wonder the Rajah did not like the tone of that letter; but it was quite true, for the loyalty showed by the Sultan and great Rajahs to Sir James Brooke was a marvel to us all. The effect of this correspondence, however, was good, as I find the Rajah writing to Captain Brooke: "Pray instruct Charley to pay to the Sultan all fines levied in his territory: it is important as a peace-offering;" and to me, that the fines should be paid over to the Sultan, which on the Rajah's return to Borneo was done.

While Mr Johnson was thus watching events abroad, he saw little of what was passing in Sarāwak. Up to that period he had been more accustomed to his Dyaks at Sakarang than he was to the Malays of Kuching; and the chiefs of Sarāwak, with that fear of consequences which is inherent in them, did not give the complete confidence which would have enabled him to unravel the plot which soon disturbed the country.

It will be remembered that after the Chinese insurrection the Rajah had pardoned Sirib Musahor and the Datu Haji. These two chiefs had never forgotten or forgiven the punishment they had received for their own bad conduct, and longed for the moment of revenge. Both Captain Brooke and the Rajah were absent—the latter reported dangerously ill—and the Government was in the hands of Mr C. Johnson. Now, the two chiefs thought, was the moment; and they formed a plot for cutting off all the Europeans. The well-intentioned gave warnings, but they were not sufficiently heeded.

The first blow struck was in the most distant station. In the Rejang were two officers of merit—one a Mr Steel, well versed in native ways and language (the Mr Steel whose conduct in 1849 had been so great a distress to Mr Gladstone); the other Mr Fox, a man of considerable abilities, but inclined to push reforms too fast, and very excitable, though he was at the same time one of the most amiable of beings.

In June, he and Mr Steel were at the Kanāwit fort, Mr Steel sitting in a chair in the Court-house, and Mr Fox superintending some alterations in their garden, when a party of Kanāwits,[1] accompanied by some of Sirib Musahor's followers, a mongrel breed, came in apparently on business, and, spreading about the fort, on a given signal fell on the two unsuspecting Englishmen and killed them on the spot. They did not, however, meddle with the fortmen.

When the news reached Sarāwak there was great excitement, but no suspicion of a plot, and the Datu Haji accompanied Mr Johnson to the Rejang to aid him. At Serikei they met Sirib Musahor, who had already taken measures to secure the fort, and who had put to death some of the

[1] The Kanāwits are a distinct tribe, and must not be confounded with the Seribas who live on the upper waters of the Kanāwit river. They are partially tattooed, and use the *sumpitan*, or poisoned arrow blow-tube.

men whom he pretended to suspect, without waiting the arrival of Mr Johnson. This created no suspicion; and the latter, listening to the advice of the two plotters, had the remaining fortmen tried on the charge of giving up their post, and they were found guilty and executed. At the same time a chief named Tani, formerly a great friend of the English, was accused, tried, and executed. On his way to death, he said, "I am innocent; but the guilty will soon be known."

The Kanāwits, after having murdered Messrs Fox and Steel, plundered the fort, and then retired to a village they had built in the woods. The Dyaks, who had asked permission to attack them, were easily defeated by the poisoned arrows of the enemy. Mr Johnson then moved up a strong force with a gun, and surrounded them. He summoned the tribe to give up the actual murderers: they refused. He then said, "If you will fight in defence of these murderers, at least let me place your women and children in safety;" but this they also refused. After giving them time for reflection, the place was attacked, burnt down, and the tribe dispersed with great slaughter, the murderers unfortunately cutting their way through the attacking force.

Mr Johnson's party suffered heavily. I heard that as many as thirty Dyaks were killed by poisoned arrows. The effect of this poison is singular: the wounded man feels drowsy; if he be allowed to give way to it, he invariably dies. One of the Malays who was wounded was kept awake, in spite of his requests to be permitted to sleep, and, after a glass of brandy, was walked down to the boats, and recovered.

Mr Johnson then returned to Kuching, and he soon became convinced that Sirib Musahor was the real instigator of the murder. He subsequently attacked him, and drove him out of Sarāwak territory, and the Sirib fled to Brunei. His flight brought to light the Datu Haji's guilt, and he was only banished from the country, though he fully deserved death; but his family connections being coextensive with Sarāwak Malay society, he escaped.

Mr Johnson throughout these transactions showed wonderful energy, activity, and courage; in fact, almost the whole burden rested on his shoulders.

The effect of the discovery of the plot on the Sarāwak officers was curious. The gentlemen, to a man, stuck to their posts with firmness, and rode out the storm—can we speak too highly of these officers?—the second class lost all courage; while the Bishop and some of the missionaries left, the former taking home news that it was a Mohammedan plot, with the Datu Imaum (the rival Mohammedan Bishop) at the head of it—whereas the Datu Imaum showed himself, as ever, the true and faithful friend of the English. The story apparently spread the panic to England, for I notice the Rajah writes to Brooke: "The day you want protection against your faithful subjects, that day the government of Sarāwak should end."[1]

Still confidence was in general shaken, and although the Governor of the Straits had sent over the steamer Hoogly and a detachment of marines, yet had not England authoritatively declared she wished to have nothing to do with Sarāwak? Hereupon the Rajah turned to Holland, and instructed Captain Brooke to open negotiations with the Dutch Minister in London. Brooke, however, little liked the task, as he felt himself unfit for

[1] Extract from the Bishop's Report for 1859: "I must again, through you, beg to call the attention of the Society to the insecurity and inconvenience of this place [Sarāwak] as our central missionary post. The Rajah's illness, the deficiency of revenue, the withdrawal of all British protection from English subjects here, render our position more unsafe than ever. I shall, I fear, be obliged to state that this is no longer the place where an establishment like ours, with women and children, can be safely and permanently kept up."

negotiations. The Rajah vainly endeavoured to make him an astute diplomatist, which, in dealing with Europeans, the Rajah himself was not, and urged him "to rise in tone to the dignity of the representative of a free people." I may at once say that Holland declined to enter into negotiations.

The Rajah was in great distress of mind as to the future of Sarāwak. He was incapacitated for hard work, both of his nephews were suffering, and he added, "Not a single rising man in the service—not a man fitted to rule."

Although no officer in Sarāwak could admire Bishop McDougall's management of the mission, yet they all were of accord as to his great medical skill, and his unfailing kindness in sickness. They therefore got up a testimonial to him to mark their sense of his conduct, and the Rajah wrote, "The Bishop deserves more than we can afford to give, for his kindness in sickness to each and all of us."

Although the Rajah in his writings often spoke disrespectfully of the climate of his mother-country, he could not make up his mind to live out of it; and now that a sum of money had been collected to present to him as a testimonial, he thought no better use could be made of a part of it than to buy a small property to which he could retire. He found in the wilds of Dartmoor a pretty place called Burrator with which he was enchanted, and having secured it, he felt for the moment cheerful and satisfied.

The least satisfactory event to my mind of this year was the Rajah's correspondence with his nephew Brooke about his resignation of authority. There is a want of precision in it which was the cause of all subsequent misunderstandings. One day he announced that his surrender of authority was final, the next all this was forgotten; but the fact was that the Rajah could not resign—he had entered into engagements with other parties which positively precluded him from handing over the government until these engagements were satisfied. Those not conversant with the Rajah could not understand the correspondence of this period. It ended in this question, "Do you acknowledge my authority or not?" The answer was, "Yes." So that all previous correspondence was annulled by this question and answer. One thing was, however, clear; the Rajah, even during the height of his despondency and illness, never intended to retire until the financial arrangements were definitively settled.

I notice that the Rajah was pleased with the following extract from one of my letters, and that he several times refers to it in his correspondence: "You say you shall never recover, but it is quite wrong to think so. Though you may abandon public work, you have many happy years to look forward to, if happiness can be found in the remembrance of unmixed good performed. I have lived with you many years, and know you well, and I am assured no man can have less to reproach himself with; and it must be a pleasing thought to know what kindly remembrances you have left in the minds of all who ever knew you out here. I assure you that when I heard the name of Tuan [Mr] Brooke mentioned with interest by the Dyaks of the far interior, in spots where even the Malays had never penetrated, I was struck with the vast influence you exercise over the native mind, and regretted the neglect which has prevented your being unable to turn that influence to still greater good."

The year 1860 opened dully. Holland had refused to enter into negotiations; England was indifferent; so the Rajah turned to France. Here, too, there was failure: no nation would touch what was refused by England, as it was known that England would be displeased by such interference. Uncle and nephew were also in disaccord, and although outward peace was established, confidence was for ever gone. Added to this, the Rajah was troubled by false news from Sarāwak, where in their panic they had mistaken a slave of Sirib Musahor's, who had appeared at Muka on some intrigue, for the Pangeran Tumanggong, the most powerful nobleman in Borneo, and son of the late Sultan. How such a blunder could have occurred is past all understanding.

In the spring Captain Brooke returned to Sarāwak, and I met him there on my way home. Hearing of the panic which existed, I went over from Singapore in a merchant-vessel, and could scarcely recognise Kuching, where once native and European were as one. I soon came to the conclusion that there was more panic than danger, and in that the native chiefs confirmed me. This I wrote to the Rajah, and I notice he says—"It delights me to hear St John's strong opinion of the love of the people to our Government." And there could be but one opinion as to the conduct of Mr Johnson in this crisis. He had acted as vigorously as it was possible for any man to act. It was clear that the conspiracy had been confined to those whose position had been rendered less important and lucrative by the advent of civilised men into the country, and that the mass of the chiefs and the people were as loyal as ever.

The confusion of relative position between the Rajah and Captain Brooke still continued. The Rajah talked of abdicating but still governing, and Brooke of not being able to afford to pay the Rajah's allowance, and of his presence being invaluable in order to restore confidence in Sarāwak.

In April, Bishop Wilberforce had an interview with Lord Palmerston—that thorough "*English* Minister," as he calls him—on the subject of Sarāwak; and this led to a renewal of intercourse between the Premier and the Rajah, which subsequently bore good fruit.

The Bishop's letter is perhaps worth inserting:—

"26 PALL MALL, *April* 4, 1860.

"MY DEAR RAJAH,—I have had an opportunity to-day at Windsor of talking with Lord Palmerston over Sarāwak affairs. He says there is a great difficulty, 1st, in the purchase or acceptance of your sovereign rights by our Crown, *because* they are not absolute, but held under the suzerainty of another—a position which our Crown could not occupy. 2d, That there was a difficulty in recognising by any formal act a subject as a sovereign. But if you wanted, as I understood, no such formal acknowledgment at present, but to be allowed to call for the aid of British ships of war to maintain your position if attacked, and if, waiving other questions, you applied for that, he plainly intimated that he would grant it, and recommended that such appeal should be made to him as the head of the Government. He had a due feeling about *you* personally, and with all his faults he is an *English* Minister; and I cannot, therefore, but hope that if, before you go further with the French, you will make this appeal to him, England will be spared the disgrace of such a transfer."

At this time that generous hand which had been stretched forth to save the Rajah from the Borneo Company now came again to the rescue, and the money was found to fulfil the dearest wish of his heart—the purchase of a steamer for Sarāwak. He soon found a suitable vessel in Glasgow, and christened her the Rainbow. Now he was happy; what a change of tone in all his letters! He "will nail his colours to the mast." In truth, no better or more welcome succour could have been found. As he said, "Sarāwak was saved."

Before I arrived in England, the Rajah addressed a letter to Lord Palmerston asking for protection for British subjects in Sarāwak.

Now came another trouble. I have mentioned that in 1859 he had cause to suspect an intimate friend's conduct towards him, and now he found that the same individual had been going about suggesting that he (the Rajah) was mad. Another had aided him in these calumnies, but as he signed a written retractation and apology, I will not refer further to him. When the Rajah first heard of his friend's ungenerous conduct, he wrote to him for an explanation, but got only an evasive reply. A long correspondence ensued, which Sir James Brooke submitted to his lawyer, who wrote back, saying "that anything more mean and shuffling he had never known," and "anything is better than a hollow pretence of friendship." Thus ended a long friendship; for although years afterwards there was a reconciliation, it was but nominal.

In August the Rajah went down to Glasgow to start the Rainbow. I accompanied him. Never was he in better spirits about Sarāwak: he might talk of abdicating, but that was indefinitely postponed.

Everything was now looking prosperous, when news arrived of a most unexpected event. The Governor of Labuan, whom I had charged with the Consulate-General during my absence, had suddenly interfered in the most "untoward" manner.

When Sirib Musahor fled from Mr Johnson he called in at Muka, and from thence went to Brunei and Labuan. Of course the Borneon Government knew with whom they were dealing, and the Sultan had given me information before I left the capital, which left no doubt on my mind that Musahor had instigated the murder of—had, in fact, by his paid agents, murdered—Messrs Fox and Steel; but Mr Edwardes, Governor of Labuan, believed nothing of this. An able, active man, was this governor, capable by his vigour to infuse life into a colony if a spark of vitality was left in it; but he had strong prejudices, and one of these was against Sir James Brooke and Sarāwak. When, therefore, he heard Sirib Musahor's plausible tale (and the Sirib was both plausible and taking in manner), Mr Edwardes, against his better judgment, decided that the man was innocent, and had been unjustly dealt with. Accordingly, when he met the Sultan, and heard his complaints respecting the encroachments of the Rajah's nephews,—for his Highness had neither forgotten nor forgiven Mr Johnson for fining his envoy—Mr Edwardes determined to interfere. He was delighted to have a chance of giving a blow to Sarāwak: and against the strong advice of his experienced officers, he started in the Indian steamer Victoria, to effect his object.

When Captain Brooke arrived in Sarāwak in April 1860, I had talked a good deal with him on the subject of the relations between Brunei and Sarāwak, and he had determined to abandon the high-handed policy, and try to live in peace with his neighbours. But the disorders on the frontiers had reached such a height that he found it impossible. Pangeran Dipa, the Governor of Muka, had taken part with Sirib Musahor, had collected the latter's dispersed followers, had fortified his district, fired on the Sarāwak flag, and driven its traders from the river. A Sarāwak Government envoy was treated in the same manner. Captain Brooke, uninfluenced by fiery counsels, went to Muka with a small force, decided to endeavour by conciliation to bring about such an arrangement as would permit the trade to continue. But no sooner did he attempt to open communications with the chiefs than his boats were fired into, and he was compelled to pause at the mouth of the river. He threw up a stockade there, and decided to wait for reinforcements. When these arrived he made a brisk attack on the forts, and his brave brother Charles, under a heavy fire, passed the defences in the gunboat Venus, with a crew of only twelve Europeans, and took up an advantageous position in rear of the stockades. At this moment Mr Edwardes arrived in the Victoria, and commanded Captain Brooke on his allegiance to cease the attack.

Captain Brooke remonstrated, but in vain, and found to his surprise, nay, horror, that Mr Edwardes had encouraged Sirib Musahor to come down to Muka. Captain Brooke need not have paid any attention to such a summons, and it is probable that had he refused to listen to it, Mr Edwardes would not have dared to interfere by violence. But Captain Brooke took the wise step of withdrawing his forces, and appealing for justice to the British Government. For this conciliatory and prudent step he received Lord Russell's thanks. I will not enlarge on Mr Edwardes's conduct, but his constant association with the murderers of his countrymen was very adversely commented on.

Captain Brooke retired within the frontiers of Sarāwak, and had a difficult game to play to hold his own under this renewed loss of prestige. Some of the chiefs became anxious, nay, frightened, at the idea of a conflict with the British Government.

But there was little danger of that. No sooner did the despatches reach home than Lord Russell saw where justice lay, and very soon decided as to the steps to be taken.

CHAPTER XVII.

LAST VISITS TO BORNEO—RETURN TO ENGLAND—QUIET LIFE—DEATH IN 1868.

1860–1868.

AT that time, October 1859, I was down at Burrator, staying with the Rajah, and suffering from fever and ague. A note from the Foreign Office recalled me to town, and, in spite of my illness, on hearing of what had occurred I could not do otherwise than volunteer to return to Borneo, and resume my duties as Consul-General. As I left Downing Street, I stepped into the old telegraphic office at Charing Cross and sent the following message to the Rajah: "Edwardes disapproved. Return to Borneo November mail. Will you come?" "Yes, certainly," was the reply.

We decided to sail by the next mail. I was confident of settling the affair without bloodshed. My instructions were elastic, and rather startled my friends, who thought the responsibility was thrown on my shoulders. I cared little for that, but as a matter of curiosity inquired at the office. I was told, "The meaning of your instructions is this: if you succeed, all right; if you don't, Lord John Russell will pull you through." I was satisfied, and required no more.

In the meantime, the same generous hand that had twice before come to the rescue, placed means at the

Rajah's disposal, and guns and powder, and shells and rockets, were sent out by the ton.

We started from Southampton the 20th November, and had a pleasant voyage. We enjoyed the Turkish bath in Egypt, and when we got into the Red Sea we settled down to chess. The Rajah was in great force. The return to Sarāwak excited his spirit, and the work to do made him brace up his nerves and conquer lassitude. Little of politics, but much of chess. We had fortunately pleasant passengers. One lady, a connection of the Rajah's, and a missionary (formerly captain in the Inniskilling Dragoons), joined our party, and a pleasant party it was. The missionary (I forget his name, but do not forget his look, manner, and appearance, which impressed both the Rajah and myself) we vainly tried to persuade to give up the idea of being a missionary among the Santhals and come to Borneo. There was a field worthy of his exertions.

On our arrival in Singapore we were eager for news, and were relieved to find that only loss of trade had followed Mr Edwardes's action. We found also that he himself had begun to doubt the wisdom of the course that he had pursued. I now started for Brunei in H.M.S. Nimrod, Commander Arthur, calling in at Sarāwak on my way, where I officially informed the Sarāwak consul that the Government did not approve of Mr Edwardes's doings. I went on to Labuan, relieved my substitute of his position as Consul-General, and established myself in the capital, to find the Sultan sulky at the failure of Mr Edwardes's promises. I remained quiet a few weeks, when I found his Highness gradually coming round; but it was long ere I was again established first adviser to the Crown, for Mr Edwardes's promises had either been great or had been misunderstood, and they thought that the British Government were about to remove the English from Sarāwak and return the country to them. I could promise the Sultan, however, that all the fines levied in his territory should be remitted to him, all arrears of revenue paid, and that some satisfactory arrangement would now be made to relieve him of all further trouble in the sago rivers.

And this was really all the Sultan cared about. The revenue hitherto raised in the sago rivers was quietly absorbed by the agents, and little ever reached the Borneon Government. They therefore listened eagerly to the idea that this uncertain item should be exchanged for a fixed payment.

The Rajah went over to Sarāwak in his own steamer, and was received as one risen from the dead, for all had heard of his serious illness, and many believed that the news of his death had been concealed from them. This idea also was very prevalent in Brunei.

In April the Rajah came to the capital, and, as I expected, all the clouds of doubt and distrust vanished. The Sultan and his ministers received him most cordially, and agreed to all he required, which was to disavow all complicity with Sirib Musahor (whom we found that the Sultan had intended to arrest, but he fled in time), and to order a general disarmament at Muka. As I had been directed by our Government to do my best to see the affair settled without bloodshed, I undertook the part of mediator, and decided to go down to Singapore and obtain a ship of war in which to visit Muka, and convey the commands of the Sultan to the chiefs.

The Charybdis, a magnificent 21-gun corvette, was there with the Hon. Captain Keane as senior officer. He soon settled to take me over, and in a few days more we were at the mouth of the Muka river. I had given notice of my coming; but as precautions were necessary with such a desperate set, Captain Keane manned and armed all his boats, and we pulled in. When we saw the fort crowded with men, we half expected to receive a broadside from their guns; but no opposition was made as we passed the boom, and 200 Englishmen were soon in a position to command all their defences. Pangeran Dipa received me well, and I produced the *chops*[1] from the Sultan advising a cessation of hostilities, and that Sirib Musahor and his men were to leave the country. While these were being read Sirib Musahor himself came in, naked to the waist, which looked anything but peaceful, with his *kris* ready for action. He came and sat down by my side, and I had leisure to look at my old chess opponent. He was a fine man, with a heavy, though in general a smiling, sympathetic expression; but now it was all gloom. But after he had heard the Sultan's orders read, he burst into protestations of his innocence. I quieted him as well as I could. We had been old and familiar friends in days gone by, and then I told them that her Majesty's Government had directed me to return to Borneo in order to explain to them that they did not approve of Mr Edwardes's interference; but they wished peace and order to be established, and all questions to be settled without bloodshed. I pointed out to them how useless it was to endeavour to cope with Sir James Brooke's forces, and that the best thing which could happen would be to receive him peacefully, and enter into a final settlement.

Both Dipa and Musahor promised to obey the Sultan's orders, and they faithfully kept their word. I pitied Dipa, as he was not a bad man, and had been pushed into hostilities by the action of Sarāwak in supporting Matusin, the murderer of his father. On returning to Sarāwak, I informed the Rajah of the result; but to make sure, he went with a strong force to Muka. Some silly story was told him about resistance if his steamer could not cross the bar, but that was all nonsense. Dipa and the people

[1] Literally "seal," but really meaning credentials, or orders under seal.

of Muka were weary of fighting; and when I pledged my word to Musahor that his life and property should be respected, he made up his mind to go away. The appearance, too, of such a vessel as the Charybdis, and Captain Keane's judicious conduct in taking an overpowering force up the river to the middle of the town, showed them that Mr Edwardes's support was no longer to be relied on.

As soon, therefore, as the Rajah reached Muka, everything was at his feet—the fort was disarmed, and handed over to him; and though fierce and severe in words when opposition threatened, he was disarmed by submission, and treated all with as much kindness as possible.

Faithful to the promise I had made, the Rajah permitted Sirib Musahor to leave the country with his property and those members of his family who elected to follow him; and as the Sultan had positively forbidden him to go north, his *prahu* was towed to Sarāwak, and he and his belongings were shipped to Singapore. That he was guilty of plotting against the rule of the English in Sarāwak, and that he actively incited the Kanāwits to murder Messrs Fox and Steel, I have no doubt—in fact, the chief murderers, Sawing and Talib, afterwards confessed that they acted by the direct orders of Sirib Musahor; and the Dutch authorities had already sent from Pontianak, *en bon voisin,* information which distinctly implicated the Sirib. Yet I could not help pitying the man. He had been such a good fellow when we first knew him; and he had been so severely treated without a sufficient allowance being made for his education—for his having been brought up in the idea that his will was law. Then he had been very unfortunate in the English with whom he had been brought in contact. Mr Steel, though able and admirable in his knowledge of native languages and customs, was an ignorant, hard man: he had been an apprentice in a merchant-ship. Mr Fox was efficient; but having been accustomed to deal with Chinese, was brusque in manner with natives of rank; and the others were an ex-valet and ex-lawyer's clerk, one who would have done credit to the Club of "the Glorious Apollos," so pleasantly described in the 'Old Curiosity Shop' by Dickens. I believe the Sirib still lives somewhere in the Straits Settlements.

Pangeran Dipa was sent away to explain his conduct to the Sultan, and the Rajah took up his residence in the fort to endeavour to settle matters, for constant civil war had introduced the most extraordinary confusion. Here the Rajah was in his element: from morning to night he was there listening to the complaints and wishes of the inhabitants, reconciling enmities, settling quarrels, undoing injustice, and aiding the poor. The Sarāwak traders now flocked to the open fort, and brought comfort in their train, as the population, half starved, could now obtain supplies in return for their abundant sago.

The Rajah expressed an opinion at this time that the Sultan and his ministers were cognisant of the plot to murder the white men in Sarāwak; but this is only a repetition of his nephew Charles Brooke's unsupported suspicions. He had an idea that Brunei intrigue was omnipotent, and saw its hand in every unexpected event that occurred in Sarāwak, which should rather have been placed to the account of the sadly imperfect government of so extended a territory with so few competent officers. When Sawing confessed that he had been ordered to murder Fox and Steele, he added that Sirib Musahor had declared it was *titah*, or by the order of the Sultan. On hearing this, Mr Johnson felt convinced that the Brunei Government were responsible, forgetting that nothing would be more natural than for Sirib Musahor to say this, however false it might be. Mumein, the Sultan, was perfectly incapable of such conduct. When the Bisayas of Awang killed the Sultan's favourite, Makota, they too shouted

Titah! titah! but no sensible person believed that the Sultan had ordered these men to kill him.

Whilst this was going on, I was living quietly in Sarāwak. We had a party of anxious ladies—Mrs Brooke (for Captain Brooke had just married his second wife), and Mrs Welstead, her sister-in-law; but we soon heard that there was to be no fighting, and calm was restored.

Captain Brooke arrived with the bulk of the Sarāwak forces, and then we set to work to consider the plans for the future. The Rajah soon joined us, and our party was all again united, but the old confidence did not exist. One day in July, whilst reposing in my room, I heard my name called, and, getting up, I opened the door, and found Captain Keane with another gentleman: "Let me introduce you to the Governor of Labuan, and your successor." "Then what am I?" "Oh, something in the West Indies." And on opening my letters, I found that I had been promoted to be *Chargé d'Affaires* to the Republic of Hayti. I was delighted, for I began to weary of my solitary life in the capital.

I accompanied Mr Callaghan, the new Governor, to Labuan in the Charybdis, in order to hand over to him the archives of the office, and introduce him to the Sultan; and a few days after the Rainbow came into Brunei, bringing the Rajah and Mr and Mrs Crookshank. The Sultan was pleased at the Muka settlement, and though sorrowful to see his districts vanishing from him, was very well satisfied with the Rajah's liberal offer to pay him an annual sum instead of his previous precarious revenue. He and his ministers soon drew up the documents which handed over the coast as far as Kidurong Point to the Rajah and his successors, for the annual tribute of a little over £1200.

The Rajah was delighted with my spacious consulate, with the cool rooms and breezy situation, and said if he were rich he would purchase it of the British Government for himself. We soon left, and I bade adieu for ever to Brunei. I could not leave it, however, without much regret. In spite of their faults, I liked the Brunei Rajahs, I liked the life, and I liked the people. I particularly liked those long exploring excursions which I had made into the interior.[1] But I felt that I was doing nothing, and I was pleased at my advancement, for which I had to thank Earl Russell.

We called in at Labuan, and then stopped at Sarāwak, as the Rajah, having decided to return to England with me, wished first to pass a short time in settling the future of the country. How wisely he settled every question, eighteen years of peace have shown; but he scarcely believed in the great stability of the edifice he had erected.

Whilst we were waiting for the return of the Rainbow from Singapore, Captain Brooke told me that he wished the Rajah publicly to install him as Rajah Muda, or heir-apparent to the Rajah. I suggested to Brooke to write a note to the Rajah, and I would deliver it. I knew the Rajah would do it without hesitation, but why or wherefore I took the precaution to have the wish expressed in writing I do not know. The terms, however, settle the status of the two relatives beyond dispute:—

"*September* 16, 1861.

"MY DEAR RAJAH,—I shall be very much gratified if you will publicly install me as Rajah Muda before you quit the country. If you will do so, it will not only be a pleasing sign of your confidence in me, but will strengthen my hands in carrying on the government.—Yours, &c.,

"J. BROOKE BROOKE."

After this letter there can be no question of their relative positions. All Captain Brooke asked was, that he should

[1] *Vide* 'Life in the Forests of the Far East.'

be publicly acknowledged as the Rajah's heir. I always told Brooke that as long as the Rajah lived he would never abdicate: he might talk loosely about it, but he would never in reality hand over his power and authority to another. And why should he have done so?

All the Europeans and the native chiefs were summoned to a public meeting. I was late, and did not enter the Court-house, but from the outside I heard the speeches, and the Rajah addressed the assembly in a manner which I can never forget. He spoke of the past, the present, and the future: introduced Brooke as the Rajah Muda, on whom would now fall the burden of government, as he was becoming old, and could no longer bear its fatigues; but he would return among them if necessary, and whatever danger threatened they might count on him. He then addressed his farewell to the audience in such feeling terms that all of us were affected. It was a splendid speech, in that choice Malay of which he was a master.

The Rajah determined before he left to come to a settlement with his nephew about future negotiations with foreign Powers and with England, and these terms were reduced to writing by myself and accepted in writing by Brooke. I had seen enough of the misapprehensions which had already arisen from uncertain talk and random notes, so I had every proposition reduced to writing, as I was sure to be mixed up with the negotiations in England.

In order to have a quiet and long talk together, Brooke and I started for a short trip up country, when we settled clearly what was to be my line of conduct in these negotiations, and what I was authorised on his part to support.

Everything now being settled, and the Rainbow having arrived, the Rajah started with a large party for Singapore.

We were detained there for several weeks on account of the non-arrival of a mail, and the inhabitants showed Sir James Brooke that all the divisions of the past were no more, and that he was now truly appreciated as the great pioneer and champion of civilisation and commerce in the Indian Archipelago.

What a pleasant voyage we had home! The Rajah was thoroughly happy. He had succeeded in all his plans beyond his hopes; nŏt a drop of blood had been shed. He had restored the prestige of Sarāwak, and shown that English agents worked in unison with him. We now again became devoted to chess; and one day he proposed a match, in which he and I were to play a consultation game against two good players on board. They were getting the best of it, when the Rajah said, "Let us dash at them with this move." I easily showed that though bold it was too risky, and quickly pushing a pawn said, "That is our move." At first he did not see the effect of this apparently innocent proceeding, which looked like an effort to gain time; but it soon flashed on him, and he whispered, "The best move you ever made." Our adversaries treated it, as we thought they would, as a sign of uncertainty, and it was not till the next move that they saw their danger. It was fun to hear the Rajah's gay laugh as he pushed our advantage and triumphantly won the game.

We had a very quarrelsome set on board, and in their troubles all came to him for advice, and many must remember with satisfaction the way in which he prevented anything serious arising from these misunderstandings. He was in private life the most conciliatory man I ever met, which would make me surprised at the tone of his recent public correspondence, if one did not remember how deep and indelible a wound the Commission had inflicted on him. As I have said, he never did get over the mortification, wounded pride, and the sense of deadly injury which this wanton act inflicted. But Muka was the last kick of the dying beast.

The Rajah spent the year 1862 in endeavours to bring

Sarāwak under the attention of Ministers, and fortunately he left all communications in the hands of his friends. Among the communications which he received this year was one from Philarète Chasles. The following is an extract: "Votre grande entreprise de civilisation orientale, si puissament et si prudemment soutenue, me touche le cœur et élève mon esprit; elle me plait et me ravis comme une des belles choses de notre temps."

Belgium, in the meanwhile, appeared inclined to treat for Sarāwak, but as nothing came of these negotiations, I need not enter into them. Belgian statesmen evidently neither understood how to acquire a colony or to develop it had they acquired one.

The Rajah thoroughly enjoyed this year. He passed it in alternate visits to his relations and friends, including a trip to Paris, and varied it by long stays in Burrator, where he was endeavouring to bring up two young cubs for the Sarāwak service. But, as usual, these cubs remained cubs to the end, and were a source of trouble and mortification until they disappeared from the scene. Strange infatuation to believe that he could do anything with such materials, when gentlemen cadets were to be had by the score.

Whilst the Rajah was making an unusually long stay in Burrator, I thought he was moping, so I tried to induce him to come up to London and enter general society, but partially failed, as the following playful note shows:—

"BURRATOR, 21*st June* 1862.

"MY DEAR NESTOR,—Why should not I enjoy the dregs which life has left in my cup? Why should that sentimental and sympathetic harpy Society devote me to a routine which I dislike, under a delusive expectation of doing good to Sarāwak? Have I not sacrificed taste, feeling, ease, and independence in a vain pursuit after a substantial good to come out of this shadow?

'Now my weary eyes I close,
Leave, oh leave me to repose.'

Pleasant are country sights and sounds, spite of rainy weather. I love retirement, I love mine easy-chair, I love my bed at half-past ten at night. Now, how can Society make amends for this loss? Let her only promise, and I will devote myself to the yawning chasm of finery and false pretence, as Marcus Curtius did of old to the gaping earth,—only Sarāwak would be none the better, as Rome was supposed to have been. It is a delusion and a snare prompted by the substance of thirty to the skeleton of sixty according to ordinary chronology; but by my reckoning six hundred and sixty years have I been upon this earth. So it is, though I do not know exactly how. You must believe, but not inquire. Now, my dear boy, go to church twice every Sunday because you owe it to society; wear light duck trousers this cold weather because society obliges; sit up late at night, eat and drink too much, listen to twaddle and praise it. Set up a tabernacle on the mount of fashion, bow down to it and worship it. Duty to society commands it. But do ask yourself in your sober moments what society has ever done for us that we should do so much for society. Now, having said all this, I beg to tell you I am coming up in July, just to touch the hem of society's garment."

The news from Sarāwak continued good. Sadok was taken easily, through the defection of Rentab's followers, and this closed the series of mismanaged expeditions. The Rajah's nephew Charles returned to England, and was a comfort to him.

In May 1862 I published two volumes of travels in Borneo, entitled, 'Life in the Forests of the Far East,' which were well received. In them I had the chapter on missions, which, in consequence of a conversation with Captain Brooke, who had strong views on the subject, I had agreed to write. I thought I was stating the opinion of all,[1] particularly of Mr Chambers, the present Bishop, with whom I had had long conversations on the subject of the non-success of the mission; but my chapter raised a storm in a teapot, and I was attacked both publicly and privately, particularly by my good friend Mr Chambers,—but this I could readily forgive, as he was now connected by marriage with the Bishop of Labuan. The assailants were not satisfied with attacking my public statements, but one went so far as to send private letters to my friends attacking my personal character.

The Rajah was very angry,—much more angry than I was, for I knew with whom I had to deal—and his letters are full of the subject: "The slanders propagated make my blood boil, and I shall authorise Brooke and Crookshank to express my opinion publicly." "The letter is weak and wicked." "I cannot stand by to see St John driven to the wall for saying what is strictly true."

I was not much moved by all this excitement, and intended to have been contented with adding a note to the chapter on missions in my second edition; but the Rajah was not so easily satisfied, and used his literary talent and energy in writing a reply, which was called a vindication, and to which I put my name. Competent authorities considered it "crushing;" at all events it put an end to the discussion. One satisfactory result arose from my chapter: the management of the mission was completely changed, and I heard that most of my recommendations were put into practice; and when, later on, Bishop McDougall

[1] In pointing out the defective management of the mission I was only conveying my own ideas, and, as I believed, those of the Rajah and Captain Brooke, though neither knew what I had written until published.

retired from Sarāwak, the series of reforms may be said to have been completely carried out.

While these things were fermenting, news arrived of a successful action having taken place under the personal direction of Captain Brooke between the Rainbow and some Lanun pirates, and the Bishop of Labuan sent a very unfortunate account of the transaction to the 'Times,' which greatly excited and scandalised the religious world, and caused many erroneous impressions to get abroad.

In the autumn I went down to stay with the Rajah in Burrator, and, while there, received a letter from Captain Brooke, in which he complained of discouraging prophecies sent him by his uncle. I answered, September 22, 1862: "I should not much trouble myself about the Rajah's gloomy views; they have been gloomy for the last five years, and none of his prophecies have come to pass; and after all, they will not in any way affect the country. I myself cannot conceive any one desiring to change the status of Sarāwak—it is a unique thing in the world. I hold that as Sarāwak has gone on for twenty years of times full of trouble, it has a good chance of going on another twenty, with increased prospects of success. Don't be depressed by any one, and I am sure there is a bright future for Sarāwak."

My prophecy appears to have been a true one.

During this time the negotiations with the British Government were making progress, though very slowly. Mr John Abel Smith, afterwards M.P. for Chichester, had charge of them, and he opened the campaign by going down to Earl Russell's at Richmond, where the two questions of protection or cession were discussed. It was evident that Earl Russell's opinions were for protection; but the difficulty was to interest the Ministers, and this Mr Smith undertook to do. At length even the question of protection was dropped, and it all centred in the recognition of Sarāwak as an independent State. During these negotiations I was asked to furnish documents on the three questions, and sent in a memorandum on each. But the cautious Government wanted to know more, and at length referred the whole question to Lord Elgin, Governor-General of India, and he instructed the Governor of Singapore, Colonel Cavanagh, to make a private visit to Sarāwak.

I left England just as the Rajah received some intelligence from Borneo which induced him to return there. I had been very ill, and rose from a bed of fever to embark for the West Indies; but my leave was up, and go I must. On the point of starting I received the following affectionate notes, which show the Rajah's tender nature: "*Dec.* 30, 1862.—I was greatly relieved by your letter of yesterday, and care will restore you,—care of your health in details, and nothing else. God speed you on your way! You will understand how I regret parting from you, but duty must be done." "*Dec.* 31.—God bless you, and farewell! remember me as your true friend, as I shall remember you."

Captain Brooke, who had heard from Lord Elgin on the subject of Governor Cavanagh's visit, received his guest well, and offered him every aid in his power to further his inquiries: in return, the Colonel thought he could not do less than show some of his papers, among others one of my reports which was marked "Secret and confidential." It would appear that this had an extraordinary effect: perhaps it was only the last hair that broke down an exhausted patience. Captain Brooke thought that his rights were being overlooked, and in a moment of excitement sent a defiant letter to the Rajah, and protests to her Majesty's Government against a transfer of the country without his consent and that of the council. At that time Captain Brooke was scarcely himself: he had lately lost his second wife, and previously his eldest boy, and was impatient under these repeated losses, and perhaps under the Governor's inquisitiveness. He did not consult any of his officers, and it was thought that he made confidants of those whom, being the Rajah's enemies, he should have distrusted: but a diligent examination of all the papers that have been intrusted to me, make me doubt whether he consulted any one. I think that the letter was scarcely written and sent ere it was regretted, as but one subsequent communication refers to the subject, and the others are as calm or calmer than any previous ones. I may notice that the Rajah never saw the report referred to, as I made it a practice not to show anything official to him, for we differed greatly in our views, and he would have asked me, perhaps, to alter or to modify them.

The Rajah did not receive this defiant letter quietly—in fact he could not. He found his negotiations with the British Government rudely interrupted, and he at once determined to go back to Borneo and resume his position as Rajah, confident, I believe, that defiance would vanish at his approach. As I have said, Captain Brooke no sooner had written the letter than he regretted it, and to avoid every appearance of divided counsels, went over to meet his uncle in Singapore, and affairs were speedily arranged. The Rajah insisted on submission, and Captain Brooke took leave of absence and returned to England on an allowance, while the Rajah continued his course to Sarāwak in a British ship of war. What a change! And he could not help writing, "The Ministers were very kind on the recent occasion." There he found every one surprised at his return, as few, if any, had suspected the true cause. I may here remark that at that time the only real question before the Government was recognition, as both cession and protection had been put on one side, and Lord Russell wished only for an independent opinion as to whether there was an established Government in Sarāwak acknowledged and obeyed by the people.

The Rajah found the country prospering—indeed, Captain Brooke could manage admirably everything pertaining to the details of the home administration, and there was nothing to change in the system. As I anticipated, not a single person had a suspicion of what had occurred; and the native chiefs when informed begged that nothing public should be said, as it would be calculated to do harm.

The chiefs must have rather wondered at this time to find seven English ships of war anchored in the Sarāwak river, ostensibly to look after pirates; but I do not doubt that there was another motive, and that they were sent to strengthen the Rajah's hands, and to show all that the days of estrangement were past, and that the English Government were again friendly to Sarāwak. In fact, since this great demonstration every shadow of suspicion that the Rajah was abandoned by England vanished from the native mind, and no more has been heard of conspiracies among any class of natives.

In England affairs were progressing favourably, and Lord Russell sent a message to Sir James Brooke to the effect that, should his authority be undisputed, he was now ready to propose to the Cabinet that Sarāwak should be recognised as an independent State, under his rule and government.

On his arrival in England, Captain Brooke fell into injudicious hands, and openly attacked his uncle. The dispute ended in the Rajah's disinheriting his nephew: and there I might take leave of the subject, but in fairness I should add that I was, and ever have been, a warm partisan of the Rajah, and strongly supported and approved the measures he took; but I cannot but confess that, after reading all the correspondence which passed between Captain Brooke and his uncle, I am not surprised at the former showing considerable irritation, though in this instance he acted under a very mistaken view. Both the Rajah and Brooke were my intimate friends; perhaps I was most intimate with the younger man, who was of my own age, but I was devoted to the interests of both. Had mutual friends both in England and Sarāwak been more conciliatory, the estrangement between uncle and nephew would never have gone so far as it did.

Knowing that his health would only permit a short residence in Sarāwak, the Rajah did his utmost to give an impetus to the country. The only enemies that were then troubling its peace were the powerful Kayan tribes who dwelt in the far interior. Their warriors would come down the great river Rejang and attack the outlying Sarāwak villages: and as they had recently destroyed several, and murdered their inhabitants, the Rajah determined to put a stop to it. He therefore directed his nephew, Mr Charles Johnson (now Mr Charles Brooke), to organise an expedition, and this he did on a great scale. He displayed wonderful energy in conquering the obstacles presented by a mighty river, dashing over innumerable ledges of rocks, and foaming rapids and cataracts of a formidable character, in spite of which he carried his force of 15,000 men into the far interior, where, however, the enemy fled at his approach. This expedition effectually cowed the Kayans. They were a cruel set, as the following account of their conduct, taken from the report of the Resident of Kanāwit, will show :—

"The Kayans killed all the Dyak captives, to the number of seventeen. After torturing them in the most fearful manner, cutting them about and taking out their eyes, they then cut their throats. There were some women tortured and killed."

After the expedition, the Rajah, seeing that all was quiet and peaceful, and being convinced that he could rely on all his officers, whether native or European, decided to return to Europe, leaving his nephew, Charles Brooke, in charge of the government.

In one of my letters I slightly reproached him with having the appearance of forgiving one who had deeply injured him and remained impenitent. His answer is characteristic: "True it is he injured me, and deeply, and perhaps what you say is true that he will injure me again; but in Sarāwak I *cannot quarrel* or feel resentment towards any one, however great the evil done to myself." He could be superior to the petty feelings which sway too many men.

One officer alone[1] in Sarāwak left the service, as his friendship for Captain Brooke would not allow him to make the pledges the Rajah thought necessary—Mr Hay; but of him the Rajah wrote, after detailing the motives of his leaving, "He is a man of honour and a gentleman."

In September 1863 the Rajah left Sarāwak for the last time. He fully intended to return, but that hope was never realised. He left Sarāwak tranquil and prosperous; without an element of discord, without a single native chief on whom suspicion could rest; prosperous too in finances,—a prosperity which has but increased as time has passed away.

Whilst in Port au Prince I received in December 1863 a letter from the Rajah announcing his safe arrival in England in health and good spirits.

During the time the Rajah had been away in Borneo, his friends in England had not been idle, and the question of the recognition of the independence of Sarāwak was kept constantly before the Ministers, with a result which was highly gratifying to all those who felt an interest in

[1] Mr Grant had previously left Sarāwak, and was confidential agent of the Government in England, which appointment he resigned at this time, in consequence of Captain Brooke's deposition.

the success of Sir James Brooke's great experiment in Borneo.

The following extracts from a series of letters in my possession give a curious and interesting account of the negotiations, and do credit to all the statesmen whose names are mentioned in them. The writer, Mr John Abel Smith, M.P., addresses these letters to the Rajah, and as the events are now all past, it is a pleasure to be able to record the opinions of such men as Lord Palmerston, Lord Russell, Sir Henry Layard, the Duke of Argyll, Lord Granville, Lord Grey, and Lord de Grey (Lord Ripon), &c.

"*June* 30, 1862.—I now proceed without preface to give you more in detail a report of my interview with Lord Palmerston on Saturday.

"I told him that I had come to express the interest I and many others felt in your work in Sarāwak, which I believed deserved the support and good wishes of all who felt interested in the wellbeing of their kind; and that we wished to ascertain whether he would feel justified in giving effect now to the favourable opinion he had long since entertained and expressed of your conduct and objects. I then, as briefly as I could, told him all I knew of your position and prospects, including the application from Belgium, and all that I have myself been doing lately with Lord Russell, &c.; and concluded by saying, that after offering upon the shrine of Sarāwak your own fortune and the contributions of your friends, and devoting to that soil the best years of your life and all the spring-time of your health and strength, you were naturally now bent on securing, as far as you could, the future of a people for whom you had made such costly sacrifices, and that you could not reconcile yourself to the chance of their returning to barbarism in the event of your death happening before you had made arrangements to secure the permanency of the system of government which you had established, and of that gentle rule by which you had succeeded in winning them over to the gradual introduction of religion and civilisation.

"I said that although you felt that the sacrifices you had made should not fall wholly on yourself, if others reaped the benefit of them—in which opinion I thought public feeling would strongly concur—still, that I honestly believed that personal and pecuniary motives were with you wholly secondary to the great object of securing the future permanency of a civilised and civilising government in Sarāwak; and that, trusting to the justice of your country, you would not allow the money question to stand in the way. I explained to Lord Palmerston the extent and population of the extensive territory now ceded to you, its fertility and resources, its increased and increasing trade, its present revenue, &c. . . . I ended by telling him that I did not come to ask him to pledge himself at that moment to the details of any particular plan or scheme; that from what I knew of the views and feelings of some members of the Cabinet, I was aware how difficult it might be to assume the immediate sovereignty on the part of the Queen, and at once add Sarāwak to the Colonial Government of England; and that even if this difficulty did not exist, I entirely concurred in an opinion you had expressed to me, that it would be most desirable in the interests of humanity and civilisation that the present form and manner of government should not be at once abandoned, and that you were able to do more at present than any one else could, and much less expensively, and that I was disposed to believe that a system of avowed and unequivocal protection would be better for all parties than immediate occupation. That in exchange for the support, sanction, and open recognition of the British Crown, you would not be indisposed to make a conditional agreement as to the future, and that your views of the protection necessary were most moderate, involving no troops or costly effects, and embracing only such support as would be given by the presence of such a naval force as might be thought desirable or necessary. That what I asked *then and there* only was, a declaration of Lord Palmerston's general views and feelings, and permission, in concert with any subordinate of his Government in whom he might have confidence, to lay before him such details as might appear best calculated to carry out the object in view, for the purpose of being afterwards submitted through Lord Russell to the Cabinet.

"Lord Palmerston, in answer, went at some length over your past proceedings, and what had occurred at various times in reference to your establishment and government in Sarāwak. He spoke of you in the handsomest terms, and seemed thoroughly to appreciate your character and objects. He admitted the value and capabilities of Sarāwak, and never attempted to depreciate its importance in order to advance his own views. He said, 'I understand you to mean that you want to have the English Government at the back of Sir James Brooke distinctly and openly, and declaring to the whole world that they were prepared to stand by him and protect him.' He said the great difficulty has been, and is, that of the Queen recognising one of her subjects as an independent ruler, but 'there were probably the means of getting over this.' That with the strong adverse feelings of Mr Gladstone and others, immediate assumption of Sarāwak with the colonial system of Great Britain was most difficult; but that he was well disposed to forward your views as far as he could, and he used the words, 'I will do what I can to help you.'

"He told me to go to Layard, to repeat to him what had then passed, and to see if a scheme could be put on paper which would be within his power to carry out, while it would also satisfy your wishes and secure the objects you had in view.

"I had a good deal of discussion with him as to the line which would be taken in a matter of the kind by various members of the Cabinet, especially Sir George Lewis, the Duke of Newcastle, and Sir Charles Wood. He spoke very hopefully of the Duke of Newcastle. I asked him plainly and directly, if such a matter as this was strongly and unreservedly supported in the Cabinet by himself and Lord Russell, whether there would be probably any determined opposition on the part of his colleagues, and he replied with a decided *negative.*

"I cannot pretend to recollect all the details of the lengthened observations made by Lord Palmerston, but I think I shall have enabled you to understand their general purport, and the amount and extent of his agreement with us. I should be most ungrateful if I omitted to state that I have rarely had a pleasanter interview on public business; that he was frank, cordial, and unreserved in manner and expression, and seemed entirely disposed to enter into my feelings as to yourself and your claim on the admiration and consideration of your countrymen. He gave me leave to come to him again whenever I had anything to say, and was altogether most obliging and gracious.

"I find that I have omitted one remark of Lord Palmerston's which struck me at the time as significant of his meaning to enter into my views. He said, 'What can we do to declare to the world our interest in Sir James Brooke unless we take possession of his territory?' 'By the by,' he added (answering himself), 'we could write a letter to the Sultan of Borneo saying how great a friend the Queen was to Sir James Brooke, and that she meant to protect him heartily.' This means little, but shows he appreciated and understood my application and statement."

The Rajah's friend continued his negotiations, but as Colonel Cavanagh had been directed to proceed to Sarāwak and report thereon, everything was in abeyance for the moment, and the result of the Governor's mission was impatiently expected. At length it reached England, and Mr Abel Smith continues:—

"*Feb.* 24, 1863.—Very shortly after my last to you, Colonel Cavanagh's long-expected report arrived, and Lord Russell at once permitted me to read it. It appears to me to do little credit to Colonel Cavanagh's powers of observation or discernment, and to be totally silent on the future to be expected from the country, if placed under the direction of a power able to secure its internal and external security, and an impartial administration of law. All this is comparatively unimportant, inasmuch as I am able to inform you that Lord Russell is satisfied with it to a certain and (for you) sufficient extent, and considers it establishes the fact that there is a regular government carried on in Sarāwak in your name, acknowledged by a willing and obedient people—that the ordinary forms of government are observed—and that the administration of justice is careful and regular. He also admits that the statements made by me previously, under your direction, as to the population, trade, and resources of the country, are substantially borne out by Colonel Cavanagh's returns. . . . I have the present satisfaction of informing you that Lord Russell has authorised me to inform you that if your authority in Sarāwak is undisputed, he is ready at once to propose to the Cabinet the recognition of Sarāwak as an independent State under your rule and government."

The disparaging remarks on Colonel Cavanagh's report may not perhaps be deserved, as the Colonel may have only followed instructions, perhaps a little too strictly.

News having reached England that the Rajah had arrived in Singapore, and had, as I have previously stated, met

Captain Brooke, who unreservedly submitted to his uncle's authority, the correspondence continues as follows:—

"*April* 24, 1863.—I called on Lord Russell last Sunday. I delivered and saw him read your letter transmitted through me, and I also read out to him slowly and carefully your letter to myself accompanying the despatch to Lord Russell. I informed him that you had proceeded to Sarāwak from Singapore; that your authority in Sarāwak had never been for a moment questioned or disputed; that I had conveyed to you his message, 'that if all was quiet in Sarāwak, you were prepared to propose the recognition of the Rajah's territory as an independent State;' and that the time was now come for him to determine when and how he would proceed to carry his intentions, as expressed through me, into effect. Lord Russell remained silent and thoughtful for a minute or two, and then said, 'What I promised to do was to send out a consul to Sarāwak, and ask for an *exequatur*. That I am now ready to do, and I will at once make inquiries at the Foreign Office of the steps to be taken, and the requirements for which I am bound to ask.'

"I said, I conclude that what you say still implies a communication with the Cabinet, and a reference of the whole matter to the decision of that body; to which he replied in the affirmative: and I asked at once if he wished me to see any members of the Cabinet before the question was brought forward; and I mentioned the names of Lord Granville and the Duke of Argyll. He interrupted me with the suggestion of the Duke of Somerset's name, who he thought was likely to take a strong interest in a question of this kind, both personally and officially; but he subsequently expressed a wish that I should see also the Duke of Argyll and Lord Granville. Lord Russell said he would give me ten days or a fortnight to prepare the members of the Cabinet for the discussion.

"I told Lord Russell that the more I heard of the state of matters in the China seas, and of the present and future probable progress of Sarāwak, the more I felt disposed to believe that a closer connection with that 'State' than was implied in simple recognition was desirable for England, and that it would naturally be more economical and more easily arranged now than if postponed to a later period, and I pressed the prudence and propriety of his consulting others besides myself as to the best mode of dealing with the question. He was again silent for some time, and then broke out with, 'I suppose Lord Palmerston will be favourable to Sir James Brooke.' I recalled to his recollection what had passed between Lord Palmerston and myself, and the strong message he had sent you, and told him that, as far as I was able to judge, simple recognition would fall short of Lord Palmerston's wishes. He made no reply to this, and the matter dropped."

"*May* 9, 1863.—The Duke of Argyll, I am sorry to say, has gone to Scotland; he will not be here to give us any help, which he is well disposed to do. . . . Lord Granville is very friendly.

"Previous to his [Mr Fairbairn's] going abroad, he had at my request an interview with Lord de Grey, and he also was so kind as to write me a letter expressive of his feelings towards you, and of his opinion and recollection on one or two points on which I desired to have the support of his evidence. I think his letter so good a one—it breathes so strongly the spirit of a true and generous friendship, and I have found it so useful already both with Lord Russell and Lord Palmerston—that I have ventured to send you a copy of it, in the hopes that you will share the pleasure it has given me.

"Since I last wrote I have seen Lord Stanley and Lord de Grey. The former was most cordial and earnest in his expression of respect and confidence towards you, and in the unreserved declaration he made that the time was come for the recognition of the independence of Sarāwak, and that he was ready to support this opinion either in or out of Parliament. Lord de Grey, as a new member of the Cabinet, was more guarded; but he spoke of you most kindly and favourably. I am persuaded Lord de Grey will not oppose, I hope he may warmly support, the recognition of Sarāwak, when proposed by Lord Russell."

"*May* 19.—I had a long and detailed conversation with Lord Russell at Pembroke Lodge. I recalled to him all that had occurred in reference to yourself, and did my very best to make him master of the situation in all respects. I am happy to add that he declared himself perfectly satisfied, and expressed his willingness to propose the recognition of Sarāwak to the Cabinet on the first practicable occasion, instructing me to send certain papers, and especially Mr Fairbairn's letter (which I read to him), to the Foreign Office, for future reference, and asking me to call on Lord Palmerston and give him such information as I could on the various points alluded to by me in my recent interviews with him (Lord R.), and ascertain whether Lord Palmerston was willing that the question of recognition should at once be submitted to the Cabinet. In obedience to Lord Russell's suggestion, I saw Lord Palmerston at Cambridge House last Friday, and am happy to say that I found him as warmly prepossessed in your favour as ever. He authorised me to convey to Lord Russell his hearty assent to recognition *at once;* and he repeated what he said in a former interview —viz., that it would be a pleasure to him to find himself able to do anything agreeable to you."

"*June* 10, 1863.—I have had a long chat with Earl Grey this morning on Sarāwak matters, and found him most kind and friendly. Lord Grey's disposition, in short, is entirely satisfactory."

"*Aug.* 17, 1863.—When I wrote to you on the 10th, I was much disappointed and disturbed at not having heard from Lord Russell. He had seriously promised me not to leave London without attempting to settle the question of recognition, and as he has never in his life deceived me, I was utterly perplexed at hearing of his departure from London without any communication to me.

"I was more than angry, and was meditating a strong remonstrance, when I luckily met Layard on Thursday last in the street, who came up to me and said,—'I suppose you know that the recognition of Sarāwak was settled in the last Cabinet, and that the Government have agreed to appoint a consul as the most direct and least formal method of recognising it as an independent State. I told Lord Russell that I would tell you of it; but I have had my mind full of other important things, and I forgot it. I hoped that you might have heard of it in the interval from Lord Russell himself.' I told him that I had not heard from Lord Russell, and after expressing my joy and thankfulness, agreed to call on him on Friday to go more into detail."

"*Sept.* 9, 1863.—I have the satisfaction to inform you that recognition is at last a *fait accompli.* Since I last wrote to you, the Foreign Office have applied to the Treasury to sanction the salary of a consul to Sarāwak, and I have reason to believe that the terms of the letter conveying this application are most satisfactory and honourable to Sarāwak, as the appointment of a consul there is justified on the ground of public policy, and its advantage to the general interests of England.

"This application to the Treasury is an invariable and necessary form, but after the consent of the Cabinet, only a form. The reply of the Treasury is not yet received, in consequence of all the higher officials being out of town. I urged Layard strongly, as this decisive step had been taken, not to delay longer the writing an official communication to the Rajah of Sarāwak of the intention of the English Government to ask his *exequatur* for a consul; but he replied, first, that it could not be done until the Treasury had sanctioned the salary; and secondly, that he felt it to be of the greatest importance for you that the Office should treat Sarāwak exactly as they would any other independent power, and that the course always pursued was not to write until they had selected the consul, when your *exequatur* will be asked for that particular person. He gave me, however, leave again, in precise terms, to let you know in confidence how matters stand, and he (Layard) has behaved with such friendly earnestness in your behalf, that I cannot doubt the sincerity of his motives." (Mr Layard's was the only official course that could be pursued.)

"*Oct.* 26, 1863.—I have seen Lord Russell since his return from Scotland. He was full of kindness about you and your concerns, and quite prepared to name a consul at once."

Thus was Sarāwak recognised in the fullest and most generous manner, and the hand that had inflicted the deepest wound on the Rajah by appointing the Commission in 1853, now healed it, by according to him the most important sanction of his policy that it was possible for the Government to give. The statesmanlike view taken by Lord Russell proved how fully alive he was to the importance of the north-west coast of Borneo.

The accounts of the interviews with Lord Palmerston are very interesting, and show how frank and generous was his nature.

Mr Ricketts was appointed consul, and went to Sarāwak in May 1864, and his presence there was valuable as a public recognition of the Rajah's Government by that of England. But as there was nothing for a consul to do in Kuching, it was not found necessary to keep up the appointment. He could only repeat in his reports what had been published fifty times over. At his departure an unpaid vice-consul was named, but even that has disappeared—why, I cannot imagine, as there must be some one in Sarāwak unconnected with the local Government who could hold the appointment.

During the whole of 1864 the Rajah's correspondence shows a return of health and strength; in fact, he spent the summer in a round of visits—a fair proof of improved condition,—and in the autumn he went out partridge-shooting.

The following brief extracts from letters to a friend show an altered tone from the gloom of previous correspondence: "31*st August* 1864.—I know your flight for Italy is not far off, and before you go I want to tell you of my innocent doings, and to hear yours. At Fairbairn's I caught fish and saw races, then to H. L., and from H. Lodge to Keppel's cottage at Basingstoke, a *bijou* after your wife's own heart and yours. A tiny green lawn, bounded to the east by the drawing-room windows, and on the west by the clear gushing stream of the Itchen; a pretty country, social neighbours and county magnates within reach. . . . From the cottage I moved to my cousin Charles Stuart's, close at hand, where we played at croquet till I was *fit to drop*. From Basingstoke I went to West Sytherley, where I had my regular exercise at croquet, and one day was devoted to an inspection of John Day's racing stable—well worth seeing. At last, on Saturday I returned to Burrator, and was pleased to find how fresh and green it was by comparison to other places. A fine summer is very enjoyable, and I have found the open air, gentle exercise, and relaxation, good for my health and spirits."

"*23d October* 1864.—For myself I get on well, and nothing shows it so much as *hope* restored to my mind. I met with a clever remark in a novel called 'Julia Malatesta,' by T. A. Trollope: 'The surest mark of the intensity of suffering is the limitation of the sufferer's desires to absolute repose.' I have felt the truth of this, and have recently experienced new sensations of life and hope."

The Rajah was, as I have noticed, exceedingly pleased with the recognition of Sarāwak as an independent State under his rule and government, but he soon began to wish that Lord Palmerston would add protection to it. His news, however, from Sarāwak were highly satisfactory—telling of peace, commerce, and rising revenue; and despising the petty annoyances to which he was subjected, he appeared fairly happy.

The news of the progress of Sarāwak, of the indirect revenue alone rising £4500 in one year, of the steam gunboats being built, made all those who had separated from the Rajah when things looked less prosperous regret the estrangement; and I was not surprised to hear that one, of whom I had the worst opinion, was anxious for a reconciliation, but that the Rajah should have listened to these hollow advances is rather remarkable. He gave me his reasons for this reconciliation, saying that as he who had injured him had retracted what he stated, and confessed himself wrong, "I let bygones be bygones." "As for ——, I have shaken hands with him, and I do desire peace and goodwill, and forgiveness and charity but there is a limit to these things. We may pardon without weakness, and recollect without revenge, but he can never possess my confidence again. When you write next, assure me, thou bad heathen, that thou dost not hold hatred, malice, and uncharitableness to be cardinal virtues." I had to rest satisfied with these assurances; and glad was

I to learn afterwards that the Rajah had not fallen under an influence which had been so pernicious in former days.

A great change was now noticed in all transactions with the Dutch officials. I have referred to their prompt and generous conduct in sending help to Sarāwak at the time of the Chinese insurrection, and their friendly warnings concerning the conspiracy of the Datu Haji and Sirib Musahor, but naturally they were puzzled how to deal with a Government which was unacknowledged. One result of the sending of a British consul to Kuching was immediate. The Dutch officials met the Sarāwak officials on equal terms, and quickly and amicably settled all border questions, and expressed not only a desire for frequent and friendly intercourse, but Mr Kater, their principal agent, spoke of their warm appreciation of the Rajah's labours.

The American Government began now to turn their attention to Borneo, and a Mr Moses was sent to Brunei as consul; but as he had no money, and was unsupported from home, his grand schemes came to nothing. He obtained from the Sultan a great cession of territory for certain sums to be paid annually; as this stipulation was not adhered to, everything fell through. But it showed that money could do anything in the capital.

In 1865 the Rajah wrote, "I have improved wonderfully in health and spirits." Friends also brought about a formal reconciliation between the Rajah and his nephew, Captain Brooke, but it never went beyond. This, however, I am sure, was not altogether the Rajah's fault. On September 15th he wrote to me: "The Tuan Muda" (Charles Johnson, now[1] Brooke) "has unreservedly consented to become my heir, and I have left it with him to adopt his nephew Hope[2] as his successor if he wishes to do

[1] Younger brother of Captain Brooke.

[2] Captain Brooke's only surviving son.

so. I would not injure the poor boy's prospects if I can help it."

In June 1866 I reached England, and during my ten months' stay at home I saw a great deal of the Rajah. He was now full of obtaining further concessions from the British Government—either protection or a gradual transfer of the country; and at one time I thought he had a good chance of success, as the present Lord Derby was favourable, but it all fell through. The Rajah was in very fair health, and spent a good deal of the summer and autumn in visiting his friends.

In September, however, he received what he thought was such bad news respecting the finances of Sarāwak that it made him ill. He thought that certain transactions there would be looked upon as a breach of faith by some of his friendly financial supporters, and he would not be confuted. We did our best to point out to him that this financial *imbroglio* was nothing, simply arising from the stupidity of one officer and the carelessness of another in not pointing out to his nephew, Charles Brooke, that he was undertaking too many improvements at a time, and that a cessation of expenditure on these public works would soon set matters right. The sale of the steamer Rainbow, however, in order to pay debts, was an error; but by resolutely setting himself to work out the problem, the Rajah soon raised the necessary money, and ordered another steamer to be built, the Royalist, so named after his famous yacht. In October I went down with the Rajah to stay with Mr Fairbairn at Brambridge House, and there he signed a document offering to hand over Sarāwak to the English Government without any personal gain, but with the fullest guarantees for the rights of the natives and of the debts contracted, whose payment he was most anxious to effect during his lifetime; but there was then the cold fit in England with regard to colonial extensions.

The following is the Rajah's letter, addressed to the Right Hon. Lord Stanley, the present Earl of Derby :—

" MY LORD,—I have the honour to state to your lordship, for the information of her Majesty's Government, that I am willing to cede the State of Sarāwak and all my rights therein to the British Crown. I would merely stipulate—

" 1. That the religion, laws, and customs of the people be respected.

" 2. That the State debts, amounting to a sum not exceeding £75,000, be paid or guaranteed. Regarding the interests involved, I venture to urge upon your lordship the importance of an early consideration of this proposition. (Signed) J. BROOKE."

In November we went to Burrator. I never enjoyed a visit more. We were nearly two months almost alone, and we could talk over everything which interested him, and it was the last time that he had an opportunity of unfolding his views. He was uneasy about the future government of Sarāwak. He had decided to make Charles Brooke his heir, but he felt uncertain about the result should his brother Captain Brooke present himself in Sarāwak and claim the succession. I myself thought there could be little doubt about the result, and that Captain Brooke would have been received by all, as the natives in general could not understand why he had ceased to be heir to the Rajah.[1] As between the Rajah and his nephew there could be no doubt, but between the two brothers there could be little question as to the result. The Rajah felt this, and did not know what steps to take. Many combinations were proposed; but as they came to nothing, it is not worth while to refer to them.

[1] Captain Brooke died in 1868, leaving one surviving son, Hope Brooke.

The Rajah used every day to take walks and rides, and as his mind became more satisfied as to the financial future of Sarāwak, he grew more cheerful. We went occasionally to the churchyard, and he pointed me out the spot where he intended to be buried. "At all events," he said, "I shall have rest there." He used to dwell on the future, and say that his successors, he was sure, would keep Burrator in remembrance of him; and that he should like it devoted to the purpose of supplying the officers of the Sarāwak Government with a refuge where, during sickness, they might come to recruit in the bracing air of Dartmoor. Alas for the vanity of human wishes!

It was during this visit that I more especially noticed the Rajah's kindness to all his neighbours, and the respectful affection with which he was regarded. He would take me into the cottages, the farmhouses, the school. He was everywhere at home. He seemed to remember each face, and no detail of their absent relatives was forgotten by him. It was really touching to see the confidence shown him by the little children, who would approach him and quietly touch his hand.

The Rajah interested himself greatly in all that concerned the parish, aided in restoring the old church, keeping up the school, and doing the duties of a country gentleman.

Poor Rajah! In his anxiety about the finances he offered me £1000 a-year to go out to be treasurer in Sarāwak; but that I could not accept, and he felt that it would not be fair to press me to leave the Government service for so uncertain a position. Curiously enough, I received a letter written about the same time by his nephew, Charles Brooke, asking me to come out and report for the Rajah's satisfaction on the condition of Sarāwak.

As Christmas approached I prepared to leave Burrator, as I had promised to dine at home that day. He accompanied me as far as Plymouth, where we passed the evening with his friend Dr Beith, an old Dido man, and there he sat down to a game of whist and played till midnight. How cheerful he was, and how full of fun! I left him on the 22d December. He returned to Burrator, and I went up to London.

On the morning of the 24th December I received a telegram from Dr Beith saying that the Rajah had had another attack of paralysis. I decided to go back immediately, but went first to inform some of the Rajah's friends of what had happened, and after calling at the Foreign Office I had barely time to reach the Great Western station and catch the afternoon express. There I found two of his warm friends ready to go too, and we started on our melancholy journey. What a sad Christmas eve! We flew along through the snow, arriving at Plymouth about midnight, and after a delay waiting for a carriage, started again to the moors. What a bleak night it was! how the wind blew as we drove across Dartmoor, to find ourselves about 4 A.M. at Burrator!

I choked as I asked the question, "How is the Rajah?" He was nearly speechless, but when I entered the room he recognised me. As soon as he was struck down by the attack, Dr Beith had been sent for. He came, and after a short time asked, "Shall I telegraph for St John?" The Rajah shook his head and said, "No; Christmas time." His thoughtful kindness so apparent even in his own great danger! He had soon two of his Sarāwak officers with him—his nephew Mr Stuart Johnson, and Arthur Crookshank, who was worth more than all in a sick-room. We agreed to divide the night into watches, and sit up each so many hours.

Dr Beith, Dr Willis, and the clergyman of the parish and his wife (Mr and Mrs Daykin), were unremitting in their kind attentions.

The attack was a severe one, for although in a month he could get down into the drawing-room, make his will, look over and destroy papers, yet he never really recovered sufficiently to use his hand. I made a temporary will for him, and he began, "I leave to my dear friend, Spenser St John——" I looked up and said, "No, dear Rajah, don't let there be any money between us; leave me your papers, as you promised." A shade of disappointment appeared to pass over his face, and then he went on and clearly dictated the rest.

In February the Rajah was sufficiently well to be removed to the Baroness Burdett Coutts's house at Torquay, and I left him in good hands while I went up to London to prepare for my departure to the West Indies.

I ran down to Torquay once more before leaving, and in the beginning of April 1867 I saw him, and as I bent over him I felt it was for the last time. As I neared the door he called me back, and I saw the tears falling, and then I could see how he also felt that it was our last adieu. Although more than ten years have since passed away, its remembrance affects me deeply.

The Rajah somewhat recovered from this attack, but was never able to do any work again. He occasionally dictated a letter, but the exertion was a great tax on his little remaining strength. I was, however, carefully kept informed of all his movements. He returned to Burrator for the last time in May 1868; and in June, whilst coughing violently, he suffered from another paralytic stroke, and he never again recovered consciousness. Mr Crookshank, who hurried down to the aid of his old chief, watched him during these last painful days, and was present when the spirit of that grand old man passed away.

His funeral was considered quite private, though there were present Arthur Crookshank, Sir James's nephew Stuart Johnson, Mr Knox, Mr J. A. Smith, General

Jacob, the Rajah's old friend and follower, Charles Grant, and many others who came to pay their last tribute of affection, besides, of course, all the warm-hearted parishioners. He lies buried in the spot he had himself chosen. His memory is and will be long cherished among his neighbours, for though they know not what great deeds he had done, yet his sympathetic kindness was felt by all who approached him.

CONCLUDING REMARKS.

For many years I have had little intercourse with those residing in Sarāwak, but I have heard enough to be well assured that the country is advancing in prosperity, though slowly. The best account that I have seen of the condition of the interior is that given by a Mr Denison, formerly an officer in the Sarāwak service; and the impression which the reading of his account gives is, that the land Dyaks, about whom the old Rajah felt so much interest, are deteriorating. They appear to be now abandoned in favour of tribes farther from headquarters. No one, not even the Government, feels the old interest in them, and the abuses of the Malay rulers have been gradually creeping back. They never, however, forget the old Rajah, and their noble-minded chief is ever a subject of profound respect and tender affection, on which they love to dwell. The missionaries appear to despair of making an impression on these poor savages, and, like the Government, devote most of their energies to the sea Dyaks.

I should have been glad to have been able to give a favourable account of their work among the energetic Dyaks of Seribas and Sakarang, but I have not the materials.[1] Indeed, last year I met accidentally a superior

[1] Since the above was written, I have been informed that the missionaries amongst the sea Dyaks have had considerable success—particularly

officer of the Sarāwak Government, and he spoke most despairingly of the work of the missionaries and their second bishop. But he allowed that as they extended education they freed the Dyaks from the influence of the Malays, and thus, politically, were an important element in the cause of progress.

The discovery of *cinnabar* (quicksilver) added to the riches of the country; but the principal profits appear to be absorbed by the Borneo Company, and little finds its way into the coffers of the Government. But the real progress of the country has been quite recent—in fact, since the present Rajah has turned his attention to agriculture. Now I hear the Chinese are cleaving down the forests, and *gambier* and pepper are beginning to be important exports.

It is a satisfaction, however, to know that, on the whole, the work of the old Rajah is being so well carried on; that peace and security reign both in the exterior and interior; that piracy is a thing of the past; and that English influence is extending, however slowly.

I should have been pleased to have been able to visit once more the scene of my early life, to make an extended comparison between the past and the present, and to be able to bring to the notice of one's countrymen the gallant work that is being quietly but surely carried out in Borneo, to fix the impression that the old Rajah did not live in vain,—that the seed he so wisely sowed is bearing wholesome fruit. That it is so, I am well assured.

with the Seribas, some of whom, both men and women, have taken to teaching Christianity amongst themselves. If Mr Gladstone has been unable hitherto to discover any justification for the policy which found expression in the battle of Batang Maran, perhaps he may now have the candour to see it in this; and other results I have mentioned have not been altogether uninfluenced by that action and policy.

APPENDIX.

From the Consular Reports, 1877—*presented to both Houses of Parliament by command of her Majesty. February* 1878.

BORNEO.—SARĀWAK.

REPORT BY CONSUL-GENERAL USSHER ON THE PRESENT CONDITION OF SARĀWAK.

TERRITORY.

Sarāwak, or the territory now included under that general appellation, has a coast-line direct of about 220 miles in length, taking it from Tanjong Datu, in latitude 2° 5′, to Tanjong Kidurong, in 3° 10′ north. Owing to its irregular character, deep bays, and indentations, its actual extent may be calculated at 400 miles. The territorial area is supposed to comprise about 28,000 square miles, and extends a great distance into the interior, in easterly and north-easterly direction, probably as far as the head waters of the Rejang river. Its capital and seat of government, with a population numbering 20,000, is called Kuching, and is on the Sarāwak river, about 16 miles from the sea.

POPULATION.

It appears to be difficult to calculate the population of this large territory. Many tribes and races are found within its limits, Malays as well as Dyaks. The estimate arrived at by the Rajah's Government is as follows :—

	Number.
Malays, ...	60,000
Chinese, ...	7,000
Milanaus, ...	30,000
Sea Dyaks of the Batang Lupar and other rivers,	90,000
Land Dyaks, ...	35,000
Total, . .	222,000

The sea Dyaks once formed the famous piratical tribes who, in common with the Balagñini and Illanuns, once devastated the Archipelago. Since their reduction and subjugation by Sir James Brooke, they have proved to be amongst the stanchest and most loyal supporters of the Brooke dynasty, and are principally relied on as a local militia in case of trouble and danger. It was these tribes, once so sternly dealt with, that a few short years afterwards flew to the assistance of the late Rajah, when he was surprised and forced to fly from Kuching, the capital, by the Chinese rebels, whom these wild warriors pursued with relentless ardour until the miserable remnant of the mutineers was enabled to seek a refuge in Dutch territory.

The Chinese in Sarāwak are now an orderly and well-conducted community, and although many times more numerous than they were in those days, would never again dream of such a rash experiment as their outbreak of 1857. This conspiracy was supposed to have been incited principally by exaggerated news of English disasters in China, and to have had extensive ramifications elsewhere. The Government of Sarāwak has since then instituted a heavy penalty against those taking part in a *huey* or secret Chinese society, the members of which in Sarāwak, as in Singapore and Penang, are generally the instigators of riotous and rebellious conduct on the part of the Chinese. The recognised leader and active chief of a secret *huey* is liable to capital punishment by the law of Sarāwak.

The Malays of Sarāwak struck me as being a superior race to those of Brunei, although the latter set store by their purity of descent, and the former must be derived from the same stock. Many of the Sarāwak Malay chiefs have pleasant faces, and exhibit an intelligence that has probably been called into existence by the active part they are permitted to take in the government of the country. The Malays are also traders, and engaged in industrial pursuits.

The Dyaks appear to be steadily improving; their country, once a terror to strangers, as the land of head "hunters," is now orderly and safe for the most part; and their chiefs, as will be seen eventually, take an active part in the management of their own local affairs, and are subsidised and recognised by the general Government.

Government.

The Government of Sarāwak may be termed a mild despotism, its arrangements being in their general features and effect not unlike the constitution of a Crown colony. The Rajah is of course the absolute head of the State, and he possesses the power analogous to but superior to that wielded by a colonial governor in a Crown colony of spontaneous and independent action. This power is, however, rarely exercised, and for all practical purposes of local and general government, he is assisted by a legislative council composed of two Europeans and five native Malay chiefs.

While the current business of the country is carried on by this body, a larger assembly is periodically held. This council is composed of the principal representatives, native as well as European, of the various districts, and in the ordinary course sits once in three years, except when specially summoned to discuss important and pressing business. It numbers between fifty and sixty members, nine-tenths of whom are natives. Any important change in the law or modification of native custom would be considered by the General Council, and rejected or confirmed as occasion might suggest. Doubtless also all matters of general importance to the State would be discussed by this body.

The Executive Government is carried on by the Rajah and his European officers, assisted by natives, members of both councils, and would appear to partake of the nature of a privy council.

The government of the various districts and out-stations, forts, and rivers, is intrusted mostly to European officers, who are termed Residents; these are generally assisted by subordinate or Assistant Residents. There are also employed native or Eurasian and Chinese clerks, who act as writers, accountants, interpreters, &c., at each station.

The number of the European staff is as follows :—

	Number.
Divisional Residents,	2
2d class ,,	4
Assistant ,,	3
Magistrates,	2
Commandant at Kuching,	1
Treasurer,	1
Medical officer,	1
Junior and subordinate officers in training for higher posts,	5
Total,	19

There are also the commander and officers of the Rajah's gunboat Aline.

The native staff of paid chiefs and members of council consists of twenty-two in all, who are distributed as under :—

	Number.
At Kuching,	5
Ladong and Lundu,	3
Batang Lupar,	3
Rejang,	3
Muka,	3
Bintulu,	2

There are no fixed regulations as to promotion, and the salaries are of a modest description.

Laws and Customs.

The greater portion of the Sarāwak law, especially as regards social matters, such as divorce, inheritance, marriage, &c., is, as might reasonably be expected, unwritten. It is administered principally with the assistance of native authorities properly versed in the traditions governing their social code, and handed down to them under a patriarchal system, but altered from time to time by the Europeans and natives governing the country, to suit, as well as may with safety be attempted, the increased scale of civilisation progressing steadily among the native inhabitants of Sarāwak.

The criminal law is framed and generally administered

upon the basis of English law. Special enactments are in some cases made to meet the particular circumstances of the country and people, and considerable freedom is necessarily permitted to those administering it, avoiding for the present the technicalities of a regularly framed code; and as a rule, the discretion thus extended to the administrators of the law does not appear to be abused. Moreover, most decisions of importance come under the notice of the Rajah, who himself administers justice in the Supreme Court of Kuching in patriarchal fashion, assisted and backed by European and native assessors.

The courts of first instance are three in number: one for criminal cases, presided over by the Resident of each district; one for social matters, as divorce, matrimonial disputes, &c., presided over by native judges (members of council); and a petty debt court, or Court of Requests. From the two first of these courts appeals lie to the Rajah in the Supreme Court in Kuching, or, when the Supreme Court is held in the districts, to the Divisional Residents, who are also the presiding judges thereof.

This Supreme Court appears also to take initiatory cognisance of all serious criminal offences, such as murder, forgery, arson, &c.

No sentence of death may be carried out without the confirmation of the Rajah, to whom the evidence is submitted.

Appeals from the Court of Requests lie to a full bench of magistrates of not less than four in number.

Besides these regular centres of jurisdiction, the distant tribes of Dyaks and others are permitted to elect their own chiefs of villages, or "long houses," who may settle petty cases, receive trifling fines, which are limited in amount, and do other little acts of justice, for which they are held responsible; but any case of importance must be referred to the established courts.

The most important of the native customs, which is against our ideas of humanity and justice, the institution of slavery, has been reduced to the narrowest limits that can at present be safely reached. The Sarāwak Government appears to act wisely under the circumstances in which it finds itself situated with regard to this custom. Unable summarily to abolish it, but prudently recognising its existence, instead of pretending to ignore it, the Government has grasped the nettle, and by gradually surrounding it and hedging it with prohibitory and doubtless obstructive regulations, is taking the best means to effect its eventual extirpation. The export and import of slaves is peremptorily forbidden under severe penalties, and is conscientiously checked. But lately the Rajah fined and drove out a Brunei chief of importance, who was convicted of an attempt at this serious offence.

The first important feature of the slave regulations is this, that under ordinary circumstances any slave may work out his own freedom, by paying to his master a comparatively low amount, which any bondsman, I should think, could raise if he desired his liberty: £6 sterling seems to have been the sum originally fixed by the law, but I do not know if this is the limit; and female slaves and children are less weighted. One good point resulting from the general tenor of the slavery laws is, that a man who has so worked out his freedom will in most cases highly value it, and prove a steadier and more beneficial member of society than one of a ruck of idle, dishonest, and ignorant savages, suddenly set free without a proper appreciation of the value of liberty, with no sense of responsibility, and not possessing the power of utilising his precious possession. Without controverting the great principle of the right of freedom inherent in every man, the results of a sudden and total abolition are, as we have reason to know, generally disastrous for a long period.

The regulations respecting immoralities between masters and female slaves are of the strictest description. Any woman slave with whom her master has had intercourse becomes *ipso facto* free. Special exemptions also are made with regard to the offspring of such connections, although in this respect, and as regards the children of bondsmen generally, some improvement will, I trust, be effected as soon as prudently may be.

No slave can be transferred without the full privity and consent thereto of the local court, before which the transaction must take place and be duly registered. Certain classes of slaves become *ipso facto* free by lapse of time, or from neglect of the owner to claim them.

A distinction is made between outdoor and indoor slaves, in favour of the former; these may be looked upon as partaking of the character of the serfs in feudal times, and would appear to be specially benefited by the rule as to lapse of time mentioned in the preceding paragraph.

Maltreatment of slaves is severely punished, at times to the extent of manumission by the court having the power of liberating such maltreated slaves.

To quote from the works of an officer of the Sarāwak Government:—

"In Sarāwak any sudden steps to abolish slavery could not have been carried out without giving offence to the native chiefs, on whose goodwill the Government, especially at the outset, had much to depend. However, the system relating to slavery that was then organised, and which has been steadily pursued, has been successful in leading to a decrease of the evil, especially in preventing masters from holding and wielding unjust and cruel power; and the natives are finding out that slave labour is not equal to free labour, and the latter is very perceptibly taking the place of the former."

And again, I may cite the words of the present ruler of Sarāwak, addressed to his Residents in an exhaustive circular on this subject in 1868, before his accession to the Raj:—

"The Tuan Muda wishes to express his opinion that this is a very important question, which involves much that might raise the prejudices and antipathies of the chiefs and all those who possess slaves; he thinks if the custom were discussed publicly, with a view to bringing about reforms to ameliorate slavery, that such a discussion would have a contrary effect, and cause masters to bring claims (in most cases just ones) against people who have been slaves, but who have been living comfortably and to all intents and purposes independently, for many years past.

"Such reforms as are requisite the Tuan Muda thinks had better be quietly and gradually brought about in conjunction with the chiefs themselves, rather than permit the question to become a public one; as in the majority of cases the master would be a gainer and the slave a loser, were the question so raised as to cause the masters to put in their claim."

To summarise the general tendency of the Sarāwak laws and regulations respecting slavery is to abolish the system gradually and effectively without disturbance, to face a social evil, and by recognising it to reduce it within the narrowest limits, pending its total abolition.

The power of England has it always unquestionably within its means summarily to abolish such an evil within its dependencies; nevertheless, the system of pawns and domestic slavery was allowed to exist for many years on the Gold Coast, under the flags of her forts. It is therefore not to be wondered at that Sarāwak has to be cautious, depending as she so much does upon native goodwill, and being without the reserve of strength and prestige of a strong power, before she attempts any wholesale legislation in direct contravention of all present and past local feeling and custom; and she acts wisely, in my opinion, in gradually inducing her subjects, by an appeal, or rather by frequent appeals, to their own interest and right feeling, of their own freewill to allow this ancient but indefensible custom to drop quietly out of her records.

Much of the local legislation is marked by sound sense; for instance, the treatment of imprisoned debtors appears to be right and sensible. Every debtor confined in prison can, if he likes, work for the Government, receiving a fixed monthly wage for his labour. Half of these earnings may go to the creditor, the other half he retains for his personal use. In case of his leaving a wife and family on the hands of the public, the latter half is transferred towards their support. This arrangement cannot take place without the joint consent of debtor and creditor.

I believe that the Contagious Diseases Act, or one akin thereto, is in force in Kuching.

The laws regarding the habit of the Chinese in seeking intercourse with girls of immature age are also strictly enforced.

All native marriages, to be lawful, must be effected before a civil registrar.

Questions of divorce, marriage, inheritance, &c., are referred to native courts competent to deal with such matters.

Coroners are appointed in the principal centres.

Capital crimes are tried by a mixed jury; British or other European subjects guilty of heinous crimes are tried by a jury of Europeans, Her Britannic Majesty's Consul having the right to be present. In cases of Europeans seriously maltreating or murdering a native, one-half, or at least one-third, of the jury is to consist of natives.

The land regulations are useful, and present no peculiar features.

The law of bankruptcy appears to have been in some measure assimilated to the English law.

Military Force.

The permanent military force of Sarāwak consists of a well-drilled and effective body of men termed the "Sarāwak Rangers." They number about 200, and are well made and of powerful *physique*, being mainly recruited from the Dyaks, whose beauty of form, united with their strength and activity, is rarely to be surpassed. They are neatly dressed in white tunics and trousers, with black braid ornaments, and are armed with the Snider carbine.

The forts at the out-stations are massive wooden structures armed with a few smooth-bore guns of old pattern, and each fort is manned by about a dozen "fortmen," whose duty it is to garrison and defend the work. They are mostly occupied by the European officers of the districts, and are fourteen in number. They are quite efficient against native attack. Some of them are placed in sole charge of natives.

The militia, which constitutes the real force of Sarāwak in the hour of danger, is composed of all the able-bodied men in the tribes of the Batang Lupar, Seribas, and other sea Dyaks—of the same tribes in fact, the subjugation of whom, by the late Sir James Brooke, and their transformation into defenders of their country, from their former occupation of bloodthirsty pirates, was the subject of so much misunderstanding in England, and the cause of much undeserved animadversion on a high-spirited and humane ruler.

These people could turn out about 25,000 warriors, who are

ready to assemble at the summons of the Government, and devote themselves to its defence.

They are in various ways specially favoured, in consideration of their services, such as receiving whole or in some cases partial exemption from the capitation tax.

The naval force of Sarāwak comprises the gunboat Aline, a fine vessel; a screw steamer of about 250 tons burden; and two heavy river steam-launches, the Ghita and the Firefly.

Public Works.

Besides the forts, the most striking buildings are the "Astana," or residence of the Rajah at Kuching, a handsome and well-ordered dwelling, replete with every comfort, and surrounded with tastefully-laid-out grounds.

The court-house, a solid and handsome building in plain style, is also at Kuching, as well as a fort, barracks, and a prison. Of these the fort is being rebuilt and enlarged, and the prison is undergoing alterations and improvements. Several new roads, of good construction, are being cut through the forest in different directions from Kuching; one of these will open up a communication with the province of Upper Sarāwak. The principal means of communication, however, lie in the numerous rivers and streams intersecting the country.

Shipping.

Besides the gunboats already mentioned, there are two trading steamers belonging to the port of Kuching—the Rajah Brooke and the Royalist—of 254 and 151 tons burden respectively.

The remaining tonnage of Sarāwak, which is principally native owned, is composed of schooners and small coasting craft, 130 of which belong to Kuching; the others belong to other ports, and are apparently not registered at the seat of Government. They belong to people at Muka, Bintulu, Sibu, and other places.

Exports and Imports.

The chief native products of Sarāwak are as follows:—

Raw sago, sago-flour, pearl sago, antimony (sulphide, regulus, and oxide), quicksilver, gold, coal, timber, gutta-percha, india-rubber, cocoa-nuts, rice, *dammar*, diamonds, canes, and dye-woods.

The imports consist mostly of opium, salt, tobacco, cloth, crockery and brass ware.

On comparing the returns attached to this report, some of the exports for 1876, especially for jungle produce, are considerably in diminution of those of 1871. This is especially noticeable in the articles under the head of foreign exports, of gutta-percha, india-rubber, birds' nests, and antimony. These jungle products are getting gradually worked out in the neighbourhood of the province of Sarāwak Proper; and the natives are now turning their attention, owing to the special encouragement of the Rajah, to agricultural industry. Pepper and *gambier* are now being largely cultivated, and the results are not yet sufficiently large to show the improvement expected, but which a few years will probably develop. The natives will then, while not neglecting the various sources of wealth lying easy to their hands, not have to rely solely on them for their prosperity, but on the more stable and solid fruits of their own industry.

But the corresponding statements of 1871 and 1876 nevertheless point to a steady improvement in trade, and the territory can show exports during the past year, foreign and coastwise, amounting to more than £250,000 sterling—the imports being but little under that amount.

Revenue and Expenditure.

I have not been furnished with any returns of the expenditure of Sarāwak, and it is possible that the Government may not have been in a position to supply accurate information under this head, owing to a radical change which has lately taken place in the method of keeping accounts of the Raj.

The revenue amounted in 1871 to 157,501 dol., and in 1876 to 183,182 dol., showing a decided although not large increase during the five years.

It is principally derived from farms, such as opium, arrack, pawnbroking, &c. Gambling farms are still permitted.

An important item is furnished by the royalties on minerals, antimony bringing in the largest share.

The exemption tax is of 2 dol. per man, paid by the Malays; those serving in the militia are free.

The Dyak tax is a capitation tax of 3 dol. per family; those liable to military service are exempted in part. Bachelors pay half the tax. It is probable that this tax can be but partially collected.

The remaining revenue is made up from customs duties, both export and import, the former being a small impost on raw jungle produce. Manufactured produce is not weighted. A portion of the revenue is also derived from land and township lots.

Public Debt.

In the absence of specific information on this point, I can only state that I believe the public debt to consist of a considerable sum of money, which was advanced from his private fortune by the late Rajah, Sir James Brooke, and which now forms a mortgage or first charge upon the public assets of Sarāwak.

Ecclesiastical.

There is but one mission in Sarāwak, the English Protestant Mission, sent out by the Society for the Propagation of the Gospel. It is presided over by the Right Reverend Dr Chambers, who is Bishop of Labuan, Sarāwak, and Singapore. It numbers about six members, scattered over the territory. Its efforts do not appear to have been attended with marked success, the number of converts being stated as but few. Each mission-station has a school attached to it.

Educational.

The Sarāwak Government has three Government schools at present—two in Kuching and one in Upper Sarāwak. I have no statistics concerning them.

General Geographical and Physical Features.

Sarāwak may be described for the most part as an extensive and dense forest, intersected in every direction by rivers and streams, and traversed in some parts by lofty mountain-ranges. These streams form natural waterways, and in great part take the place of roads, and thus they afford access to the most distant tribes, who would otherwise be unapproachable. Some of these rivers are powerful and rapid, such as the Rejang river.

The timber of Sarāwak, especially in the south, is boundless in quantity: valuable woods of many descriptions abound, and are generally used in native house-building, and for other purposes. Good shingles are made in Sarāwak from an almost indestructible wood named *bilian*, and brick-making is carried on to some extent in the neighbourhood of Kuching.

Sarāwak is rich in minerals. In common with many other parts of Borneo, it possesses valuable coal-mines. In one portion of the territory an expert sent out from England calculated that a supply existed in one spot of more than 4,000,000 of tons, not difficult of access. A small mine is now worked at a trifling cost by the Government in the Sadong district, and sufficient is obtained from it to supply the three Government steamers, and to leave a surplus, the sale of which about recoups the Government its working expenditure.

Quicksilver has been worked in various places, and undoubtedly considerable quantities yet exist. Antimony has been found and worked by the Borneo Company in large amounts, and at great profit. Gold exists in small quantities, and is principally washed by Chinese; it would probably not repay European labour. The same may be said of diamonds; they are not numerous, but I have seen some good specimens.

Copper, manganese, and plumbago have also been discovered in Sarāwak, but not yet in sufficient quantity to warrant their being worked.

The aspect of the country, especially in the south, is beautiful. The landscape in the neighbourhood of the Matang and Santubong ranges, in the vicinity of Kuching, are particularly striking and romantic. The flora is numerous and interesting, and magnificent orchids abound. The fauna of Sarāwak is varied and extensive, and has been worked from time to time; but there is doubtless much to be done yet. In common with the rest of the forests of the island of Borneo, it is the chosen home of the *mias*, or orang-outang, of which two species exist.

The climate is wet, 182 inches of rain having fallen in 1876. It, nevertheless, appears to be fairly healthy for Europeans, who all seemed to be in good health and spirits during my stay there. The average temperature is rather higher than that of Labuan, and may be stated at about 85° Fahrenheit. In the mountains a sensible diminution is perceptible.

General Remarks.

The position and prospects of Sarāwak cannot fail to be of considerable interest to Englishmen. It is not too much to say that Sarāwak presents one of the few remaining chances of existence to the enervated and indolent race of Malays. Under such a Government, which appears to strive to impress them with a sense of their duty to the State, as well as with a feeling of self-respect, by inducing and encouraging them to take an active part in the administration of public affairs, the Malays of Sarāwak ought to prosper; and they have, moreover, continually before their eyes the example of the misgovernment and anarchy existing in the wretched kingdom of Borneo Proper, which is apparently hastening to ruin and decay. The comparison between Brunei and Sarāwak cannot but be gratifying to the inhabitants of the latter; and if any spark of pride and energy yet lingers in the breasts of this once powerful people of Brunei, the reflection upon their misgovernment and apathy, plainly forced upon them by the aspect of the prosperity of Sarāwak, when compared with the decadence of Brunei, should stimulate the latter to some attempt at reviving their ancient prestige. But I fear that nothing will now arrest the general decay and ruin sensibly attendant on the Malay races of the Peninsula and Archipelago. To their inherent vices of sensuality, rapacity, and indolence, the youth of Brunei, and doubtless of other Malay countries, are superadding the deadening effects of opium; and the action of this drug, injurious as it probably is to the active-minded and industrious Chinese, must in time prove positively fatal to the enfeebled and nerveless constitutions of the modern Malays.

The policy of the Sarāwak Government appears to me to be just and equitable toward the native Dyak and other races. It may fairly be assumed to be so, if we take as a test the fact that extensive tribes of savages have been transmuted from lawless head-hunters and pirates into comparatively peaceful agriculturists. The crime of head-hunting is now scarcely known in Sarāwak. Indeed, I regret to state that it appears to be more common in the territories of his Highness the Sultan than elsewhere; and so bold did I find these gentry on my arrival in Labuan, that two cases were absolutely before me of head-taking within the British colony of Labuan itself, where a panic on that account had existed for some time. The same remarks apply to the crime of piracy, a pet offence with the old marauders, and a venial one in their eyes. What little piracy exists on the western coasts of Borneo is not to be found within the dominions or seaboard of Sarāwak. It is rather to be looked for in the territories on the north-west coasts of Brunei, again partially within the nominal jurisdiction of the Sultan, and to a greater extent on the north-east coast.

There are doubtless to be found in the rule of Sarāwak many defects, some of which might be at once amended—others, again, that time only can efface. In criticising severely any special legislation or custom at present obtaining within the country, it would be necessary that all the attendant circumstances should be thoroughly elucidated and considered, before arriving at a sweeping and condemnatory judgment upon matters which, to the eyes of the most civilised colonists in the world, appear anomalous or even wrong. Sarāwak is yet not forty years old, and has time before her to amend and improve any defects of government. As long as the main objects constituting the welfare of the community be kept in view, and the people are generally contented and happy, the objections to which I have referred can practically have but small weight in the balance.

One of the principal recommendations attaching in the eyes of the native to European rule in Sarāwak is the honesty of its administration, especially in pecuniary matters. The object of the Malay nobles in olden times, and indeed now in the territories of Brunei, was to squeeze as much as might be from the wretched aborigines; the principal aim of the European appears to them to be, to solve the problem of how to carry on an effective Government at the lightest possible cost to its subjects. This difficulty has met with a creditable solution in Sarāwak: a taxation of about £40,000 per annum, distributed amongst a population of 200,000 souls, and giving a statistical average of from 5s. to 6s. sterling per head, can scarcely be called oppressive. (In calculating this average, I strike out about 50,000 frontier natives, who probably escape paying taxes at all.) Another recommendation in the eyes of the native is the possibility of obtaining even-handed, if rough, justice. It is not necessary, as they see and admit with satisfaction, that litigants should enter into a pecuniary competition with their opponents to purchase the favour and countenance of their judges.

Education and progress will safely and surely eradicate many of the evils remaining in the State, which may be viewed as legacies, fortunately diminishing, of a barbarous *régime* long since extinct. In the meantime, natives, Mohammedan as well as pagan, will be best managed and improved by showing a proper degree of respect for their usages and customs, especially by a complete tolerance of their forms of worship; and the only real blot in Sarāwak, that of slavery (although existing in a modified form), may be trusted to die out with the gradual extension of European rule, and the increased intelligence of the coming generation.

The occasions requiring the employment of armed force are becoming rare, and disturbances are strictly local.

The real power of Sarāwak is based upon the remembrance of, and gratitude due to, the late Rajah, Sir James Brooke, as well as upon the firm administration and even-handed justice of the present Government. No one visiting Sarāwak can fail to observe the respect and affection in which the present Rajah and his family are held by the entire community. The fact is as noticeable among Europeans as among the natives; and I may observe that the moderately paid but fairly efficient European staff is socially on a par with the officials of the generality of our colonies. The mode of life amongst the European body is quiet and unostentatious; but of hospitality there is abundance, and no visitor leaves Sarāwak without pleasant reminiscences of his stay.

A further noteworthy feature is to be found in the results obtained with so little money. The civil list of the Rajah is, I know, modest in all respects; and it is not every Government that, on a yearly revenue of £40,000 sterling, would be enabled effectively to rule 25,000 square miles of territory, with a population of over 200,000 souls, to keep up a respectable standing military force, to garrison and maintain fourteen forts, to pay a competent staff of European officers and native authorities, to maintain three gunboats, to protect commerce and agriculture, and generally to guarantee safety to life and property within its limits.

Sarāwak is on good terms with its Dutch neighbours, who have lately by vigorous measures repressed and put an end to the disturbances caused by their frontier Dyaks of Kapuas.

In conclusion, I may observe, that although *de facto*, and, as he contends, I believe, *de jure* absolute ruler of Sarāwak, the Rajah clings to his English nationality; and "British interests," to use his own words, "are paramount" within his dominions. In spite of the anomaly of the position, this fertile country should only be looked upon as an item of the great colonial empire of Great Britain, and, I trust, as virtually under her shield and protection.

I append a list of printed instructions for young officers from the Rajah.

They appear to be wise and practical, and offer an indication of the spirit in which the Government of Sarāwak is carried on.

Hints to Young Out-station Officers from the Rajah.

An officer, to be efficient, must have regularity in his habits; and to possess this, he should tie himself down to do things at stated times and intervals. He should never give an order for anything to be done except he sees that his order is properly carried into effect. No out-station in the country is so extensive that it cannot be guided and governed by one man who possesses an active mind combined with discipline or regularity. Such an one can not only superintend all things pertaining to public interest, but devote a certain amount of time to social and friendly intercourse with those he has to govern, and this is necessary in order to obtain some knowledge of the character of the people. He can do all this, and yet find plenty of spare time for leisure or study, and would enjoy the latter the more by keeping his mind in a state of discipline. Such occupations also tend to preserve health.

Fortmen.—Men entered to take service should be free men, and not slaves or debtors—and, so far as can be ascertained, bear an honest character. The recruits should go through some kind of drill, even to make them hold themselves upright and march up and down when on duty. To keep themselves clean is something towards making soldiers of them. They must be made to obey with alacrity, and be useful. If other good men are over and with them, it will not be long before they show whether they are good for anything. Besides their watches, they should turn out at seven o'clock every morning and do at least one hour's work in clearing or cutting grass, or in other ways making the grounds tidy.

Arms.—Besides a regular inspection once a-week, an officer, if he has an eye practised to see things in order, could never walk through or enter his fort without observing if anything was out of place: a trial of a lock or two, and a feel with the finger in the muzzle to see if rust exists, will show him the state of the arms. An officer who has not been brought up to manage arms can soon put himself in the way of becoming accustomed to them, by trying a few experiments. Seeing how others use them will teach him something. Seeking information and facing difficulties till they are mastered, is the only road to efficiency in every profession. Everything to the smallest item should be in eye of an officer. Arms, number and kind—ammunition—cartridges, big and small—shot, shell, caps, wads, rammers and sponges, priming wires and horns, vent lists, &c.; how stowed; how to be got at quickly; safe from fire; properly arranged. There should be a regular inspection of all these things once a-week, the officer examining for himself, and not trusting to the reports of others.

Some difficulties may arise at times in obtaining supplies from Kuching: there would, however, be no difficulty in making good the deficiencies, if there were a stock of fine and coarse grained powder, a bullet-mould, and some lead. Native fortmen can make as good small-arm cartridges as those supplied from Kuching. The charges for big guns can be put into *blachu* bags, made the size of the bore of the gun. Wads can be twisted into *grummets*, the size of the different bores, of rattan, or *akar*, as good as rope. Torches can be made of the sheath of cocoa-nut blossom, or old manilla rope, well-dried cocoa-nut husk, or many barks of trees. The length of any dispart is one-sixth the difference between the circumference of the base and centre rings—on the latter the dispart is placed; a temporary one of wood, or wax tied or stuck on, will prove as true as the best.

One-fifth, or even the sixth, of the weight of the shot is used as the weight of an ordinary charge; with light brass guns even less will be found sufficient in firing round-shot. Swivels and light guns are apt to kick dangerously when overcharged. In loading big guns, the captain of the gun should always stop the vent—and the man who sponges should do it by turning it round quickly three or four times, well forced into the base of the gun. In ramming home the charge, the loader should only leave his arms and never his body exposed before the muzzle of the gun, and, after striking it sharply twice, should spring back, while the captain pricks the charge with the priming-wire to find out whether it is home: if so, the shot is then forced in with a wad on it, to keep it from rolling from its place. The wad should fit tight, as, if the shot gets misplaced, it is apt to burst the gun. In marching or walking with natives who are carrying loaded muskets, it is very difficult to make them understand that it is necessary to keep their arms at half cock. An officer had better see that they do this, or keep out of the way himself, as there is almost as much danger in letting the hammer rest on the nipple with a cap as carrying it on full cock.

On Cleanliness.—It is a mistaken idea that natives who are fortmen cannot be taught to keep a place clean and tidy. After the necessity of this is pointed out to them for a short time, they will see to doing it as well as any Europeans, who also require to be taught at first in a similar way. It cannot be too carefully attended to, as it looks well to visitors, and is a good example to the whole country. An officer has little pride who allows any untidiness, however mean his habitation may be. Morning and afternoon sweeping of every part is necessary. A brushing away of cobwebs, and the floor kept clear of oil-spots and other dirt.

On Watching.—The men who watch should be kept from sitting down: if they are once allowed to do this they become slovenly, will soon lie down, and sleep will be the result. The best and surest test of a good watch is to hear the steps of a sentry as he marches day or night. If these are silent, ten to one if he is not asleep, or at any rate not watching. Fortmen, especially new ones, are very liable to be insolent in their bearing to the inhabitants of the country, and sometimes hail boats to stop them for amusement or for private purposes—such as to buy fish, *sirih*, or fruit. Such anomalies must be prevented, or the fort gets a bad name, and is looked upon more as a hindrance than a benefit. The people will hold themselves aloof, through fear of being insulted, and the use of such a building as a centre point, where all parties can bring their complaints and seek protection, is lost. The fort, or officer of the fort, must make the rules obeyed, and in doing so he may often, or sometimes, have to resort to severe measures, such as firing on a passing and suspicious-looking boat, or apprehending troublesome characters—but this is only in accordance with his duty, and can be understood as such by all parties. It will only tend to make him respected, and not in any way bring about discontent or a bad name.

Four muskets, or rifles, should always be kept ready loaded

and capped—and discharged once a-week. More than four, only when the country is in a disturbed state.

Cash and Revenue.—Any transaction which has to do with money must necessarily be of importance, more especially when the money is not the individual's own property.

One of the most important branches of duty connected with an out-station is that proper supervision be kept on the Revenue and Treasury departments, and if these are not properly attended to, an officer is far from having his charge in an efficient state. If one part be faulty, however trivial, it more or less affects the whole.

Out-station cash and revenue business is of so simple a character, that any one who has no knowledge of methodical book-keeping and double entries can keep all straight and clear if he pays proper attention to it.

There never can be any excuse for extravagance nor forgetfulness in monetary concerns: twenty minutes a-day of supervision, or two hours a-week, or eight hours a-month, would prevent mistakes arising. The above times given could enable one to examine everything, to see wages and bills paid, and, if necessary, to copy out accounts or cash statements. However much an officer may be called away on other services, he can always spare eight hours a-month in attending to so important a duty, and no officer is fit to hold any post of importance unless he does so.

General Remarks. — Proper deference should always be shown to the chiefs, of whatever class they may be; and in any case of difficulty, more especially when an officer feels he has not sufficient experience to decide with safety, and even oftentimes in trivial affairs, it is as well to ask the advice and opinion of the head-men. It satisfies them and strengthens you, and a great object is to throw as much responsibility on the shoulders of the native chiefs as possible.

You are not obliged in all cases to follow their advice, but the fact of asking it is a compliment to them, and gives them an interest in what is being done by the Government. There is no doubt, when uninfluenced by prejudice and relationship, the decisions of natives are very sound and sensible. Europeans too often give them credit for knowing too little: this is a mistake; and a man should never be above taking advice, however old he may be—and when in a strange country it is trebly necessary and useful. In everyday concerns—jungle-walking, marks denoting danger from traps and animals, management of boats in river or on coast, judging the state of weather, and in many other ways—the knowledge of natives must necessarily be superior to a European.

There are certain personal dangers to be avoided by those who have to occupy positions of trust when young, who are thrown much on their own resources, and hold authority over others without the check of senior officers and public opinion always immediately over them. The danger is, in allowing any relapse of right principle or sense of duty, which an English gentleman is supposed to have instilled into him from childhood. Also, the avoidance of becoming overbearing and despotic when left so much with subservient natives. This topic might be much enlarged upon, but any one can think it out best for himself, and will come to the conclusion that the manner of life in this country should not be different from what it would be in England.

Manner with Natives.—The best manner in the long-run with natives is to be thoroughly natural, and in no way patronising. A mixture of kindness and freedom, with severity when required, without harshness or bullying. Joking to be limited to the comprehensions of the people; if they cannot understand a joke, they are liable to misinterpret and gather wrong impressions. Never put natives on a familiar footing. They hold their position in society, and you yours. They are not inferior, but they are different.

Always be guarded against giving way to sentimental likings for particular natives in calling them by familiar terms, and admitting them into intimate friend- or relationship. They cannot understand or appreciate this behaviour, and in the end will dislike it. It injures and often ruins them, as a petted horse is spoiled. An officer who tries such an experiment with the best intention, and in the hope that it will raise the tone of the native, labours under a mistake, and will lose influence by such acts of misplaced kindness. An officer on duty, as one is in an out-station, is not as a private individual who can do as he likes. He should recollect that he is in harness, and that to hold himself steadily at the collar determines not only his future, but the future of the community among whom he lives. Nothing artificial or extraneous in the shape of gilt or tinsel will help to gain the confidence of the natives. They are too matter-of-fact, and only admire and respect strength in its entirety.

An officer working for the general good in his profession, with a healthy tone of mind and body, doing his duties with earnestness, strength of purpose, and tact, is marked, and leaves an impression on the multitude. A reality is one thing, a shadow another—and by the sum total of his many acts the result is shown to be sound or otherwise.

COMPARATIVE REVENUE RETURNS.

	1871.		1876.	
	Dol.	c.	Dol.	c.
Farms—				
Opium	42,800	00	46,300	00
Gambling . . .	12,402	50	13,338	00
Arrack	7,601	50	5,703	72
Pawn	452	00	860	00
Antimony royalty . .	16,054	10	8,888	88
Quicksilver „ . .	4,444	44	4,444	44
Exemption tax . . .	19,500	00	19,343	70
Dyak revenue . . .	11,650	00	21,338	23
Import duties—				
Tobacco tax . . .	4,320	00	6,033	00
Salt „ . . .	2,850	00	5,402	00
Excise „ . . .	536	00	...	
Matches „ . . .	54	00	522	00
Spirits „ . . .	...		441	00
Jars, steel, &c. . . .	...		1,538	00
Guns	...		226	00
Export duties—				
Sago tax	135	00	5,200	00
Gutta-percha . . .	...		2,950	00
Camphor	...		300	00
Bilian	...		1,898	00
Bee's-wax . . .	...		130	00
Birds' nests . . .	...		120	00
Rattan	...		1,990	00
Guliga	...		17	00
Miscellaneous . . .	34,702	00	36,199	00
Total . . .	157,501	54	183,182	97

TRADE RETURNS FOR 1871-76.

Imports—Foreign.

Articles.	1871.	1876.
	Dol.	Dol.
Treasure	113,337	94,985
Gutta-percha	7,260	722
Rice	89,699	129,601
Gold	2,359	1,527
Cloth	212,359	250,165
Brass-ware	23,260	18,108
Fish	8,335	14,466
Opium	15,358	34,820
Tobacco	52,928	52,467
Tea	2,517	3,101
Wines	10,531	7,874
Sugar	12,305	11,491
Jars	15,571	8,417
Iron-ware	20,263	32,802
Crockery-ware	18,318	7,236
Cocoa-nuts	19,868	14,978
,, oil	20,901	25,813
Salt	11,125	10,612
Gunnies	6,200	8,621
Paddy	280	1,715
Bêche de mer	200	3,110
Raw sago	673	3,033
Birds' nests	240	580
Sundries	116,481	108,884
Total	780,368	845,128

TRADE RETURNS FOR 1871-76—*Continued.*

EXPORTS—FOREIGN.

Articles.	1871.	1876.
	Dol.	Dol.
Treasure	109,532	64,487
Cloth	23,103	10,975
Tobacco	2,690	4,365
Brass-ware	2,631	4,512
Fish	7,364	26,366
Rice	20,622	20,226
Paddy	3,570	2,317
Sugar	1,849	1,137
Cocoa-nuts	936	327
„ oil	1,215	2,714
Gold	4,952	2,749
Timber	635	35,220
Sago-flour	164,935	313,559
Diamonds	1,085	4,130
Bee's-wax	19,621	7,675
Gutta-percha	132,694	75,558
Dammar	4,024	6,934
India-rubber	99,948	41,402
Birds' nests	32,200	22,651
Fish maws	6,027	2,777
„ fins	837	1,965
Quicksilver	24,992	108,050
Antimony	51,690	45,958
Canes	6,382	1,088
Rattans	769	47,301
Camphor	3,920	11,221
Raw sago	11,752	20,781
Bêche de mer	2,295	3,390
Sundries	26,706	40,677
Total	769,026	930,542

TRADE RETURNS FOR 1871-76—*Continued.*

IMPORTS—COASTING.

Articles.	1871.	1876.
	Dol.	Dol.
Treasure	114,847	85,912
Gutta-percha	136,483	50,391
India-rubber	96,930	32,674
Bee's-wax	17,795	4,849
Birds' nests	13,686	7,339
Rice	24,665	10,536
Paddy	10,292	3,871
Raw sago	130,743	174,131
Rattans	639	7,867
Gold	4,048	5,748
Timber	2,673	2,365
Cattle	500	289
Sago-flour	26,993	900
Camphor	8,860	22,554
Fish maws	614	516
„ fins	68	261
Dammar	1,209	459
Galiga	1,374	1,183
Canes	6,260	62
Vegetable tallow	...	1,987
Fish	90	17,898
Coal	...	5,335
Sundries	48,786	35,410
	647,555	472,537
Add foreign . .	780,368	845,128
Total	1,427,923	1,317,665

TRADE RETURNS FOR 1871-76—*Continued.*

EXPORTS—COASTING.

Articles.	1871.	1876.
	Dol.	Dol.
Treasure	154,905	138,857
Cloth	159,835	158,249
Jars	7,232	3,549
Tobacco	29,778	36,625
Salt	7,138	6,765
Brass-ware	21,283	12,961
Iron-ware	4,866	5,005
Crockery-ware	5,185	3,869
Fish	902	5,220
Rice	16,659	33,699
Paddy	1,039	3,311
Sugar	5,549	4,342
Opium	19,064	25,054
Cocoa-nuts	6,210	3,965
,, oil	6,317	4,548
Gold	10,475	6,499
Timber	737	984
Wines	670	1,418
Gunnies	800	500
Sundries	40,667	37,177
	499,311	492,597
Add foreign	769,026	930,542
Total	1,268,337	1,433,139

PRINTED BY WILLIAM BLACKWOOD AND SONS.

Oxford in Asia Hardback Reprints

Indonesia

Dance and Drama in Bali
BERYL DE ZOETE and WALTER SPIES

The History of Java: Plates
THOMAS STAMFORD RAFFLES
Preface by John Bastin

The History of Sumatra
WILLIAM MARSDEN
Introduction by John Bastin

Memoir of the Life and Public Services of Sir Thomas Stamford Raffles
LADY SOPHIA RAFFLES
Introduction by John Bastin

Travels in the East Indian Archipelago
ALBERT S. BICKMORE
Introduction by John Bastin

Zoological Researches in Java, and the Neighbouring Islands
THOMAS HORSFIELD
Memoir by John Bastin

Malaysia

The Blockade of Kedah in 1838: A Midshipman's Exploits in Malayan Waters
SHERARD OSBORN
Introduction by J. M. Gullick

The Expedition to Borneo of HMS *Dido*
HENRY KEPPEL
Introduction by R. H. W. Reece

Life of Sir James Brooke The Rajah of Sarawak
SPENSER ST JOHN
Introduction by R. H. W. Reece

Oriental Silverwork: Malay and Chinese
H. LING ROTH
Introduction by Sylvia Fraser-Lu

The Pagan Tribes of Borneo (2 vols.)
CHARLES HOSE and
WILLIAM McDOUGALL
Introduction by Brian Durrans

Sketches of Our Life at Sarawak
HARRIETTE McDOUGALL
Introduction by R. H. W. Reece and A. J. M. Saint

The Straits of Malacca, Siam and Indo-China: Travels and Adventures of a Nineteenth-century Photographer
JOHN THOMSON
Introduction by Judith Balmer

Singapore

One Hundred Years of Singapore (2 vols.)
WALTER MAKEPEACE, GILBERT E. BROOKE, and ROLAND ST. J. BRADDELL (Editors)
Introduction by C. M. Turnbull

One Hundred Years' History of the Chinese in Singapore
SONG ONG SIANG
Introduction by Edwin Lee

Thailand

Journal of an Embassy to the Courts of Siam and Cochin China
JOHN CRAWFURD
Introduction by David K. Wyatt

Journal of a Voyage to Siam 1685–1686
ABBÉ DE CHOISY
Translated and Introduced by Michael Smithies

The Kingdom of Siam
SIMON DE LA LOUBÈRE
Introduction by David K. Wyatt

The Mission to Siam and Hue 1821–1822
GEORGE FINLAYSON
Introduction by David K. Wyatt